EXPERIMENTS ON COSMIC DUST ANALOGUES

ASTROPHYSICS AND SPACE SCIENCE LIBRARY

A SERIES OF BOOKS ON THE RECENT DEVELOPMENTS
OF SPACE SCIENCE AND OF GENERAL GEOPHYSICS AND ASTROPHYSICS
PUBLISHED IN CONNECTION WITH THE JOURNAL
SPACE SCIENCE REVIEWS

VOLUME 149
PROCEEDINGS

EXPERIMENTS ON COSMIC DUST ANALOGUES

PROCEEDINGS OF THE SECOND INTERNATIONAL WORKSHOP
OF THE ASTRONOMICAL OBSERVATORY OF CAPODIMONTE
(OAC 2), HELD AT CAPRI, ITALY, SEPTEMBER 8-12, 1987

Edited by

EZIO BUSSOLETTI

Istituto Universitario Navale, Napoli, Italy

CARLO FUSCO

Istituto Universitario Navale, Napoli, Italy

and

GIUSEPPE LONGO

Osservatorio Astronomico di Capodimonte, Napoli, Italy

KLUWER ACADEMIC PUBLISHERS

DORDRECHT / BOSTON / LONDON

Library of Congress Cataloging in Publication Data

Osservatorio astronomico di Capodimonte. International Workshop (2nd
  : 1987 : Capri, Italy)
    Experiments on cosmic dust analogues : proceedings of the 2nd
International Workshop of the Astronomical Observatory of
Capodimonte, Capri, Italy, September 8-12, 1987 / edited by Ezio
Bussolotti, Carlo Fusco, Giuseppe Longo.
        p.   cm. -- (Astrophysics and space science library ; v. 149)
    Includes index.
    ISBN 9027727775
    1. Cosmic dust--Congresses.   I. Bussoletti, Ezio, 1947-
II. Fusco, Carlo. III. Longo, Giuseppe. IV. Title. V. Series.
QB791.O85 1987
523.1'12--dc19   88-17581
                                                          CIP
ISBN 90-277-2777-5

Published by Kluwer Academic Publishers,
P.O. Box 17, 3300 AA Dordrecht, Holland.

Kluwer Academic Publishers incorporates
the publishing programmes of
D. Reidel, Martinus Nijhoff, Dr W. Junk and MTP Press.

Sold and distributed in the U.S.A. and Canada
by Kluwer Academic Publishers,
101 Philip Drive, Norwell, MA 02061, U.S.A.

In all other countries, sold and distributed
by Kluwer Academic Publishers Group,
P.O. Box 322, 3300 AH Dordrecht, Holland.

TABLE OF CONTENTS

INTRODUCTORY REPORTS

CONTRIBUTED PAPERS

REPORTS OF THE WORKING GROUPS

EDITORS' FOREWORD

Astrophysical analysis relating to solid matter requires data on properties and processes. Such data, however, expecially appropriate to space conditions are mostly lacking.

It appeared then very tempting to gather together experimentalists, observers and theoreticians working in the field of cosmic dust and in related areas. The Workshop held in Capri (Italy) from September 8th till 12th 1987 gave the participants a unique opportunity for exchange of ideas and discussions of problems and experimental procedures.

Introductory reports were prepared with the aim of giving the state of the art about single subjects; contributed poster papers presented, on the contrary, very recent results in the various fields.

According to his specific interest each attendant has also contributed to three Working Groups respectively on a) carbon, b) silicates, c) ice and related topics.

Scientifical and technical problems about these items were discussed in great detail. Though no definite answers were given, useful indications come out which will be of some help for future works.

In addition to the scientific efforts, the Capodimonte Observatory and the Istituto Universitario Navale sought to give a warm welcome to the participants. Thanks to several sponsors, the LOC could organize some excursions and shows to entertain people during their spare time.

In particular, the LOC wishes to thank for their contributions, the Italian Ministry of Education, the Italian Council for Scientific Research (CNR), the Regione Campania, the Italian Group for Cosmic Physics (GIFCO), the Department of Physics of the University of Lecce, the University of Naples and the Polaroid Company.

Needless to say, to organize an international workshop requires an enormous amount of work which could not be accomplished without the help of many people. A special thank goes to Mrs G. Iaccarino, Mr G. Cuccaro of the OAC and to Mr M. Scalzo of the University of Lecce for their help during the meeting. We also thank Mr S. Marcozzi, Mr R. Trentarose, Miss T. Ievolella and Mrs L. Sorvillo for their cooperation.

A very special thank is due to Mrs A. D'Orsi and E. Acampa for their unvaluable help in preparing these proceedings. Without their dedication and accuracy, the retyping of the papers and their final formatting would have taken much longer than it did.

Ezio Bussoletti
Carlo Fusco
Giuseppe Longo

Scientific Organizing Committee

L.J. Allamandola - USA
S.H. Bauer - USA
E. Bussoletti - Italy
L.B. d'Hendecourt - France
B. Donn - USA
W.W. Duley - Canada
R.H. Giese - FRG
D.R. Huffman - USA
W. Krätschmer - FRG
M. Rigutti (director OAC)
G. Strazzulla - Italy

Local Organizzing Committee

A. Borghesi
C. Fusco
G. Longo (contact person)
A.A. Vittone

LIST OF PARTICIPANTS

AIELLO S., University of Florence
ATZEI A., ESA/ESTEC, Noordwijk
BAR-NUN A., University Geophysics, Tel Aviv
BERETTA F., Istituto Ricerche Combustione, Naples
BLANCO A., University of Lecce
BOHME D.K., York University, North York
BORGHESI A., University of Lecce
BUSSOLETTI E., Istituto Universitario Navale, Naples
CHLEWICKI G., Laboratory for Space Research, Groningen
COLANGELI L., University of Lecce
CORADINI A., Istituto Astrofisica Spaziale, Frascati
CORADINI M., ESA Headquarters, Paris
D'ALESSIO L., University of Naples
DE MAIO A., Istituto Universitario Navale, Naples
DONN B., NASA/Goddard Space Flight Center, Greenbelt
EBERHARDT P., University of Bern
FEDERICO C., Istituto Astrofisica Spaziale, Frascati
FULLE M., International School for Advance Studies, Trieste
FUSCO C., Istituto Universitario Navale, Naples
GRIM R., Laboratory Astrophysics, Leiden
GRÜN E., Max-Planck-Institut für Kernphisik, Heidelberg
HUFFMAN D.R., University of Arizona, Tucson
JONES A.P., U.M.I.S.T. (Mathematics Department), Manchester
KNEIßEL B., Ruhr-Universitat Bochum
KRÄTSCHMER W., Max-Planck-Institut für Kernphysik, Heidelberg
KRUEGER F.R., c/o MPI Kernphysik, Heidelberg
LONGO G., Astronomical Observatory, Naples
McDONNELL J.A.M., University of Kent, Canterbury
NASTRO A, Istituto Universitario Navale, Naples
NUTH J.A., NASA-Goddard Space Flight Center, Greenbelt
OROFINO V., University of Lecce
PIRRONELLO V., University of Calabria
RIGUTTI M., Astronomical Observatory, Naples
RUSSO A., Istituto Universitario Navale, Naples
SAKATA A., The Univ. of Electro-Communications, Tokyo
SCHMITT B., Laboratory Astrophysics, Leiden
SCHWEHM G.H., Space Science Department of ESA, Noordwijk
SMITH R.G., Max-Planck-Institut für Extraterrestrische Physik, München
STEPHENS J.R., Los Alamos National Laboratory
STRAZZULLA G., Istituto di Astronomia, Catania
TEMI P., Università "La Sapienza", Roma
VAGLIECO B. M., University of Naples
VITTONE A.A., Astronomical Observatory, Naples
WÄNKE H., MPI für Chemie, F.R.G.
WDOWIAK T.J., University of Alabama, Birmingham

ADDRESS OF THE DIRECTOR OF THE OSSERVATORIO ASTRONOMICO DI CAPODIMONTE TO THE PARTICIPANTS

M. Rigutti
Osservatorio di Capodimonte Napoli, Napoli, Italy

It is a honour and a great pleasure to me to warmly welcome you on behalf of the Osservatorio Astronomico di Capodimonte, who, together with the Istituto Universitario Navale of Naples and the Dipartimento di Fisica dell'Università di Lecce, organized the present workshop on cosmic dust analogues experiments which has been well liked by so many distinguished scientists from all over the world.

This is the second international workshop of the Capodimonte Observatory. As a matter of fact the Capodimonte Observatory's Management Committee decided to organize every year a meeting like the present one, to be devoted to topycal subjects of great interest for the development of astrophysics. Last year, important problems we faced when treating Fe II in astrophysical objects were discussed; next year, together with the Kiepenheuer Institut of Freiburg, we will organize a very nice workshop on solar physics problems. Proceedings of all these meetings will be published by Reidel. With this well known Publisher we had very cordial and constructive contacts, and came to a quite interesting arrangement. Let me thank here the Reidel Company for the nice way they do their job. The initiative of organizing annual workshops must also be seen in the frame of many activities the Capodimonte Observatory is performing for the development of astrophysical studies and research in the South of Italy. Until recent years the scientific activity of the Capodimonte Observatory, founded in 1819, was essentially devoted to classical astronomy, and in this field many important results were achieved throughout the 19th and part on the 20th century, particularly in the fields concerning small planets and the polar motion. However, since about fifteen years, astrophysics officially arrived to the Gulf of Naples. Its arrival was anyway made easier also by the fact that the few heirs of the classical tradition still operating on the spot had practically got exhausted. A very good deal of efforts were then devoted to educate and train a new and modern generation of astronomers, i.e. of astrophysicists, and to give a new look and organization to the old observatory. This work was much more difficult than we expected. We experienced hard times, some bitter disappointments, but also very nice and good moments. A peculiar kind of difficulty we met was the practical impossibility of getting new astronomers not already living in this area. Of course, the main reason for this is that to be a sort of pioneer is not exactly the kind of work one wants to do. But another good reason to explain that difficulty lays in the fact that in Italy, for a number of different and good reasons, people want to live their whole life through, from the cradle to the grave, in the same spot. All this means that, with very few exceptions, the new astronomers had to be found here in Naples and its surroundings. This means, also, of course, that the average age at the Capodimonte Observatory is rather low, and I think that this is a very important detail to be considered when we have to judge the kind of effort we are doing in order that our Observatory may develop with a good gradient and produce science and scientific culture at a good international level. In short, let me say that not many years ago, in a certain sense we refounded the Observatory, and that since then we are working hard to establish also in this for many aspects not very easy area a modern and efficient centre for theoretical and a practical astrophysical studies and related laboratory activities both from. I think we succeeded in realizing our program and nowadays we can perform research at the right level in three main fields, namly, solar physics, particularly dynamics of the photosphere and low chromosphere, stellar physics, with a particular stress on early and late stages of evolution, and physics of galaxies, with special attention to the morphological and dynamical properties of early type galaxies. But this year, for the first time in the history, we had the fortune to have in the Neapolitan area two new full-time professors of astronomy. One of them is professor Ezio Bussoletti, you all know very well. This made me particularly happy because Ezio is not only an old friend of mine, he is also a source of know-how in a field new for us. We then already made nice plans to make up a fourth group devoted to studies concerning some topics of research in the infrared, also from a technological point of view, having in mind research in space. Then, we all were very happy to organize the present workshop. It is a way to make science in an ever growing and exciting field such as infrared astrophysics, to get new friends, and to say to Ezio that we are so glad he is among us, to work with us. Of course, at the same time, this is also something like a trap for him, because it put in great evidence his future responsabilities. In fact, the idea of having here an operating infrared group will be carried out only with his unquestionably necessary contribution. Of course, your help will also be necessary and the Capodimonte Observatory will be grateful to you for any kind of scientific co-operation.

I do not want to lengthen out my speech, but before leaving you to your work I have the pleasant duty of thanking some persons and public bodies who made possible this meeting. First of all, I must emphasize the enthusiastic co-operation of Istituto Universitario Navale and the Dipartimento di Fisica dell'Università di Lecce. Then I have to thank the Ministero della Pubblica Istruzione, the Regione Campania through the Assessorato Istruzione e Cultura, the Università degli Studi di Napoli, the Consiglio Nazionale delle Ricerche, the Società Italiana di Fisica, the Gruppo Nazionale di Fisica Cosmica under whose patronage this workshop takes place. I also have to thank the Comune di Anacapri and the Azienda Autonoma di Cura, Soggiorno e Turismo for their kind hospitality which will certainly make more pleasant your staying here and help to have good memories. In this connection, I want to remember also the kind contribution from the firm Polaroid which offered the nice development machine you all will certainly use with satisfaction and the bag where to put all the tools for working. Then, let me thank the journalists and the TV-men who came here to meet you. We are grateful to them because we are sure not to work only for our intellectual satisfaction. We believe to work for the tomorrow world even if in a not very showy way. Of course, we are not alone in doing that, but we too are among those who live looking also to the future of humanity. And it is good that people know what we are doing, also because, after all, people pay taxes where money for our work comes from. Finally, I would express my sincere warm thanks and appreciation to the scientific organizing committee and the local organizing committee for the beautiful job they did and to the staff of the Observatory who worked so nicely to give us the opportunity to attend a smooth meeting. And last, but not least, I thank you colleagues and friends for coming, and wish you a pleasant stay at Capri.

COMET NUCLEUS SAMPLE RETURN MISSION

A. Atzei
ESA/ESTEC, Noordwijk, The Netherlands

1 Introduction

The Rosetta/CNSR mission is one of ESA's four Cornerstone missions to which ESA has committed itself in its approved long term plan, Horizon 2000. Additionally, this mission has been categorised in NASA Solar System Exploration Committee's augmented programme as a mission of high scientific merit. The Rosetta mission constitutes the next logical step in cometary exploration after fly-by and rendez-vous missions. It can be regarded as one of the most ambitious space endeavours imaginable with present day technology and it is therefore an ideal candidate for international cooperation. Within this spirit a Science Definition Team was appointed in September '85 by ESA and NASA to study and define a "cooperative ESA/NASA mission which is scientifically excellent, technically feasible and financially and programmatically acceptable to NASA and ESA". At the same time several technical studies have been accomplished in NASA and ESA, the objective of which was to conduct an assessment of candidate mission concepts and to define key design, operations and technology requirements. These studies have by now provided a technical baseline to sufficient extent to generate performance, schedule planning and cost data. Initial steps are being undertaken to establish a joint mission implementation plan for a mission to start around year 2000.

2 Mission objectives and requirements

The Rosetta mission shall allow earth-based sample analysis of a comet nucleus, to initiate the scientific study of the presolar environment and possible sample materials from interstellar and galactic regimes. It is expected that a comet nucleus sample will consist of ice phases, non-volatile presolar mineral grains and organic compouds. These compounds are probably contained in fragile, porous aggregational structures and may also be present in coherent, rocklike masses. The definition of the type of sample to be collected and conditions under which they should be taken and preserved were given during a CNSR workshop held in 1986 at Canterbury.

The nominal mission should be designed to comply with the following requirements:

1

E. Bussoletti et al. (eds.), Experiments on Cosmic Dust Analogues, 1–16.
© 1988 by Kluwer Academic Publishers.

- rendez-vous to an active and fresh comet

- characterisation of the nucleus by means of remote sensing and "in situ" instrumentation

- acquisition of three samples: one from the core at 1 m depth, one from the crust and one rich on volatiles

- store the samples until return to earth at temperatures below 160 K

- distribute for study cometary samples to scientists in established laboratories.

Remote sensing and "in situ" measurements should provide the target characterisation and the sub-surface properties in an area local to the sampled core. Specific scientific objectives of a model payload are summarised on table 1.

Table 1: Additional scientific objectives and corresponding instrumentation.

OBJECTIVES \ INSTRUMENTS	HIGH RESOLUTION CAMERA	MEDIUM RESOLUTION CAMERA	WIDE ANGLE CAMERA	RADAR ALTIMETER	DUST COUNTER	NEUTRAL MASS SPECTROMETER	VISUAL AND INFRARED SPECTROMETER	PLASMA WAVE ANALYSER	MAGNETOMETER	SMALL CAMERA
REMOTE OBSERVATIONS AND PRIMARY MEASUREMENTS										
SHAPE AND SPIN	XX	XX	X	X						
SURFACE MORPHOLOGY (ICE, CRUST, CRATERS)	XX	XX		XX			XX			
DUST ENVIRONMENT					XX					
PURE SCIENCE MEASUREMENTS										
GEOLOGY, MINERALOGY	X	X					XX			
SURFACE TEMPERATURE							X			
GAS ENVIRONMENT						XX				
COMET - SOLAR WIND INTERACTION								XX	XX	
NUCLEUS STRUCTURE	X	X								
ON COMET OBSERVATION										
SAMPLING SITE OBSERVATION										XX

3 Mission options

The Comet Nucleus Sample Return Mission is a high energy mission (the sum of the required velocity increments, starting from low earth orbit and returning into an earth orbit, is of the order of 15 km/s). Based on the available status of expected technology in the time frame foreseen, launch between 1998 and 2002, three major design options exist, based respectively on:

- chemical propulsion
- solar electric propulsion
- combination of these two propulsion systems (hybrid propulsion)

The corresponding trajectory modes are illustrated on figure 1.

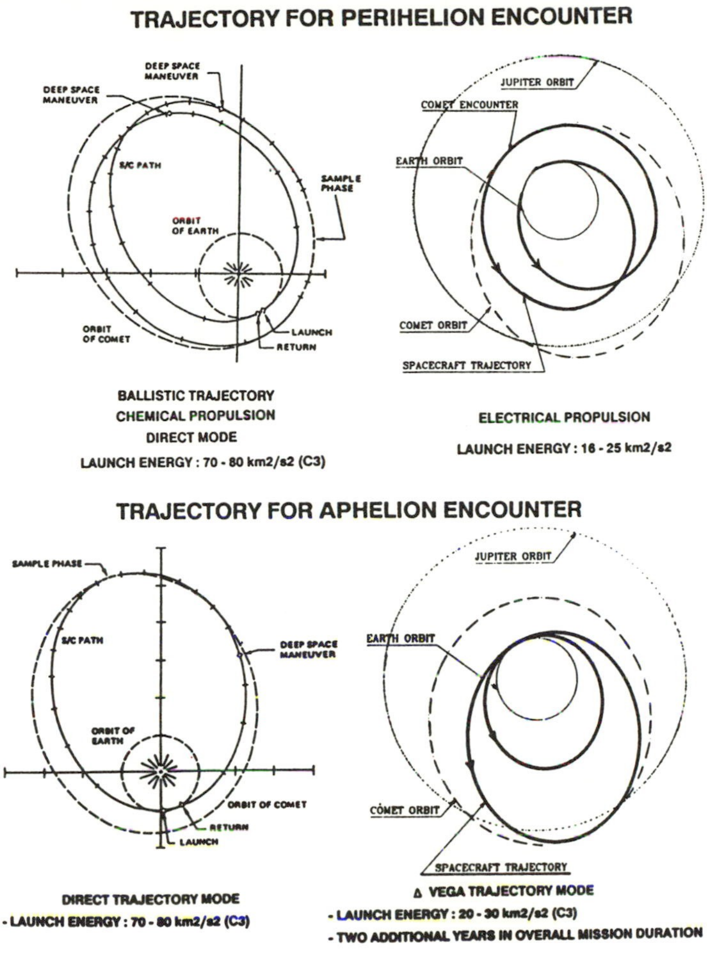

TRAJECTORY FOR PERIHELION ENCOUNTER

BALLISTIC TRAJECTORY
CHEMICAL PROPULSION
DIRECT MODE
LAUNCH ENERGY : 70 - 80 km2/s2 (C3)

ELECTRICAL PROPULSION
LAUNCH ENERGY : 16 - 25 km2/s2

TRAJECTORY FOR APHELION ENCOUNTER

DIRECT TRAJECTORY MODE
- LAUNCH ENERGY : 70 - 80 km2/s2 (C3)

Δ VEGA TRAJECTORY MODE
- LAUNCH ENERGY : 20 - 30 km2/s2 (C3)
- TWO ADDITIONAL YEARS IN OVERALL MISSION DURATION

Figure 1: Trajectory modes.

3.1 Chemical propulsion mission

A chemically propelled comet rendez-vous mission without earth gravity assist needs so high an injection energy at earth departure (C_3 80 km^2/s^2) that in orbit assembly of upper stages or in-orbit fuelling is required. This type of scenario is very complex and costly and it will not be feasible before the year 2000. An orbit injection strategy, which considerably reduces the launch energy ($C_3 \sim 27$ km^2/s^2), is the use of Delta-V-Earth Gravity Assist (Δ VEGA). In this case, the spacecraft is injected into an about two years lasting earth return orbit. Thus, at the cost of two extra years mission duration, the launch energy requirement can be reduced to the point that mission opportunities based on launchers under-development and using conventional propulsion systems are available. For chemically propelled spacecraft, comet rendez-vous is only possible near aphelion (~ 6 AU) and mission duration are around 8 years (including the Δ Vega orbit).

Solar electric propulsion (SEP) mission

For a non conventional propulsion system like the SEP working on a 10 times higher exhaust velocity (about 30 km/s instead of 3 km/s), higher energy missions can be flown ($\Delta V \sim 20$ km/s). A SEP system requires about 10 times less propellant than a stored chemical propellant system to produce a given velocity increment. The major drawbacks of SEP are:

- only low acceleration (thrust) levels can be reached;
- at large solar distances the reduced power input leads from poor to doubtful SEP performance, unless non-conventional solar arrays are developed.

As a consequence of this, typical optimum interplanetary SEP mission profiles imply a comet rendez-vous near perihelion with mission durations of 6–8 years. Missions with Ariane 5 or other expendable launchers are possible. SEP missions require injection energy around 30 km^2/s^2 if a 30 kw SEP system would be used.

3.2   The hybrid mission

A combination of both electrical and chemical propulsion is effective in reducing requirements on the launcher capability and increasing mission reliability. After earth departure the perihelion velocity (escape energy) can be increased by a SEP-stage which is separated at a solar distance at which it is not further usefull. From there the mission profile will be similar to that of the chemically propelled (aphelion) missions, but the rendez-vous can now take place at post-perihelion (2 to 3 AU). This last feature is important in minimizing the hazard to the spacecraft from cometary environment at perihelion. The most attractive mission opportunities exist when an earth gravity assist mode is incorporated in the trajectory (injection energies $C_3 = 10 - 25 km^2/s^2$ with a 25 km SEP). This mission scenario provides many mission opportunities and largest launch vehicle mass margins, making this option attractive also for the Ariane 5.

4   Mission opportunities

The number of possible mission opportunities within a given period depends on the injection energy, launcher capability, mission scenario and ultimately on the launch mass. Two major scenarios have been examined:

- a baseline mission to return core, crust and volatile samples and with on board a complement of highly desirable payload instrumentation. This mission is necessarily based on a landing scenario.
- a minimum mission based on the **harpoon scenario** and with instrumentation limited to support essential mission needs (i.e. imaging system).

Finally the study has focussed on the mission opportunities compatible with a 1998–2002 launch period but eliminating those targets which require a mission duration above 8 years or are associated with a high inclination orbit. These constraints reduced the choice of the suitable comets from about 90 short period comets to about 20. Mission opportunities have been determined taking into account three main launch vehicles, namely: Ariane 5, Titan-Centaur and the possible future family of heavy launchers. Spacecraft masses have been derived for a direct earth direct re-entry scenario and depend primarily on the trajectory type selected (low thrust, ballistic, hybrid). An overview of mission opportunities for the different possible launchers and trajectories is given in table 2 and figure 2.

Table 2: Mission opportunities.

POTENTIAL COMETS FOR APHELION ENCOUNTER

COMETS	LAUNCH DATE	NOMINAL		REDUCED WITH LANDING		HARPOONS	
		TITAN	A5	TITAN	A5	TITAN	A5
TRITTON	95/1/28						
HANEDA CAMPO 5	95/8/3						
NEUJMIN 2	96/1/24			X	X	X	X
HOWELL	97/5/28						
TEMPEL 1	98/7/11						
HONDA-MRKOS PAJDU	98/12/17						
SCHWA WACH 3	99/7/29					X	X
FINLAY	99/10/25			X		X	X
WIRTHANEN	2000/11/7			O		X	
CHURRYUMOV-GERASIM	2001/1/1	X	X	X	X	X	X
WILD2	2001/4/11						
DU TUIT HARTLEY	2001/5/19			X		X	X

POTENTIAL COMETS FOR POST-PERIHELION ENCOUNTER

COMETS	LAUNCH DATE	NOMINAL		HARPOONS	
		TITAN	A5	TITAN	A5
CHURYOMOV-GERASIM	1997/12/18	X		X	X
HARRINGTON	1998/7/30			X	X
KOHOUTEK	1998/11/28			X	X
NEUJMIN 2	1999/6/6			X	X
DUTOIT/N./D.	1999/6/9			X	
WILD 2	2000/1/16				
DU TOIT HARTLEY	2000/3/9	X	X	X	X
KOPFF	2000/6/25			X	X
TRITTON	2000/12/13	X	X	X	X
HANEDA CAMPOS	2001/8/16	X		X	X
TSUCHINAN 1	2001/12/5			X	X
TEMPEL 2	2002/7/8			X	X
HOWELL	2002/7/9			X	X

X Possible comets with a 20% mass margin

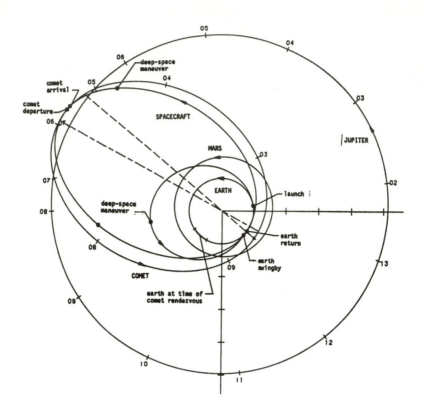

Figure 2: ΔVega ballistic mission to comet Churyomov-Gerasimenko

In conclusion it can be said that every mission type is feasible with future heavy launchers and with plenty of opportunities. As far as Titan 4 or Ariane 5 are concerned, sufficient opportunities exist for hybrid or SEP missions. Although the Ariane 5/Titan 4 class of launchers cannot support a "nominal" ballistic mission, it is still compatible with a reduced mission based on chemical propulsion and harpoons. Among the most interesting candidate targets are comet Churyomov-Gerasimenkov (Ra = 5,7 AU; Rp = 1,3 AU; i = 7,1°) and comet Tritton (Ra = 5,4 AU; Rp = 1,4 AU; i = 7°). A mission to these targets can be planned for December 2000, either using a heavy launcher and a ballistic trajectory with ΔVega (Figure 3), or using a Titan Centaur with a hybrid propulsion (Figure 4). The above missions are based on the landing and drilling scenario. A mission with harpoon to comet Churyomov-Gerasimenkov appears feasible with Ariane 5.

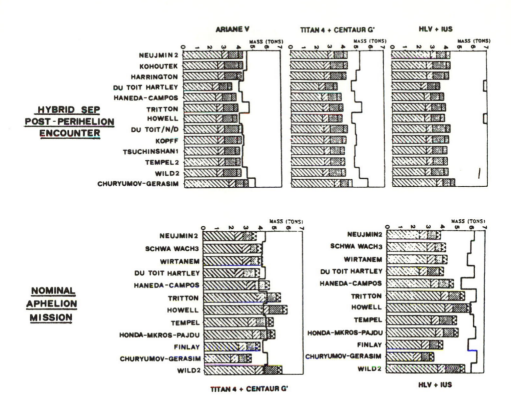

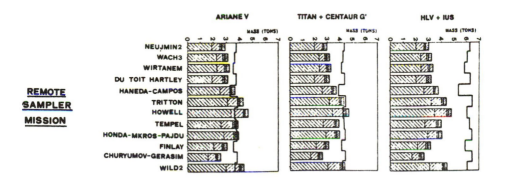

Figure 3: Launchers and scenarios.

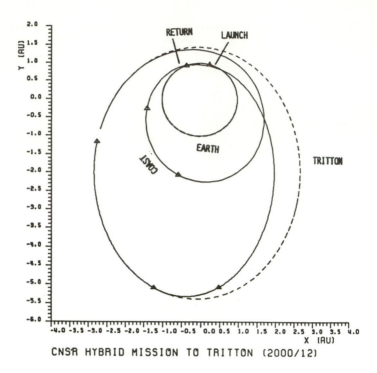

Figure 4: Hybrid mission to comet Tritton.

5   Mission scenarios

The mission is characterised by the following events and phases:
- launch and injection
- heliocentric earth-comet transfer
- near comet operations
- heliocentric comet-earth return
- sample retrieval

The essential aspects of these phases are summarised below.

5.1   Launch and injection

The preferred mode is a direct injection into Earth escape trajectory for all three propulsion options. The utilisation of a space station as a node for on-orbit assembly of the spacecraft and the upper stages or for upper stages refuelling is not essential, it imposes undue penalties, it may be very costly and there is no guarantee today that these capabilities of the station will be available in time for CNSR ($\sim$ year 2000). In the specific case of the SEP and the hybrid mission options, the launch vehicle injects the spacecraft into an Earth escape trajectory with a relatively small excess velocity ($\sim$2–4 km/s). For the high-thrust mission option, the excess velocity should be such that the resulting helio-centric trajectory has an aphelion radius about equal to that of the comet.

5.2 Heliocentric trans-comet phase

- For the electric propulsion missions, the SEP module is used to increase the heliocentric energy of the spacecraft so that its orbit matches that of the comet. This typically requires two long duration continuous thrust phases separated by a coast phase. For this mission option comet rendez-vous can only take place near perihelion because near aphelion insufficient power from the solar rays would preclude SEP operation.

- For the hybrid mission the SEP module is typically used during near Sun phases only ($\sim$ 3 AU) and it is then jettisoned.

- For the high thrust direct transfer the trans-comet heliocentric phase is basically an all ballistic phase apart from possible deep-space manoeuvres to change the inclination and refine the transfer trajectory.

The ΔVEGA trajectory mode requires injection into a heliocentric orbit with a period of approximately two years. Near aphelion of this orbit a manoeuvre changes the trajectory such that Earth reencounter takes place about two years after launch. At Earth encounter the gravity assist technique is then utilized to increase the heliocentric energy of the spacecraft. After Earth swingby the trajectory is identical to that of the direct mode.

5.3 Near comet operation phase

There are two basic scenarios and two possible strategies. The scenarios refer to the orbital locations selected for the comet rendez-vous (RDV), i.e., aphelion or perihelion; the strategies refer specifically to the possible methods to be used for collecting the sample. Two sample acquisition strategies have been considered, namely landing on the surface and drilling into it (nominal mission) or shooting an harpoon and collecting the impact sample by means of tethers. The near comet phase begins at about $2 \cdot 10^6$ km (worst case for aphelion RDV), when optical sightings of the comet nucleus become available in the spacecraft cameras. Final orbit corrections are made until space craft and comet are closed co-orbiting around the sun. An approach navigation phase will be implemented from the ground until the spacecraft reaches an altitude of a few thousands km, where the relative range measurements by the on-board radar altimeter become available and on-board autonomous guidance and navigation is enabled. The spacecraft is then controlled to drift until a station keeping point at about 10 nucleus radii is reached. This phase is characterised by an observation program, the objectives of which are primarily the identification of the nucleus kinematics and its geometrical and physical characteristics and to search for suitable landing sites. Fine mapping of suitable landing sites are obtained while coasting around the nucleus for the primary objective of determining its gravitational factor. The ground processing of these high resolution images will allow the selection of the landing site where two or three penetrators, equipped with RF beacons and special impact sensors, will be positioned. The touch-down error should not exceed few tens of meters, since these penetrators will provide the lander with navigation and landing beacons and possibly with essential information about the mechanical and physical properties of the nucleus upper-layer. This information can be used by the ground for deciding whether to proceed with the landing or to search for a new site.

Another scenario can be considered for the observation phase if the hazards around the comet are too important. This is the case especially for a perihelion encounter, for which the areas and altitudes where the environment is safe could be determined by a dust counter. From these locations, images of the nucleus can be taken, and the ground control can define the next operations. If landing is not possible, a back-up solution could be to use one or more harpoons to take the samples. This harpoon could be released at some hundreds meter above the nucleus surface and guided over a flat area; the impact velocity required to acquire the sample (which is function of the soil hardness) would be provided, after separation from the spacecraft, by a solid motor. A tether could be used to bring back the sample to the spacecraft.

5.4   Surface operation phase

In general the surface operations are autonomous, but the ground can modify and optimise the operation plans. Landing is based on autonomous navigation and guidance, with reference to the RF beacons positioned on the landing site. After deploying the landing legs and anchoring devices, the drill is positioned, the surface sampler is deployed and the samples are acquired and stored into the canister. The duration of the surface operations strongly depends on the ground hardness of the nucleus and under worst case conditions, i.e. rocky nucleus, the acquisition of 3 m core sample would last several hours. It should therefore take place in a night side region to avoid hazard from dust and gas jets. This last restriction applies specifically to a perihelion mission and affects the reliability of this scenario. A sampling scenario will typically consist of repetitive cycles, each optimised to collect a 50 cm x 10 cm sample, and consisting of the following operations:

– 50 cm depth drilling

– core bracking

– transport of cylinder with sample to the surface

– insertion of an additional extention cylinder into the hole and so on until the sixth core sample is brought to the surface.

These different core segments will allow to preserve a coarse order of stratigraphy, should the sample be of a porous structure of weak ice phases too fragile to survive the acceleration loads of the return trip to earth. The core segments can be stored by a robot arm into a sealed container, incorporating a pressure relief valve to prevent the risk of high pressure build-up and the alteration of less volatile solid phases which may be present in orther core sample segments. The volatile sample can be a small quantity (100 g) of core materials from the bottom of the drilling hole, where the most volatile rich material is expected. This will be stored in a strong, closed and unvented container. Finally a large quantity ($\sim$ 10 kg) of cometary dust and other non volatile material will be collected from the surface (mantle) of the nucleus, by a process of digging or scraping. Having completed the sampling and storage operations, a considerable part of the vehicle, called lander and sampling module, is not needed any further and will be left on the comet.

This module contains also the scientific payload and it would be highly desirable, resources permitting, to maintain this package in operation in order to monitor the comet evolution during its orbital period.

5.5   The sample return phase

This phase will last typically 3 years and will consist basically a ballistic trajectory to encounter the Earth. This is a quiet period and the major task to be accomplished is the perservation of the sample within the specified condition of temperature (< 160 K). This may require attitude manoeuvres to keep the sunshire out of the sample container radiator, which need to be deployed at the beginning of the return cruise. Under these conditions, it shall be possible to survive the Earth return phase until aerocapture.

5.6   Aerocapture

When approaching the Earth atmosphere, the Earth return vehicle separates from the aerocapsule and it will be injected in an escape orbit, to prevent the RTG entering the Earth atmosphere. The aerocapture (i.e., the aerodynamic braking of the hyperbolic excess velocity of the re-entry probe to a value below the local escape velocity) is the preferred sample recovery mode and its performance depends primarily on the lift coefficient L and drag coefficient D. The aerocapture manoeuvre might be performed in different ways, such as:

– single or multiple pass braking

– recovery to orbit or recovery to Earth

– uncontrolled or controlled atmospheric flight

– ballistic entry (L/D $=$ 0) or lifting entry (L/D$>$0).

The CNSR aerocapture simulation suggested to rule out:

– uncontrolled re-entry and recovery to orbit, due to very high sensitivity against variations of the entry angle and generally for the very small safety margin in the execution of the aerocapture manoeuvre.

– uncontrolled flight for recovery to ground due to low landing point accuracy and high structural and thermal loads

– controlled recovery to orbit due to the difficulties involved in the recovery operations, including large out-of-plane manoeuvring to allow recovery by an orbiting manoeuvring vehicle (OMV) from the space station. In addition the thermal requirements of the sample are not compatible with the time to be spent in a low Earth orbit.

As a conclusion, the baseline recovery mode for CNRS is a direct recovery to Earth, with a controlled re-entry and L/D$>$0. The aerocapsule design should have a L/D ratio of 0.3 to 0.5 to achieve good landing accuracy. Typical trajectory and acceleration profiles are shown in figure 5.

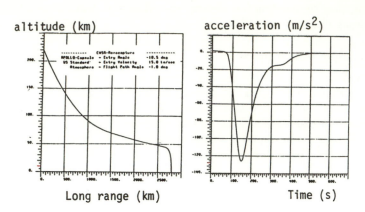

NOMINAL TRAJECTORY FOR APOLLO-TYPE CAPSULE

Figure 5: Aerocapture characteristics.

The aerobraking phase is very critical for the magnitude of heat input to the vehicle: this is largely governed by the density and velocity experienced along the trajectory and its dissipation shall not affect the cometary samples. The selected aerocapsule design uses an Apollo-type re-entry vehicle with a phase change material and high temperature resistant ablator materials as heat shield. This heat shield will be ejected as soon as possible in order to decrease heat-transfer to the inner structure. At this point a parachute will be deployed to allow a safe landing at a preselected site.

6   Spacecraft concepts

There are four major spacecraft configuration options. Two are related to the rendez-vous location; this demands a chemical propulsion for an aphelion RDV and electrical propulsion for a post-perihelion RDV. Two other options are related to the sampling strategy, which may be based on landing and drilling scenario or on the harpoon scenario. In any configuration there are basic tasks assigned to dedicated modules, such as:

– **the Earth return vehicle** ($\sim$ 400 kg - dry), which contains all conventional subsystems and controls all mission operations, except during the aerocapture phase. It is powered by an RTG and includes a chemical propulsion system for the return phase (300–500 kg of propellant).

– a large (2200 kg) **chemical FIRST stage** for the aphelion mission or

– an optional **electric propulsion stage** (2700 kg) for the hybrid post-perihelion mission. This configuration is largely dominated by a very large solar array ($\sim$ 250 m^2).

– a **lander/sampling facility** (600 kg) containing the payload and all instrumentation for the surface operation, and applicable to the baseline mission. The alternative sampling scenario does not make use of this facility, but includes instead

– one or more **harpoons** (100 kg each), specially shaped to operate in the wide range specified for the cometary nucleus (snow – soft rock)

– the **aerocapsule** (400 kg) containing the cometary samples, an autonomous guidance system, a special thermal protection and a parachute system.

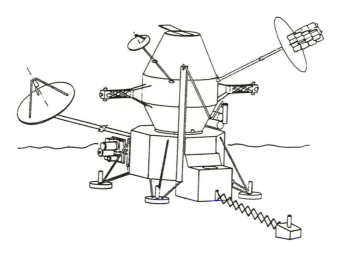

Figure: 6: Lander configuration on comet.

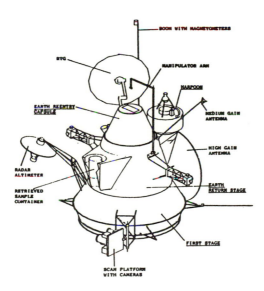

Figure 7: Harpoon configuration.

Some selected configuration concepts are shown in figure 6 (lander mission), figure 7 (harpoon mission), and figure 8 (aerocapsule).

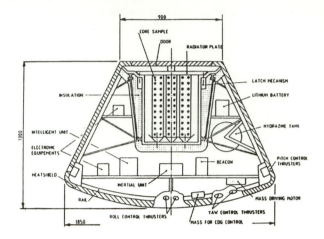

Figure 8: Aerocapsule.

7   Programme implementation

The major milestones and activities are outlined in table 3.

Table 3: Overall programme schedule.

ROSETTA/CNSR Implementation Programme

	87	88	89	90	91	92	93	94	95	96	97	98	99	00 08
ESA PREPARATORY PROGRAMME				•										
NASA STUDIES			•											
NASA/ESA COMMITMENT		•												
JOINT SYSTEM STUDY				•										
CORNERSTONE 03 SELECTION				•										
ANNOUNCEMENT OPPORTUNITY					•									
PHASE B					•		•							
PHASE C/D								•			•			
LAUNCH													•	
SAMPLE RETURN														•

The planning assumes a go ahead (major budget committment) at the end of 1991 (cornerstone N. 3 selection). A NASA/ESA Memorandum of Understanding for the Joint implementation is needed by that time. Under the above conditions, the short term essential feature of the Rosetta mission implementation is its preparatory programme. This preparatory programme, the content of which is shown in table 4, shall support the credibility and the feasibility in relation to the critical scenarios and technologies.

Table 4: The preparatory programme.

ROSETTA-CNSR PREPARATORY PROGRAMME

	86	87	88	89	90	91	92
1) SYSTEM STUDIES							
A) MISSION DEFINITION STUDY							
B) SYSTEM DEFINITION STUDY							
C) JOINT ESA/NASA STUDY							
2) SPECIAL ACTIVITIES							
A) MISSION ANALYSIS UPDATE							
B) GROUND SEGMENT DEFINITION							
C) AEROCAPTURE GUIDANCE AND NAVIGATION							
D) COMETARY MODELLING AND SIMULATION (*)							
3) TRP							
A) AUTONOMOUS AND ADVANCED NAVIGATION							
B) ELECTRIC PROPULSION							
C) SOLAR ARRAY CONCENTRATIONS							
D) SAMPLE ACQUISITION SYSTEMS							
E) ENTRY PROBE THERM. PROTECT. & DESCENT SYST.							
F) AUTONOMOUS DATA SYSTEM							
G) COMET APPROACH SYSTEMS							
H) TETHER DYNAMIC (*)							
I) TEST FLIGHT (*)							

It shall further develop the understanding and definition of the mission to a level where all cost drivers are sufficiently appreciated so that a realistic cost estimate can be made at the beginning of Phase B. This preparatory program includes studies and technology development and heavily relies on the Agency's Tecnology and Research Programme. Based on the present understanding of the baseline mission and the proposed scenarios, the key mission aspects which shall receive special consideration in the preparatory programme are primarily related to the operational aspects of the near comet phase (rendez-vous, landing, sampling) and of the Earth re-entry phase (guidance and thermal control). The autonomy aspects possibly required in the above phases is especially emphasised. Finally, the uncertainties regarding the availability of a suitable launch vehicle to support a ballistic mission with the desirable scientific return is a strong argument to increase and continue the electric propulsion programme.

8 Conclusions

There are several credible scenarios for designing, developing and accomplishing a Comet Nucleus Sample Return mission. All these scenarios rely on international cooperation, as an essential element to support the ambitious aims and significant cost of this mission. The techniques and technologies required are well achievable in the next decade and on time for the attractive mission opportunities available around the 2000–2001 launch period. This conclusion is based on the assumption that the proposed preparatory programme is adequately supported in its content and within the time frame indicated. It is anticipated that such a scientific and technological venture will stimulate new ideas, provide a ground for space application of advanced systems and in general will yield not only great scientific results but also significant technology benefits for industry. The system definition study to start in 1988 will reconsider all those factors (technical, financial and political) on which the mission feasibility depends. It is the present plan to define by 1990 the final scenario for which the Rosetta mission will be designed.

Acknowledgements

This mission definition study was supported by an ESA contract with Matra/Erno/-Fokker/Saab. The mission analysis aspects have been studied within ESA, by Messrs. J. Cornelisse and M. Hechler.

INITIAL COMET SIMULATION EXPERIMENTS AT DFVLR

E. Grün[1], H. Kochan[2], K. Roessler[3], D. Stöffler[4]
[1] *Max-Planck-Institut für Kernphysik, Heidelberg, F.R.G.*
[2] *Institut für Raumsimulation, DFVLR, Köln-Wahn, F.R.G.*
[3] *Institut für Chemie 1, KFA Jülich, F.R.G.*
[4] *Institut für Planetologie, Universität Münster, F.R.G.*

1 Introduction

The space missions to comet Halley have led to a better understanding of cometary physics and chemistry (for a review see e.g. Whipple, 1986). They produced new impulses to simulation and modelling of the nucleus and its behaviour in the solar radiation field. Since 1969 several attempts of laboratory simulation had been made, first by Russian groups in Leningrad and Dushanbe (Kajmakov and Sharkov, 1972; Dobrovolsky and Kajmakov, 1977; Dobrovolsky et al., 1986; Ibadinov and Aliev, 1987). Small samples of water ice with frozen volatiles, organic compounds, and dust particles on a cryostat finger in a vacuum cell were exposed to a strong light source. Evaporation rates, surface structures, albedo changes, physico-chemical and mechanical properties were measured. The results give a first insight into the behaviour of cometary ices, but are somewhat limited by the small size of the samples impeding an effective scaling up to the true cometary conditions. Furthermore, the lack of detailed analytical methods such as spectrometry or chromatography did not allow a detection of chemical changes induced by irradiation. Anyhow, destruction of ice samples by micro-explosions or larger outbursts and formation of a porous matrix were recognized already in an early stage of the experiments (Kajmakov and Sharkov, 1970). Further attempts to simulate cometary nucleus dynamics were undertaken by Delsemme and Wegner (1970) in snow samples. Sublimation experiments were carried out by Saunders et al. (1986), who prepared ice grains incorporating mineral components by spraying an emulsion of water with montmorillonite clay particles (size 0.2 μm) in weight ratios of 100:1 and 1000:1 into liquid nitrogen. After exposing the sample to visible light in a vacuum chamber, water ice component partially sublimed leaving fluffy filamentary sublimate residue of the mineral particles which did not exist in the starting material. The effect of dust mantles on sublimation of cometary ices has been theoretically studied by Brin and Mendis (1979), Fanale and Salvail (1984) and recently by Rickman and Fernandez (1986). It was shown that the observed sublimation history of comets can be understood by the quenching of the energy and gas flows by dust mantle which is only partially permeable for gases but highly thermally insulating. Shulman (1981) pointed out that such a mantle has not to be blown off by the gas during the comet's perihelion passage but can be retained in a fluidized

17

E. Bussoletti et al. (eds.), Experiments on Cosmic Dust Analogues, 17–23.
© *1988 by Kluwer Academic Publishers.*

state. Besides these investigations, which concentrated on the simulation of cometary nucleus materials, other studies on ice physics and chemistry provided valuable information. Smoluchowski (1981), Klinger (1981), Bar-Nun et al. (1985), their co-workers and many others, studied ice physics, heat conductivity, effect of phase transformations, decomposition of clathrate-hydrates, evolution of gas from mixtures, surface changes, thermal stress, etc. (c.f. Klinger et al., 1985). Greenberg (1982) investigated the behaviour of frozen gas mixtures under UV irradiation, in particular the chemical changes. Other investigations were devoted to the interaction with cosmic rays - especially the radiation chemical effects involved (Draganic et al., 1984) and with the interplanetary medium (Wykoff, 1983). The effect of solar wind or particle irradiation, and in particular high energy chemistry and sputtering of ice and frosts of volatiles was studied by various groups (c.f. Roessler et al., 1984, 1986; Johnson et al., 1983; Strazzulla et al., 1983). The ground truth for any comet simulation experiment is provided by the astronomical comet observations and the recent in-situ spacecraft measurements obtained by the fly bys of comet Halley. Television images of the nucleus (Sagdeev et al., 1987; Keller et al., 1986) showed that this is larger than previously assumed, mainly because its activity is limited to only a small portion ($\sim$ 10%) of its surface while the major part of the surface shows no emissions. The determination of the volume together with the mass of the nucleus, which was obtained from astronomical observations, allowed Rickman (1986) to estimate its mean density to about 0.2 g/cm^3. Direct measurements of size distribution of dust particles emitted from the nucleus showed that it ranges from at least 10^{-6} cm to 1 cm (McDonnell et al., 1987; Mazets et al., 1986). Measurements of chemical composition of cometary grains indicate that there are composed of silicates and refractory carbonaceous material in varying proportions (Kissel et al., 1986a, b). Many particles are rich in low-Z elements: hydrogen, carbon, nitrogen and oxygen (Kissel and Krüger, 1987). The average abundance of heavier elements is chondritic within a factor of two (Jessberger et al., 1986). The small sizes of many cometary grains and the presence of organic matter lends support to models emphasizing the interstellar connection of cometary materials. These tar-like substances probably cause the black color of cometary grains and of the nucleus itself. Although the smallest grains are by far the most numerous in the coma, they contribute little mass to the comet's total dust output - on the contrary most mass is recorded in the biggest particles. The total dust emission from comet Halley has been estimated by McDonnell et al. (1987) to be between 3 and 20 tons/s. In comparison about 20 tons of gas are emitted every second (Krankowsky et al., 1986). The gas is predominantly water ($\sim$80%)with admixtures of CO, N$_2$, NH$_3$, CH$_4$, CO$_2$ and other minor constituents (see also Eberhardt et al., 1987).

The new "post-Halley" approach to comet simulation presented in this paper originates from a large group of specialists in different fields connected to cometary research, ranging from comet modelling, ice physics and chemistry, mineralogy, to space simulation and space mission techniques, of which the four authors are only representatives. Initial experiments were performed in a small vacuum chamber and culminated in facility experiments in the space simulator chamber of DFVLR Cologne. These experiments shall lead to a better understanding of the Halley data, to assist the preparation of further cometary sample return missions and to provide knowledge of sample handling at low temperatures.

2 Simulation experiments in the small vacuum chamber

Preparatory experiments were performed in a small chamber which was equipped with a high pressure xenon lamp for simulation of solar radiation at an intensity of about 1 solar constant (1370 W/m^2). For a more detailed description of these experiments see Grün et al. (1987). Ice samples were produced by enjection of water-mineral suspensions into liquid nitrogen (c. f. Saunders et al., 1986). The starting material were, HD-clay, a mixture of kaolinite (80–90 Vol%), illite (5 Vol%), and montmorillonite (3–5 Vol%). Nearly all particles (90%) were smaller than 2 μm. The mineral/water ratio was 1/10 by mass. 1–2 percent carbon dispersed by some ethanol was added to this suspension. After installation of ice sample in the cooled sample holder and pump down of the vacuum chamber the radiation source was switched on. The two thermocouples at the surface and near the bottom of the sample showed rising temperatures. At a surface temperature of about $-30°$C the artificial comet became visibly active: dust ejection started. The front-window was totally bespattered in the anti"solar" direction by the dust ejected, an indication for the relativity high (some m/s) velocity of the dust particles. The temperature interval at which activity starts depends on the composition of the sample and the ratio of the sample size and the dimension of the surrounding cold shroud. Dust ejection mechanism does not consist of breaking up ice-clathrates which may have formed at the injection of the mineral suspension into liquid nitrogen. This was shown by an experiment in which sample material without minerals was irradiated. The sample proved to be inactive, no ejection was observed. The possibility of a trigger mechanism by the ethanol was excluded by an experiment with alcohol-free-sample matter. The model-comet was as active as before.

3 Experiments in the space simulator

To obtain some first experience on the operational behaviour of the big DFVLR space simulator during comet simulation experiments, a technological verification test was performed on 11–13 May 1987. A full description of the experimental results will be published elsewhere, here we give only a brief account of the major outcome. The space simulator consists of a horizontal cylinder 4.8 m long and 2.5 m in diameter (see figure 1). The inside is covered with a cold shroud connected to a liquid nitrogen system. The vacuum system has an effective pumping capacity of 3 times 15000 1/s. The solar simulator - 10 xenon lamps with 6.5 KW each and filtered radiation illuminates a reference plane of 1.30m diameter (in the middle of the chamber) with an intensity of nearly 1.25 solar constants. To receive a higher radiation intensity the sample can be moved closer to the entrance window or the cone of rays can be concentrated via special lenses. In these cases target diameter is confined to 0.8 m. In case of the technology verification test (see figure 2) a cylindrical target (TA), 30 cm in diameter and 15 cm in height was chosen. The artificial comet made out of HD-clay water suspension was tilted by 45° and was cooled from the backside by liquid nitrogen (BP).

Figure 1: DFVLR space simulator in open position. Solar radiator is left. Inside the chamber the cold shroud and in the center the sample container is visible.

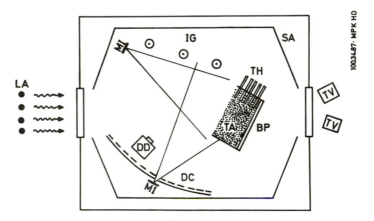

Figure 2: Schematics of technology verification test in the space simulator of DFVLR.
BP: liquid nitrogen cooled backplate, DC: dust collectors (16), DD: acoustic dust detector, IG: ion gauges (3), LA: xenon lamps (10) MI: mirrors (2), SA: liquid nitrogen cooled shroud, TA: target (water ice-minerals-carbon mixture), TH: thermistors (18), TV: television cameras (2).

The experimental setup is demonstrated in figure 2. Several thermistors (TH) measured temperatures in the target at various depths. Television cameras (TV) observed via mirrors (MI) the target and the volume in front of it. Dust collectors (DC) and a piezo-electric dust detector (DD, provided and operated by U. Weishaupt, Technical University Munich) were installed in order to observe dust emission from the target. Several ion gauges measured the pressure at different locations in the chamber. Simulation was run for approximately 13 hours at about 1.4 solar constants input. During that time, the surface of the target recessed by about 3 cm because of sublimation. Temperatures rose monotonically during irradiation. The temperature profile with depth became flat near the surface. Dust emission was observed directly on television screen and by the dust detector, which indicated that big particles ($\gtrsim$0.1mm in size) left the target. Emission speed was up to a few m/s. The gas stream leaving the target was observed by the pressure gradient in the chamber. In the later stages of the experiment a layer of condensed water on the cold shroud became visible above the target. After the simulation experiment the cold shroud was warmed up except one panel on which all the water in the chamber was collected. During all this time, the target was cooled in order to preserve its state. After opening of the chamber dust collectors were inspected. They were covered by layers of dust up to several mm thick. The target itself showed a several mm thick dry dust mantle underneath which a several cm thick ice crust was found. Only the deepest few cm of the target was still in the same condition as before simulation. Analysis of the dust showed that the majority of the particles were loose aggregates of the original micron sized particles with diameters of the order of 0.1 mm in size. The structure of the aggregates indicates that they formed already during the freezing of the droplets and not during sublimation. Several improvements of future simulation experiments have been suggested. Sample composition and preparation techniques will be varied. The sample will be characterized quantitatively by mass, density, structure, thermal and optical properties. Diagnostics of the gas and dust emission have to be further developped and most quantitative parameters will be more accurately calibrated.

Acknowledgements

The authors want to express their thanks to those who supported actively the implementation of the experiment, who contributed work, experience and hardwave to the experiments and to those who participated in the simulation experiments themselves and in the discussion of their outcome:
J. Benkhoff, A. Bischoff, H. Boenhardt, H. Fechtig, W. Feibig, B. Feuerbacher, H. Grill, F. Joo, H.J. Juraschek, H.U. Keller, J. Klinger, D. Krankowsky, P. Lämmerzahl, W. Littke, H.D. Masslow, Naehle, G. Neukum, W. Ostermann, P. Penkert, K.V. Rehmann, W. Schlosser, H.P. Schmidt, T. Schulz, G. Schwehm, W. Seeboldt, T. Spohn, W.H. Stöcker, K. Thiel, W. Weishaupt, R.H. Zerull and H. Zwettler. The Institut für Raumsimulation as a whole enthusiastically supported the simulation experiments.

References

Bar-Nun, A., Herman, G., Laufer, D., Rappaport, M.L.: 1985, Icarus **63**, 317–332.

Brin, G.D., Mendis, D.A.: 1979, Astrophys. J. **229**, 402–408.

Delsemme, A.H., Wegner, A.: 1970, Planet. Space Sci. **18**, 709–716.

Dobrovolsky, O.V., Kajmakov, E.: 1977, "Comets, Asteroids, Meteorites" Delsemme, A.H. ed. Toledo (Ohio), (University of Toledo) 37–46.

Dobrovolsky, O.V., Ibadinov, K.I., Aliev, S., Gerasimenko, S.I.: 1986, in "Exploration of Halley's Comet" ESA SP–250, 2, 389–394.

Draganic, I.C., Daganic, Z.D., Vujosevic, S.I.: 1984, Icarus, 60, 464–475.

Eberhardt, P., Krankowsky, D., Schulte, W., Dolder, U., Lämmerzahl, P., Berthelier, J.J., Woweries, J., Stubbemann, U., Hodges, R.R., Hoffman, J.H., Kliano, J.M.: 1987, Astron. Astrophys., in press.

Fanale, F.P., Salvail, J.R.: 1984, Icarus, 60, 476–511.

Greenberg, J.M.: 1982, "Comets", Wilkening, L.L., ed., (Tucson, Arizona, The University of Arizona Press), p. 131–163.

Grün, E., Kochan, H., Roessler, K.P., Stöffler, D.: 1987, "Comets", ESA SP–278, in press.

Ibadinov, K.I., Aliev, S.A.: 1987, in "Diversity and Similarity of Comets", ESA SP–278, in press.

Jessberger, E.K., Kissel, J., Fechtig, H., Krueger, F.R.: 1986, in "The Comet Nucleus Sample Return Mission", ESA SP–249, 27–30.

Johnson, R.E., Lanzerotti, L.J., Brown, W.L., Augustyniak, W.M., Mussil, C.: 1983, Astron. Astrophys, 123, 343–347.

Kajmakov, E.A., Sharkov, V.I.: 1972, in "The Motion, Evolution of Orbits, and Origin of Comets", Chebotarev et al., eds, (Reidel Publ. Comp., Dordrecht),p. 308–314.

Keller, H.U., Arpigny, C., Barbieri, C., Bonnet, R.M., Cazes, S., Coradini, M., Cosmovici, C.B., Delamere, W.A., Huebner, W.F., Hughes, D.W., Jamar, C., Malaise, D., Reitsema, H.J., Schmidt, H.U., Schmidt, W.K.H., Seige, P., Whipple, F.L., Wilhelm, K.: 1986, Nature, 321, 320–326.

Kissel, J., Sagdeev, R.Z., Bertaux, J.L., Angarov, V.N., Audouze, J., Blamont, J.E., Büchler, K., Evlanov, E.N., Fechtig, H., Fomenkova, M.N., von Hoerner, H., Inogamov, N.A., Khromov, V.N., Knabe, W., Krueger, F.R., Langevin, Y., Leonas, V.B., Levasseur-Regourd, A.C., Managadze, G.G., Podkolzin, S.N., Shapiro, V.D., Tabaldyev, S.R., Zubkov, B.V.: 1986a, Nature, 321, 280–282.

Kissel, J., Brownlee, D.E., Büchler, K., Clark, B.C., Fechtig, H., Grün, E., Hornung, K., Igenbergs, E.B., Jessberger, E.K., Krueger, F.R., Kuczera, H., McDonnell, J.A.M., Morfill, G.M., Rahe, J., Schwehm, G.H., Sekanina, Z., Utterback, N.G., Völk, H.J., Zook, H.A.: 1986b, Nature, 321, 336–337.

Kissel,J., Krueger, F.R.: 1987, Nature, 326 (6115) 755–760.

Klinger, J.: 1981, Icarus, 47, 320–324.

Klinger, J., Benest, D., Dollfus, A., Smoluchowski, R.(eds.): 1985, in "Ices in the Solar System", (Reidel Publ. Comp. Dordrecht).

Krankowsky, D., Lämmerzahl, P., Herrwerth, I., Woweries, J., Eberhardt, P., Dolder, U., Hermann, U., Schulte, W., Berthlier, J.J., Illiano, J.M., Hodges, R.R., Hoffmann, J.H.: 1986, Nature 321, 326–329.

Mazets, E.P., Sagdeev, R.Z., Aptekar, R.L., Golenetskii, S.V., Guryan Yu, A., Dyachkov, A.V., Ilyinskii, V.N., Panov, V.N., Petrov, G.G., Savvin, A.V., Sokolov, I.A., Frederiks, D.D., Khavenson, N.G., Shapiro, V.D., Shevchenko, V.I.: 1986, in "Exploration of Halley's Comet", ESA SP–250, 2, 3–10.

McDonnell, J.A.M., Alexander, W.M., Burton, W.M., Bussoletti, E., Evans, G.C., Evans, S.T., Firth, J.G., Grard, R.J.L., Green, S.F., Grün, E., Hanner, M.S., Hughes, D.W., Igenbergs, E., Kissel, J., Kuczera, H., Lindblad, B.A., Langevin, Y., Mandeville, J.C., Nappo, S., Pankiewicz, G.S.A., Perry, C.H., Schwehm, G.H., Sekanina, Z., Stevenson, T.J., Turner, R.F., Weishaupt, U., Wallis, M.K., Zarnecki, J.C.: 1987, Astron. Astrophys., in press.

Rickman, H.: 1986, The Comet Nucleus Sample Return Mission, ESA SP–249, 195–205.

Rickman, H., Fernandez, J.A.: 1986, The Comet Nucleus Sample Return Mission, ESA SP–249, 185–194.

Roessler, K., Jung, H.J., Nebeling, B.: 1984, Adv. Space Res. 4 (12), 83–95.

Roessler, K.: 1987, Rad. Effects **99**, 21–70.

Sagdeev, R.Z., SzabÂ, F., Avanesov, G.A., Cruvellier, P., SzabÂ, L., Szegö, K., Abergel, A., Balazs, A., Barinov, I.V., Bertaux, J.L., Blamont, J., Betaille, M., Dermarelis, E., Dul'nev, G.N., Endröczy, G., Gardos, M., Kanyo, M., Kostenko, V.I., Krasikov, V.A., Nguyen-Trong, T., Nyitrai, Z., Reny, I., Rusznyak, P., Shamis, V.A., Smith, B., Sukhanov, K.G., SzabÂ, F., Szalai, S., Tarnopolsky, V.I., Toth, I., Tsukanova, G., Valnicek, Varhalmi, L., Zaiko Yu, K., Zatsepin, S.I., Ziman, Ya, L., Zseinei, M., Zhukov, B.S.: 1986, Nature, **321**, 262–266.

Saunders, R.S., Fanale, F.P., Parker, T.J., Stephens, I.B., Sutton, S.: 1986, Icarus, **66** 94–104.

Smoluchowski, R.: 1981, Astrophys. J. **244**, L 31–34.

Strazzulla, G., Pironello, V., Foti, G.: 1983, Astron. Astrophys. **123**, 93–97.

Whipple, F.L.: 1986, in "Exploration of Halley's Comet", ESA SP–250, **2**, 281–288.

Wyckoff, S.: 1983, J. Phys. Chem. **87**, 4234–4242.

METHODS AND DIFFICULTIES IN LABORATORY STUDIES OF COSMIC DUST ANALOGUES

D.R. Huffman

Department of Physics, University of Arizona, Tucson, Arizona, U.S.A.

1   Introduction–Bulk-sample vs. small-particle approaches

The effort to understand the nature of cosmic dust, such as circumstellar, interstellar, and interplanetary dust, relies on the comparison of optical measurements such as scattering, extinction and emission with models ultimately derived from some sort of laboratory data. In this long-standing effort, there are two distinctly different approaches which I would like to outline in the beginning by reference to figure 1. First there is the approach in which the basic optical properties of matter, as summarized by wavelength-dependent optical constants, are used in small-particle calculations relating to the observations. This approach is diagrammed on the left side of figure 1. Although the starting point of this process is often the middle of the left column labeled $n(\lambda)$, $k(\lambda)$, which sometimes may be found in the published literature, someone had to do a lot of hard work preceding this point to furnish those optical constants. In such determinations, bulk samples of solids (or liquids) have to be prepared in slabs with polished surfaces, for example, and optical measurements such as trasmission and/or reflectance as functions of wavelength performed. From these measurements, an appropriate theory must be used such as Fresnel's equations for reflectance from a slab. Out of this process comes the wavelength-dependent "measured" optical constants. At this point it is possible to transform these bulk optical properties into small-particle optical effects by using the optical constants in an appropriate and doable small-particle theory such as Mie theory. The resulting extinction, absorption, and scattering parameters can the be compared with observed properties of cosmic dust, and parameters such as the size distribution and possibly the shape can be varied.

There are, of course, certain inherent problems in this approach. The optical constants may not be available for the materials desired, or the measurements may be suspect or have large uncertainities associated with them. The calculations may only be tractable for spherical particles or other ideal geometries such as infinitely long cylinders or spheroids, and real particles in nature are seldom no simple. Because of these and other difficulties, the dedicated experimentalist is apt to favor the direct approach of simply making particles of desired composition and directly measuring the absorption,

E. Bussoletti et al. (eds.), Experiments on Cosmic Dust Analogues, 25 42.
© *1988 by Kluwer Academic Publishers.*

scattering, or emission characteristics. This approach, which sounds quite appealing, is diagrammed on the right of figure 1.

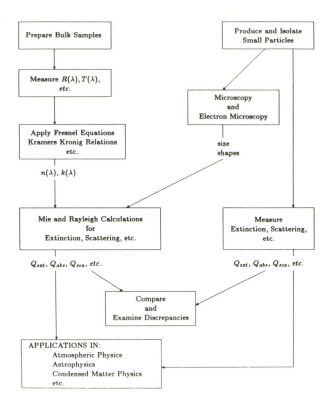

Figure 1: Overview of a program to relate measured and calculated optical properties of small particles.

The direct measurement approach is also not without serious difficulties. Life is never easy for the serious truth seeker. For example, the size of particles produced in the laboratory may not be the correct one to match a given remote measurement such as interstellar extinction. Here the first method is superior since it requires only that the worker change the size distribution on the computer input, and the calculation is easily modified. But what may have taken only a second or two on the computer may take a year or more in the laboratory if one is asked to produce a different size distribution. Certain techniques of laboratory smoke generation naturally generate specific size distributions which may be very difficult to alter. Another problem is that particles generated in the laboratory environment tend to aggregate very readly, giving rise to optical properties which, in some cases, differ markedly from the properties of isolated particles. If the interstellar particles are individual or simply in a different state of aggregation than the laboratory particles, a direct comparison may be faulty. In fact, this is the problem with a very large amount of published work on small-particle optics, and is the problem I want to discuss primarily in this paper. Before continuing this subject, however, I want to make a few more comments based on figure 1.

Please note that the bulk sample approach on the left and the small particle method on the right both arrive at the same optical properties of the small particles of interest. At this point, if both approaches were carried out, a comparison could be made which would prove very comforting for both sides if the results were favorable. In order to be able to make this comparison, it has been one of the goals of our laboratory work over the past 18 years to accomplish all of this program for a few selected materials. In fact, I have used an old diagram similar to this for quite a few years to illustrate our overall program, which in some sense is at least a lifetime's work. In cases where the comparison of the two methods has been fully accomplished, it is frequently true that THE TWO METHODS COMPARE VERY POORLY. In one sense this is unfortunate, since it does not generate great confidence in the applicability of either process to the understanding of remote cosmic dust. On the other hand, the lack of agreement is intellectually exciting since it implies that there are at least some parts of the process that we basically don't understand. As regards the modeling of interstellar optical properties, this suggests the following question. Isn't it important that we be able to model optical properties of pure, single component, small particles of known composition and size distribution before we get too confident in our ability to model the particles of interstellar or interplanetary space?

2 The dominant problem of particle clustering

I would now like to document the difficulties associated with the clustering of particles. First I will illustrate the problem based on work which is mostly unrelated to astronomy. The three examples of serious disagreement concern the infrared extinction of MgO and quartz in the infrared, the far infrared extinction of small metal particles, and the ultra-violet extinction of silver particles. As the large discrepancies are introduced, I will also put forth a very simple calculational approach which seems to permit reasonable calculations to be made for clustered particles. Following these non-astronomical examples, the ideas will be applied to a discussion of the cosmically important, solid carbon.

In general, the optical effects associated with clustered particles are greatest in spectral regions where the electromagnetic wave induces strong polarization in a particle, thus giving rise to a strong tendency for the polarization to induce a response in closely neighboring (or touching) particles. In order to treat these effects simply, we introduce the Rayleigh approximation for extinction per unit volume of material by a sphere in vacuum as

$$\alpha = \frac{6\pi}{\lambda} Im \left\{ \frac{\epsilon - 1}{\epsilon + 1} \right\}. \tag{1}$$

The complex dielectric function $\epsilon = \epsilon' + i\epsilon''$ is easily related to the complex index of refraction $N = n + ik$ by $\epsilon' = n^2 - k^2$ and $\epsilon'' = 2nk$. The approximation is valid in the limit of small size parameter, $x = 2\pi a/\lambda \ll 1$, and in th limit of zero phase shift in the particle ($|m| x \ll 1$), where m is the relative complex index of refraction of the sphere with respect to the host medium. It is quite obvious when written in this way that there is a maximum in the extinction efficiency for a sphere if ϵ' goes through -2, the height of the maximum depending on ϵ''. It had been known since the time of Gustav Mie (1908) that small metal particles in the visible and ultraviolet have strong extinction resonances which produced colors in metal colloids. Mie's solution of the general sphere problem was aimed at understanding the change of these colors with size. During the early 1970's, interest began to focus on the optical properties of small ionic crystal particles in the infrared, where the optical constants undergo strong variations giving rise to the $\epsilon' = -2$ condition. We now illustrate these infrared resonances with two examples from the infrared region.

2.1   MgO and quartz

When magnesium ribbon is burned in air and the resulting white smoke collected on an infrared transparent substrate, the measured extinction shows a broad band extending between about 400 and 700 cm^{-1}. Electron microscopy of the smoke reveals that it is composed of aggregates of sub-micron, cube-shaped particles grouped into loosely branching chains. The branching aggregation is characteristic of many smoke systems. Two among the many examples of photomicrographs for such aggregates published in the literature are found in Hunt et al. (1973) for NiO and in Stephens (1980) for carbon. An example of our comparison between measured extinction spectra of the MgO smoke and sphere calculations using bulk optical constants is given in figure 2.

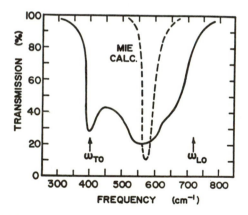

Figure 2: Measured infrared transmission spectrum of MgO smoke collected on a transparent substrate (solid curve, compared with calculations of the spectrum using bulk optical constants (from Huffman, 1977).

I choose this example of early work from my own files to avoid calling undue attention to what now appears to be an obvious deficiency in doing this kind of work, namely the use of arbitrary scaling of one curve to the other rather than determining and plotting mass-normalized or volume-normalized measurements. It is sometimes not easy to do the mass normalization accurately because the band strengths are so large that only very small quantities of the smoke should be used in making extinction measurements. Even in the arbitrary normalized spectra of figure 2, there is a large difference in widths of the band between calculated and measured spectra. In the early days, this discrepancy led to suggestions of quantum size effects and serious departures of particulate optical constants from bulk values. Later work in our laboratory, which incorporated careful mass determinations, led to the comparison of figure 3.

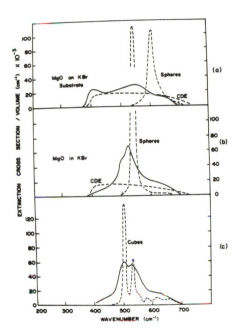

Figure 3: Measured infrared extinction by magnesium oxide smoke (solid curves) compared with calculations (dashed curves). The particles are progressively more dispersed in going from (a) to (c). From Bohren and Huffman, 1983.

The agreement was perhaps even poorer since now both the widths and the maximum peak heights were discrepant. Upon making a valiant effort to disrupt the aggregated chains by shiking the MgO sample together with infrared-transparent KBr powder before pressing into the usual KBr pellet used for the infrared transmission spectroscopy, the nature of the broad MgO absorption changed completely. In successive dispersal measured in times of minutes (first), then hours, then days, the spectrum slowly evolved into something that could be modeled by individual Rayleigh cubes using the theory of Fuchs (1975). All of this is discussed, along with a detailed treatment of such surface modes in small particles, in Chapter 12 of the monograph by Bohren and Huffman (1983). In this reference we also presented a derivation and discussion of a treatment of a wide distribution of shapes in the Rayleigh limit modeled as a continuous distribution of ellipsoidal shape parameters (CDE). By averaging over all orientations of an ellipsoid and then averaging over all possible ellipsoid shape parameters (assumed to be equally likely) the following simple expression emerged for volume-normalized absorption cross section:

$$\frac{\langle\langle C_{abs}\rangle\rangle}{\nu} = k Im \left\{ \frac{2\epsilon}{\epsilon - 1} Log\ \epsilon \right\} \tag{2}$$

In (2) Log z denotes the principal value of the logarithm of a complex number such as $z = re_{i\theta}$ and $k = 2\pi a/\lambda$. The results of this little calculation using MgO optical constants are shown in figure 4 as the dashed curve labeled CDE.

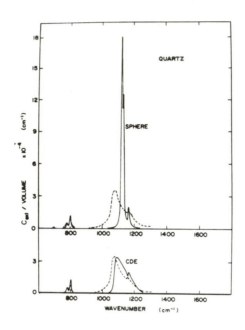

Figure 4: Measured extinction for crystalline quartz particles dispersed in KBr (dashed curves) compared with calculations for spheres (top) and for a continuous distribution of ellipsoidal shapes (bottom). From Bohren and Huffman, 1983.

Remembering that there are no arbitrary constants in this comparison, the agreement with the MgO smoke-on-substrate (top part of the figure) is rather good — much better than can be obtained with sphere calculations, also shown in the figure.

A second example of serious disagreement between measured small particle extinction and that calculated from Mie theory is for submicron particles of crystalline quartz (Huffman and Bohren, 1980). In this case the particles are irregular in shape with the sub-micron sizes having been selected by allowing larger particles to settle out before collecting and dispersing the small particles in KBr. The upper curve shows the comparison of measured extinction (dashed curve) with sphere calculations, and the lower comparison shows the distribution of ellipsoids calculations with the data. The agreement in the latter case is remarkably better, suggesting that shape effects in this case as well as quite important. Other examples of similar comparisons strongly favoring the

CDE calculation are for SiC (Bohren and Huffman, 1983, figure 12.15) and alpha Al_2O_3 (Rathmann, 1981). In all these cases it is surprising that such a naive treatment of shape effects including both irregular shaped primary particles and aggregates, is so effective, and we only partly understand why. Although we certainly claim no correspondence between any specific ellipsoid and some particular irregular shape or aggregated group, it appears that the effective *ensemble* of shape factors in a highly irregular collection of real particles is matched fairly well by the *ensemble* of shape factors in the distribution of all possible ellipsoids. It all seems to come out in the averaging process. At any rate, whether we understand it completely or not, perhaps it is worth using.

2.2 Far infrared absorption in metal particles

Perhaps the most widely discussed mystery regarding measured small particle optical properties during the past 10 years concerns the far infrared extinction of small metal particles. In a search for quantum effects on the optical properties of small metal particles it was found (Granqvist et al., 1976) that the measured extinction per unit mass of certain small metal particles in the far infrared was about 3 orders of magnitude greater than that calculated assuming spheres with bulk optical constants. This set off a small explosion of both experimental and theoretical work resulting in similar discrepancies of up to five orders of magnitudes from the work of the experimentalist and numerous theories published in Physical Review and other journals to explain the striking phenomenon. A review of the extensive literature can be found by starting with papers by Halperin (1986), Devaty and Sievers (1984), Ruppin (1979) and the Thesis of Rathmann (1981). In recent work from the laboratory of Prof. Sievers, where it all started, considerable effort has been taken to control the degree of aggregation of the particles, with the result that for the most well-dispersed Ag particles, the authors (Devaty and Sievers, 1984) report, "the far infrared absorption coefficient is enhanced by a factor of 100 at most, and the data is consistent with no enhancement whatsoever." In work done in our laboratory we have repeated the production and measurement of aluminum particles and have extended the far infrared measurements (which agree rather well with the original work) into the near infrared. These results from the Thesis of Rathmann (1981) are shown in figure 5 as the triangles with the circles for the original work. Comparison is made with Rayleigh calculations for spheres and for the continuous distribution of ellipsoids. For the particle sizes in the sample (as measured by transmission electron microscopy) there is essentially no difference between Mie calculations and the Rayleigh calculations. One clearly sees the three or more order-of-magnitude discrepancy at 100 μm and longer wavelength between measurements and calculations. Although the bump in the vicinity of 20 μm shows that there is oxide present on the particles as expected, the main features of the experimental results are much better explained by the CDE theory than by sphere theory, at least out to 200 μm wavelength. It is even more astonishing that the CDE works so well here since the effective elongations of metal ellipsoid required to contribute at 100 μm wavelength seem absurd. Nevertheless, our little one-line calculation of shape effects (Equation 2) in clustered particles seems to work rather well.

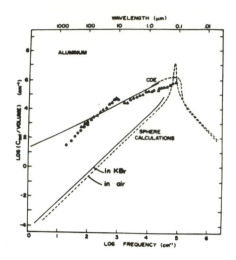

Figure 5: Measured extinction (circles from Rathmann and triangles from Granqvist et al.) for aluminum smoke particles compared with calculations for spheres and for a continuous distribution of ellipsoidal shapes (CDE). Details are found in Rathmann, 1981.

2.3 Silver

Our next example of the effect of clustering is based solely on experimental measurements and relates to clustered and unclustered particles of silver. In investigations of optical effects in small metal particles, silver has played a dominant role. The surface plasmon condition ($\epsilon' = -2$) for silver particles occurs near the visible region where spectroscopy is readily done. The extinction spectrum of small silver particles has been studied in photosensitive glasses (i.e., Stookey et al., 1978), as aggregated color centers in alkali halide crystals (i.e., Doyle, 1958), in low temperature solid argon matrices (i.e., Welker, 1978; Abe et al., 1980), and scattering spectra have been recorded for particles suspended in a flowing gas stream (Eversole and Broida, 1977). The references cited here are only representative of a considerable literature on silver particles. In most of these studies there appears to be excess broadening of the absorption peak due to particle aggregation.

In recent work we have set up a system to do both scattering and extinction as functions of wavelength for smokes of metals (including silver) produced by vaporizing the metal in an inert gas such as helium. By varying the position of the source boat with respect to the probing light beam, it is possible to sample the smoke cloud in the region where the particles are very small and have had little chance to cluster. One such experimental extinction run made on silver smoke particles is shown in figure 6. The main surface plasmon extinction peak shows up at about 360 nm, with a width similar to that expected for unclustered particles.

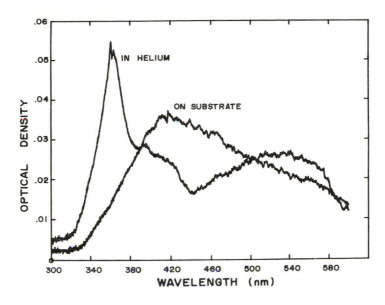

Figure 6: Comparison of extinction measurements on silver smoke made in situ in the helium gas with measurements on collected particles from the same region.

There is extra extinction in this particular run at around 400 nm and in the 500-600 nm region, which we attribute to agglomerated particles that have swirled around into other parts of the beam from earlier condensations. When care is taken to eliminate these swirls, a very "clean" 360 nm peak is obtained. I choose to show results from this particular run because, during it, we collected a sample on a transparent substrate by pushing the substrate momentarily into the smoke at the exact position where the optical beam was passing. Thus we have produced an experimental comparison of extinction from non-agglomerated particles measured in-situ, and identical smoke collected in an agglomerated way on the substrate. The differences are considerable but not unexpected (at least to those who have come to expect them). In this case there is an arbitrary scaling constant between the two curves, since we do not have accurate mass information on the in-situ smoke. Disregarding magnitude comparison, however, there is an obvious shift of the peak toward longer wavelength and a large increase in width for the collected particles. These results will be used as an intuitive guide in discussing carbon particles, on a substrate and in interstellar space, in a later section of this paper.

3 Silicate comparisons

Going almost back to the discovery of the 10-micron "silicate feature", olivine silicate, $(Mg,Fe)_2SiO_4$, has been the leading candidate to explain the band. Because of this interest, Steyer (1973) carefully measured optical constants in the infrared for all three directions in olivine, utilizing gem quality, single crystal samples that were X-ray oriented, cut, and polished. Polarized infrared light was then employed along the three axes to measure reflectance, from which the three sets of optical constants wer extracted. Based on Mie calculations from these optical constants, the extinction properties of olivine small particles could be accurately calculated. Unfortunately they caused an almost immediate loss of interest in crystalline olivine. The reason was that there occurred spectral structure in the 10-micron region that was very different from the featureless circumstellar band.

 This development prompted Krätschmer and Huffman (1975) to devise a way of producing a highly disordered layer of olivine silicate from the same single crystal sample that Steyer had used. We did this by subjecting the crystal to high doses of radiation from heavy ions produced in our Van de Graaff accelerator. Reflectance spectroscopy of the resulting isotropic layer was performed and analyzed to determine optical constants of the disordered silicate. When Mie calculations were done to determine extinction for this material, there was a much better match with the interstellar material.

 Following the amorphous olivine measurements, little interest was shown (as measured by the number of requests for the data) in Steyer's optical constants until very recently. Upon analysis of infrared measurements from Comet Halley dust, it was found that there was structure in the comet spectra — structure that is suggestive of the twin-peaked absorption feature in the spectrum of crystalline olivine particles. Hence in the continuing quest to find a connection between interstellar and interplanetary dust one might think of the interstellar dust as disordered silicates such as olivine grains which, after undergoing only very slight heating, might produce the type of infrared spectrum seen in Comet Halley.

 Since the spectra of both amorphous and crystalline olivine may now be of interest, we have dusted off the old results and present them again in figure 7 and 8 as further examples of the degree of agreement we have (or have not) achieved in matching measured extinction spectra with calculated extinction. Interestingly enough, olivine smoke can be produced in the disordered state and easily transformed to the crystalline state by gentle heating, so that small particles with the same size distribution can be made in both types. The agreement is not terrific, but it is also not all that bad compared to the problems we have shown to exist in figure 3 through 6.

4 The clustering problem in carbon

Carbon is a truly cosmic solid. It appears to be the dominant absorbing component of the earth's aerosol and, as such, it affects the earth's heat balance and the trend of global temperatures. Cosmically, carbon is by far the most abundant element forming refractory solids. The famous 2200 Å feature of the interstellar dust as long been thought by many to be due to graphitic carbon, and more recently variations of carbon have been invoked to explain unidentified absorption bands in the infrared and in the visible. Pure carbon occurs in several allotropic forms including the distinctly different crystalline forms of graphite and diamond.

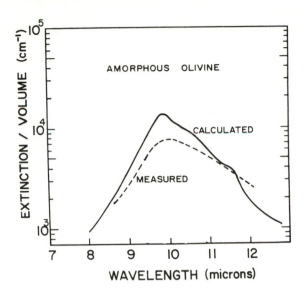

Figure 7: Measured and calculated extinction for amorphous olivine smoke.

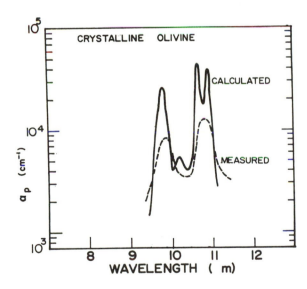

Figure 8: Measured and calculated extinction for crystalline olivine smoke.

Complicating the study of small-particle carbon is the fact that there is a continuous range in its physical properties and in its structure. Ian Spain, in his long and thorough review of the physical properties of various carbons (1981) says. "The electrical resistivity, ρ, of materials spans an enormous range of values, which is greater than that for any other material property... . Materials based on carbon have resistivity across this range." Spain further observes that, "carbon is unique in that the degree of disorder can be modified, and in some cases, controlled, over wide ranges." This variability of carbon has led to much confusion in comparisons among physical measurements done by different workers, and is likely to be an important one of the reasons why measured properties do not agree very well. An example is the optical constants of carbon, which often seems to boil down to a guessing game after looking at "measured" optical constants from various authors.

As regards cosmic dust, there are two main spectral regions upon which interest in carbon has been focused — the far infrared region and the region around 2200 Å. In the infrared not only magnitude of extinction (or absorption or emission) is of interest, but also the power law dependence on wavelength. Crystalline solids tend to give a λ^{-2} behaviour at long wavelengths, while various interstellar measurements seem to suggest a less steep falloff — perhaps λ^{-1}. The 2200 Å region is of interest because of the long-standing identification of the 2170 Å interstellar band with graphite carbon. We will discuss some laboratory measurements in each of these regions.

4.1 Far infrared absorption in carbon

Several workers have reported infrared extinction measurements for carbon particles. These include Blea et al. (1970). Borghesi et al. (1983,1985), Edoh (1983), Koike et al. (1980), and Tanabe et al. (1983). In addition, some (but not all) of these references compare the measurements with calculations based on some sort of optical constants. Draine (1985) has also calculated far infrared extinction based on his synthesis of optical constants from the published literature on graphite. When one transforms all these results to the same units and plots them on one graph there results a very large range of values. (I have made such a plot, but it is too confusing to include here). At 10 μm, the values for volume-normalized extinction range from about $1.5 \cdot 10^3$ cm^{-1} to $9 \cdot 10^4$ cm^{-1}. At 100 μm wavelength, values range from a calculated value of 30 cm^{-1} for graphite to $2.7 \cdot 10^3$ cm^{-1}. Clearly there are still problems remaining in knowing what the extinction properties of carbon are in the infrared. I think there are several reasons for the problems. One is the variability in the structural properties of carbon itself. That is, all measurements are not made on the same type of carbon. There is the remaining lack of optical constants for some type of carbon and the uncertainty in the optical constants for other carbons — graphite, in particular, for which the far infrared optical constants are highly questionable. There are also possible effects due to size distributions. If the particles are not strictly in the Rayleigh size limit, magnetic dipole terms can cause an appreciable increase of volume-normalized extinction over the size-independent value for Rayleigh spheres. Finally there is the problem of clustering, which afflicts almost, if not all, of the measurements to some extent.

Since the problem of clustering is the theme of this paper, we illustrate the magnitude of this problem in figure 9 with calculations over a broad wavelength range for carbon using the distribution of ellipsoid theory (Equation 2) and Rayleigh sphere theory (Equation 1). The optical constants chosen were those of glassy carbon measured by Edoh.

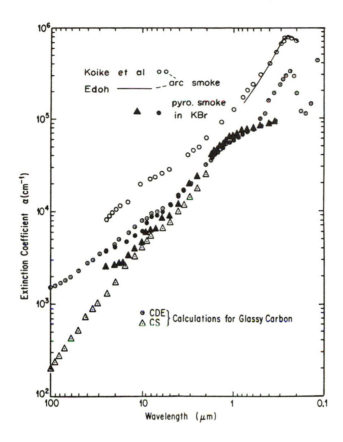

Figure 9: Comparison of various measured extinction values for carbon smoke with calculations based on optical constants for glassy carbon. Open circles are from Koike et al. (1980) on arc condensed smoke. Solid circles and triangles are from Edoh (1983) on smoke from acetylene pyrolysis, and the solid line is from Edoh's measurements on arc condensed smoke.

Glassy carbon was chosen over single crystal graphite because of evidence that small particles condensed in the laboratory and in interstellar space are somewhat disordered. Amorphous carbon was not considered because it does not show the characteristic UV band. We have also included measurements of arc-generated carbon smoke from Koike et al. (1980), which is fairly similar to the measurements that have been made in our laboratory. If one accepts that the simple CDE calculation gives an adequate indication of the effect of nonsphericity induced by particle clustering, it is clear from figure 9 that extinction for such carbon is fairly strongly affected by clustering. The magnitude of the difference becomes greater toward longer infrared wavelength, and the slope is significantly different. Upon comparing calculations with measurements one sees that most measurements result in values that are considerably higher than are produced by sphere calculations, using any reasonable optical constants that have been suggested for the carbons. Using graphite optical constants (as synthesized by Draine, 1985, for example), the differences in both slope and magnitude between individual and clustered particles, and between measurements and calculations, are even greater. In most cases, it appears that the CDE calculations are closer to the measurements in both magnitude and in slope. Our main point here is to alert laboratory workers to the possibilities of large effects due to clustering of carbon particles, with a plea for vigorous attempts to isolate the particles when one compares the measurements to single sphere calculations or to remote astronomical measurements on particles that are likely to be isolated.

4.2   The 2200 Å region in carbon

Very soon after the discovery of the 2200 Å interstellar feature, graphite particles were suggested as the cause, based on a close match of calculated extinction using measured optical constants. (The early history of both the observations and the graphite hypothesis are documented in the several review articles on interstellar dust). In 1973, Day and Huffman published direct extinction measurements for smoke particles condensed from a graphite electrode arc. A strong peak was found in the ultraviolet, although it was shifted somewhat to longer wavelengths than 2200 Å, and was broadened. Ther was also a strong upturn in the far UV for the smoke, and, in general, the direct measurements agred less well with the interstellar curve than the calculations from measured optical constants. Further work by Stephens (1980) on smoke made by vaporizing the bulk material with a focused laser beam, and further work from our lab (Huffman, 1975) confirmed the main features of the experimental work, but led to a discussion of whether the carbon was graphite, amorphous, or glassy. Calculations based on optical constants of glassy carbon seemed to give a better fit to the position and width of the measured extinction, and also gave more of a far UV upturn to match the measurements (see Huffman, 1977 for this discussion). Because of the lack of convergence between the calculate extinction for graphite (which could fit the interstellar curve) and the carbon smoke measurements (which couldn't), model makers have tended to ignore the direct measurements. This is in spite of the significant experimental fact that no additional component would be needed in interstellar dust models if carbon could produce both a 2200 Å type band and a far UV upturn. But what about the disagreement in shape and position of the carbon smoke measurements and the interstellar observations? In hopes of re-opening more discussion of carbon particles for both of these regions of the interstellar spectrum, I would like to draw from trends reported earlier in this paper that relate to possible clustering effects on particulate extinction spectra in the surface plasmon region, and to present new data on accurately mass-normalized extinction on carbon smoke.

Figure 10 shows the 2200 Å region, including both measured extinction for carbon particles condensed in helium gas from graphite electrodes, and calculations based on measured optical constants. The optical constants were taken from the widely used results of Taft and Philipp (1965) for electrical field E perpendicular to the unique (c) axis, and from Edoh's (1983) synthesis from several published measurements for $E\|c$.

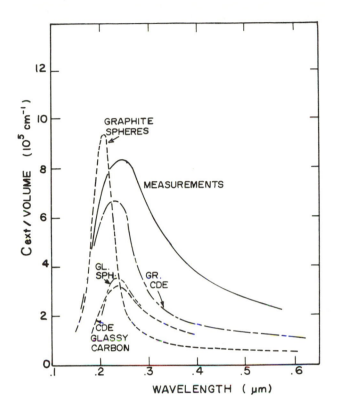

Figure 10: Comparison of measured extinction for arc-generated carbon smoke with calculations for graphite and for glassy carbon. For $E\perp c$ the optical constants of Taft and Philipp (1965) were used and for $E\|c$ those of Edoh (1983).

Glassy carbon measurements are from Arawaka et al. (1977), incorporated into Edoh's work. Only one calculated curve is shown for glassy carbon since spheres and distributions of ellipsoids are fairly similar. For graphite, appreciable differences are obvious betwen the two calculated curves. The experimental measurements have been volume normalized based on the careful mass measurements of the very small quantities of carbon used in this highly absorbing region. Details are given in Edoh's Thesis. The resuls agree quite well with the results for similarly produced carbon smoke of Koike et al. (1980; see also our figure 9). Some surprising observations follow from figure 10. With the absolute normalization used in the figure, it appears that the smoke extinction is much too large to be explained by carbon similar to glassy carbon. However, the strength is quite comparable with graphite calculations, particularly when one considers the likely clustering discussed in this paper. The graphite calculations show that clustering is expected to shift the peak toward longer wavelength and broaden it, while reducing the magnitude of peak extinction. The same trends have been shown for silver particles in our previously discussed in figure 6. Now glassy carbon has more of a short-range crystalline order than arc-evaporated films whose optical constants have been determined in this region by Duley (1984). This amorphous carbon shows no peak in this region and an even lower calculated extinction. The significance of the mass-normalized results therefore suggests that the carbon produced in the smoke state from graphite electrodes may be fairly crystalline, as measured by its optical properties in the UV region. Although it does not show well-defined X-ray diffraction patterns, it may be that tha plasma of pi electrons that resonates collectively to produce the 2200 Å band in small carbon particles may develop to almost graphite-like properties in regions that are too small to show single-crystal graphite-like X-ray diffraction.

Regardless of the truth of this speculation, the comparison of figure 10 show a very effective absorbing mechanism in small carbon grains. Further, if one reasons backwards from the measurements on the unavoidably clustered particles to what the extinction curve might be for the same particles when truly isolated, we suspect that the peak would be shortward of the measured 2400 Å, narrower than measured, and peaked at higher extinction values. The first two of these trends would bring the measurements into closer agreement with both interstellar measurements and with calculations for pure graphite. Until the isolated particle measurements can be made, these predictions are somewhat speculative, but we think they are, at least, well-supported speculation. The long-standing disagreement with calculations dating from the 1973 paper of Day and Huffman may be partly, if not completely, explained by clustering effects.

But what of the extreme disagreement in the far UV, where carbon smoke turns up very rapidly — more reminiscent of the interstellar curve than any calculations? In particular, what is the situation now that the results of figure 10 seem to argue against glassy carbon, which seemed to help explain the matter (Huffman, 1977)? We now think this is due to a certain amount of tetrahedral (diamond-like) bonding in the imperfect graphite structure. Complete tetrahedral arrangement would result in single crystal diamond, but imperfect or slightly disordered graphite will surely have some of this type of bonding between carbon atoms. As a qualitative guide to the optical changes expected, we point out that diamond shows extremely rapid absorption rise above about 7 eV (1770 Å). We therefore presume that some slight mixture of this nature could give rise to an additional far UV upturn added on to an almost graphite-like solid. We have prepared calculational models of mixtures of this kind which appear very promising for finally explaining all features of the UV extinction for graphite-vaporized smoke, and consequently interesting new possibilities for explaining both the 2200 Å and the far UV observations of interstellar dust.

5   Conclusions: some predictions and some suggestions regarding future experimental studies of cosmic dust

Two mayor difficulties have been discussed in this paper as regards comparison of direct laboratory measurements on small particles with astronomical observations. One is due to the fact that both silicates and carbon (especially) can occur in a complete range of crystalline order. In order to compare different measurements and to compare measurements with calculations, it is important that we experimentalists try to quantify and document the degree of disorder more carefully. In the case of carbon, an important tool seems to be Raman spectroscopy, which is quite sensitive to structural disorder in carbon. In regard to the clustering problem, we need to be aware of the discrepancies this can lead to and to try either to eliminate the problem by isolating the particles as individuals or find ways to properly calculate the optical properties of clustered particles.

Since I doubt that any of the experimental results published to date on carbon particles (including our own, of course) have eliminated the aggregation problem, I would like to speculate on what might be the case if and when the problem is successfully solved. I am sticking my neck out to predict the following:

1. The extinction spectrum for vapor condensed into particles from a graphite arc, which has previously shown a peak near 2400 Å, will show the band at shorter wavelengths (2200 Å?), narrower than before, with an extremely strong peak absorption (perhaps $1.2 \cdot 10^6$ cm^{-1}).

2. The optical constants appropriate to the above smoke will be fairly similar to graphite, but will contain a more rapidly rising imaginary part in the far UV due to an added contribution from tetrahedral bonding in the structure. The resulting Rayleigh peak for spheres might even occur near 2200 Å, which would solve the problem with graphite that a fairly special size distribution is required to fit interstellar observations.

3. In the infrared region I predict that isolated carbon particles will give a steeper dependence on wavelength than most measurements have shown to date. Therefore if a λ^{-1} law is the rule for the interstellar medium, carbon may not do the job. A more likely candidate might be amorphous silicates.

Acknowledgements

The author expresses appreciation for research support of this work to the Chemical System Research and Development Command of the U.S. Army at Aberdeen Proving Ground, Maryland, to the Naval Air Systems Command, and to the National Science Foundation. Many of the results have come from the work of former students Arlon Hunt, Terry Steyer, Janice Rathmann and Otto Edoh, and current students Joe Kurtz and Emmet Anderson, to whom I express my appreciation.

References

Abe, H., Schulze, W., Tesche, B.: 1980, Ber. Bunsenges. Phys. Chem. **88**, 95.

Arakawa, E.T., Williams, M.W., Inagaki, T.: 1977, Appl. Phys. **48**, 3176.

Blea, J.M., Parks, W.F., Ade, P.A.R., Bell, R.J.: 1970, J. Opt. Soc. Am. **60**, 603.

Bohren, C.F., Huffman, D.R.: 1983, "Absorption and Scattering of Light by Small Particles" (New York: Wiley).

Borghesi, A., Bussoletti, E., Colangeli, L., Minafra, A., Rubini, F.: 1983, Infrared Phys. **23**, 85.

Borghesi, A., Bussoletti, E., Colangeli, L.: 1985, Astron. Astrophys., **142**, 225.

Day, K.L., Huffman, D.R.: 1973, Nature Phys. Sci. **243**, 158.

Devaty, R.P., Sievers, A.J.: 1984, Phys. Rev. Lett., **52**, 1344.

Doyle, W.T.: 1958, Phys. Rev. **111**, 1067.

Draine, B.T.: 1985, Astrophys. J. Suppl. **57**, 587.

Duley, W.W.: 1984, Astrophys. J. **287**, 694.

Edoh, O.: 1983, Ph.D. Dissertation, Department of Physics, University of Arizona.

Eversole, J.D., Broida, H.: 1977, Phys. Rev. **B15**, 1644.

Fuchs, R.: 1975, Phys. Rev. **B11**, 1732.

Granqvist, C.G., Burhman, R.A., Wyns, J., Sievers, A.J.: 1976, Phys. Rev. Lett. **37** 625.

Halperin, W.P.: 1986, Rev. Mod. Phys. **58**, 533.

Huffman, D.R.: 1975, Astrophys. Space Sci. **34**, 175.

Huffman, D.R.: 1977, Adv. Phys. **26**, 1026.

Huffman, D.R., Bohren, C.F.: 1980, in "Light Scattering by Irregularity shaped particles", D. Scheuerman Ed. (New York: Plenum).

Hunt, A.J., Steyer, T.R., Huffman, D.R.: 1973, Surf. Sci. **36** 454.

Koike, C., Hasegawa, H., Manabe, A.: 1980, Astrophys. Space Sci. **67**, 495.

Mie, G.: 1908, Ann. Phys. **25**, 377.

Rathmann, J.: 1981, Ph.D. Thesis, Department of Physics, University of Arizona.

Ruppin, R.: 1979, Phys. Rev. **B19**, 1318.

Spain, I.L.: 1981, Carbon **16**, 119.

Stephens, J.R.: 1980, Astrophys. J. **237**, 450.

Steyer, T.R.: 1974, Ph.D. Thesis, Department of Physics, University of Arizona.

Stookey, S.D., Beall, G.H., Pierson, J.E.: 1978, J. Appl. Phys. **49**, 5114.

Taft, E.A., Philipp, H.R.: 1965, Phys. Rev. **A138**, 197.

Tanabe, T., Nakada, Y., Kamijo, F., Sakata, A.: 1983, Publ. Astron. Soc. Japan **35**, 397.

Welker, T.: 1978, Ber. Bunsenges. Phys. Chem. **82**, 40.

EXPERIMENTS ON THE FORMATION, PROPERTIES AND PROCESSING OF COSMIC DUST

B.D. Donn

NASA/Goddard Space Flight Center Laboratory for Extraterrestrial Physics, U.S.A.

1 Introduction

This report considers a variety of experiments that provide data on the formation, properties and processing of potential silicate components of cosmic dust. An earlier, similar report (Donn, 1985) included material covered at this meeting in sessions on ices and on carbon type matter. The attempt was to include all relevant processes for silicates not contained in the reviews of optical phenomena and grain irradiation. Many of the phenomena discussed here are applicable to all grain species. As consequence of the approach adopted here, this paper, is not a systematic and coherent treatment of a limited facet of experiments on grain analogs and does not generally treat the different technique in great depth. The references take care of this deficiency. In the silicate working group, particular aspects could be considered more critically and in greater detail.

2 Condensation and evaporation

One of the fundamental processes in studying grain formation, including core-mantle grains, is the evaporation and condensation mechanism. After a stable nucleus forms, subsequent growth depends upon the impact and sticking or vaporization of further refractory species. In the case of mantles, the additional species are volatile molecules, initially sticking to refractory, presumably silicate cores. After a few layers have been deposited, the process consists of evaporation and condensation of volatile molecules from or upon an ill defined mixture. We need to know the evaporation and condensation coefficient for the circumstances just described. Unfortunately, there do not seem to by any experimental measurements of these probabilities for astrophysically interesting systems involving refractories, except iron (Crittenden, 1950) and SiO on iron (Gunter, 1957).

Extensive discussions of vaporization and condensation (Rutner et al., 1964, Somerjai, 1969; Hirth and Pound, 1963; Frankl and Venable, 1970) and descriptions of condensation experiments (see e.g. Ruth et al., 1966; Sandejas and Hudson, 1968; Rosenblatt, 1976) shows that there are very complex phenomena. A full discussion is

43

E. Bussoletti et al. (eds.), Experiments on Cosmic Dust Analogues, 43–61.

out of place here. In view of the importance of these processes for the areas of grain growth and mantle formation in astrophysical settings a brief discussion is given.

First, it is noted that the condensation phenomena is very depended upon the incident flux or supersaturation.

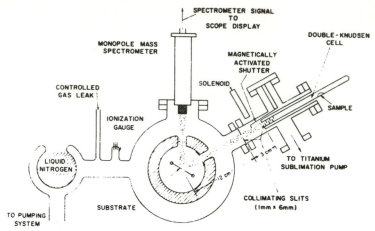

Figure 1: Experimental system to study condensation on surfaces. (Hudson and Sandejas, Vac. Sci. Tech. **4**, 230, 1967).

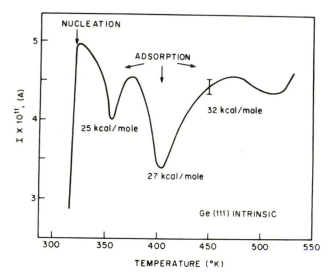

Figure 2: Adsorption and nucleation on Ge(111) surface. The error bar indicates the noise level. Mass spectrometer current plotted against target temperature starting from high temperature end. (Voorhoeve, et al., 1971), Vac. Sci. Tech. **9**, 780).

In a selected compilation of condensation coefficients, Pound (1972) tabulated values for a number of species on surfaces of the same composition. For all metals except silver the condensation coefficient is effectively unity. However, he has limited the table to experiments at high supersaturations. It has been known since the experiment of Wood (1915) that very large critical supersaturations $> 10^6$ are required for condensation. This critical supersaturation increases with temperature. At some sufficiently low temperature, depending upon the system, every impinging atom would stick. Condensation occurs preferentially at surface defects such as crystal boundaries and ledges or on pits or scratches. This last effect can be seen in the deposition of frost on windows, where a line of frost follows scratches or dirt. Impurity absorption generally increases the critical supersaturation although in some systems it enhances condensation.

The most precise and detailed studies of condensation involve the combination of molecule beams and mass-spectrometer in ultrahigh vacuum systems (Hudson 1969; Voorhoeve, 1974). This technique is illustated in figure 1 with some results shown in figure 2. Such experiments permit the determination of the lifetime of adsorbed species and of substrate atoms on itself, the critical temperature for nucleation as a function of beam flux, critical supersaturation, evaporation and condensation coefficient, heat and activation energy of evaporation and adsorption. Unfortunately, this experimental technique has been applied to very few systems.

In the experiment of Hudson and Sandejas (1967) for cadmium on clear tungsten the residual pressure was $5 \cdot 10^{-10}$ Torr. At this pressure it takes over 10^3s to form an absorbed monolayer on the surface. Under these conditions they obtained critical supersaturations under 2 compared to values between 10^6 and 10^{19} for the same combination under poorer vacuum. This illustrates the critical effect surface contamination can have.

For grain growth in circumstellar shells, the diffuse interstellar medium or mantle formation in clouds, the structure and composition of the surface and the nature of contamination are uncertain. The surface is certainly not clean, smooth and pure. In addition, we are dealing with chemical reactions producing some form of silicate or oxide layer. Experiments to study such complex gas surface interactions using the molecular beam-mass-spectrometer technique seem feasible. All the work referred to has been on bulk material. It is necessary to know how the complex vaporization and condensation process may be effected on sub-micron grains.

3 Nucleation

Nearly all nucleation experiments have been made using relatively volatile compounds (Abraham, 1977). A few, summarized in table 1 studied refractory species. In all nucleation experiments on high temperature refractory species the investigators report significant discrepancies between the experimental results and the predictions of classical nucleation theory. The Goddard experiments on nucleation (Nuth and Donn, 1982) used a bell-jar apparatus which is displayed in the poster paper by Nuth et al.. Details of the experiment are given in the references. Inside a water cooled steel bell-jar was a wire wound ceramic tube that served as a furnace. This, in combination with the heated crucible containing the SiO sample, determined the nucleation temperature. Chunks of SiO were vaporized into ambient hydrogen gas maintained at a presence of 20, 35 or 50 Torr. The SiO partial pressure was set by the crucible temperature. All temperatures were measured by appropriately placed thermocouples. The onset of nucleation was detected by extinction or scattering from the cloud which formed above the crucible.

Table 1: Summary of nucleation experiments involving refractory condensates.

Condensate	Temperature Range (K)	Method: Reactant in Background gas	Reference
Fe	$1800 < T < 2150$	Shock tube: $Fe(CO)_5$ in Ar	Kung and Bauer 1971
	$1600 < T < 1800$	Shock tube: $Fe(CO)_5$ in Ar	Freund and Bauer 1977
	$1625 < T < 2125$	Shock tube: $Fe(CO)_5$ in Ar	Frurip and Bauer 1977
	$1900 < T < 2400$	Shock tube: $Fe(CO)_5$ in Ar	Stephens and Bauer 1981
	$1000 < T < 1700$	Shock tube: $Fe(CO)_5$ in Ar	Steinwandel et al. 1981
Pb	$950 < T < 1225$	Shock tube: $Pb(CH_3)_4$ in Ar	Frurip and Bauer 1977
Bi	$755 < T < 1275$	Shock tube: $Bi(CH_3)_3$ in Ar	Frurip and Bauer 1977
Si	$1500 < T < 2800$	Shock tube: SiH_4 in Ar	Tabayashi and Bauer 1979
	$1550 < T < 2400$	Shock tube: SiH_4 in Ar	Stephens and Bauer 1981
	$1500 < T < 3000$	Shock tube: SiH_4 in Ar	Steinwandel and Bauer 1982
Yb	$500 < T < 625$	Gas evaporation: Yb in Ar	Onaka and Arnold 1981
Na	$350 < T < 400$	Gas evaporation: Na in Ar	Hecht 1979
Fe/Si	$1650 < T < 2400$	Shock tube: $Fe(CO)_5 + SiH_4$ in Ar	Stephens and Bauer 1981
FeO_x	$1950 < T < 2900$	Shock tube: $Fe(CO)_5 + N_2O$ in Ar	Stephens and Bauer 1981
SiO_x	$1250 < T < 4200$	Shock tube: $SiH_4 + N_2O(+H_2)$ in Ar	Stephens and Bauer 1981
	$750 < T < 1015$	Gas evaporation: SiO in H_2	Nuth and Donn 1982
$Mg_x SiO_y$	$750 < T < 1015$	Gas evaporation: SiO in $H_2 + Mg$	Nuth and Donn 1983b
$Fe_x SiO_y$	$1600 < T < 4000$	Shock tube: $Fe(CO)_5 + SiH_4 + N_2O$ in Ar $+CO_2 + H_2$	Stephens and Bauer 1983

Freund, H.J., Bauer, S.H.: 1977, J. Phys. Chem. **81**, 1994.

Frurip, D.J., Bauer, S.H.: 1977, J. Phys. Chem. **81**, 1001.

Hecht, J.: 1979, J. Appl. Phys. **50**, 7186.

Kung, R., Bauer, S.H.: 1971, in Proc. 8th International Shock Tube Symposium Chapman and Hall, London, paper 61.

Nuth, J., Donn, B.: 1982, J. Chem. Phys. **77**, 2369.

Nuth, J., Donn, B.: 1983, J. Chem. Phys. **78**, 1618.

Onaka, T., Arnold, J.: 1981, preprint.

Steinwandel, J., Dietz, Th., Hoeschele, J., Hauser, M.: 1982 invited paper, Flat Plate Solar Diode Arrary Workshop of J.P.L. Phoenix, Az, August 1982.

Steinwandel, J., Dietz, Th., Joos, V., Hauser, M.: 1981 Ber. Bunsenges. Phys. Chem. **85**, 686.

Stephens, J., Bauer, S.H.: 1981, Meteorities, **16**, 388.

Stephens, J., Bauer, S.H.: 1982, in Proc. 13th International Symposium on Shock Tubes and Waves, State University of New York Press, Albany, NY, p. 691.

Tabayshi, K., Bauer, S.H.: 1979 in Proc. 12th International Symposium on Shock Tubes and Waves, Magnes Press, Hebrew V., Jerusalem, p. 409.

In a following experiment a second crucible, containing magnesium, was placed next to the SiO crucible. Its temperature and therefore the magnesium partial pressure was set by the ambient temperature at the crucible.

Figure 3 shows the critical pressure of SiO to initiate nucleation of SiO and Mg-SiO. For temperatures below about 900 K, as expected, the SiO critical pressure is less when Mg was present as there were then two condensible species. For higher temperatures, the SiO pressure with Mg vapor present approached that for pure SiO nucleation and then followed the same curve.

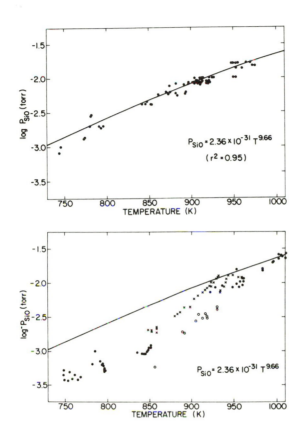

Figure 3: Partial pressure of SiO for nucleation vs ambient temperature. Upper figure, SiO-50 Torr H_2 mixture. Lower figure, SiO-Mg-50 Torr H_2 mixture. (Nuth and Donn, 1983, J. Chem. Phys. **78**, 1618).

This suggests that below 900 K Mg takes part in forming the critical nucleus but at higher temperatures only SiO is effective in creating the critical nucleus. In the experiments on grain formation to be described later, the resultant grain was the same in each case as far as we could tell. An equivalent result was obtained by Stephens and Bauer (1981) who studied the system Fe + SiO using a shock tube at higher temperatures. They found the critical SiO pressure did not depend upon the concentration of iron. Figure 4 plots the Goddard and the Stephens and Bauer results for SiO. A reasonable extrapolation of the low temperature data gives a fair agreement with the high temperature shock tube results.

   A study of the composition of pre-nucleation clusters could provide important data on the mechanism of nucleation for multielement systems.

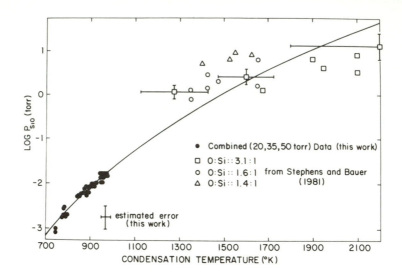

Figure 4: Comparison of critical pressure of SiO obtained in bell-jar system (Nuth and Donn) with that by Stephens and Bauer (Meteoritics, 1981, **16**, 338) in shock tube. The curve is an exponential regression best fit to the low temperature data extrapolated to the high temperature shock tube results. (Nuth and Donn, J. Chem. Phys. 1982, **77**, 2639).

Such experiments would be a definite test of the hypothesis that in metal + SiO system atoms do not contribute to the critical nucleus at temperatures above about 900 K. A system to carry out such experiments is presented in the poster by Nuth et al..

4   Grain information and properties

There has been a continual series of experiments designed to produce particles potentially representing cosmic grains. Perhaps the first of these efforts was the work of Caryel and Schatzman (1954). They dispersed colloidal graphite flakes in a cuvette filled with warm gelatin. The flakes were orientated in a magnetic field of 7000 gauss and the orientation was frozen-in as the gelatin cooled. The polarization was then measured. Further work on graphite grains was reported by Donn et al. (1968). High purity commercial graphite grains were dispersed in an oxygen carrier which flowed through a furnace ~ 1200 K. The oxygen attacked the graphite, reducing the grain size. The graphite-oxygen mixture then flowed into and was trapped in a multiple pass cell which permitted the wavelength dependence of extinction to be measured. The results are shown in figure 5 for two different grain size distributions.

   There are two points to be made. The first is that the structure in the extinction curve marked (a) for the smallest size grain distribution appears to be an artifact of the experiment. It appeared in later experiments in which the small size distribution could not be obtained as well as in experiments by others including measurements on sodium particles by Hecht (1981).

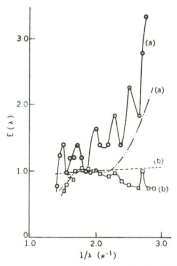

Figure 5: Extinction curves for graphite grains normalized at 540 nm. Solid curves represent experimental values. Broken curves are from Mie theory calculations using measured particle size distribution. (Donn, Hodge and Mentall, Ap. J. **154**, 135, 1968).

Thus, there is no basis for the conclusion in that paper that structure in the spectrum of small graphitic grains has been detected. The second point in the general agreement between the experimental curves and the extinction calculated by Mie theory for the measured size distribution. This indicates that the experimental procedure for obtaining small graphite grains destroyed the clustering almost universally present in smokes. The problem of grain clustering was emphasized by Huffman in his presentation at this workshop.

A different procedure was used by Lefevre (1970). He produced particulate clouds of the desired material by striking arcs between electrodes of an appropriate composition in argon or air. The extinction of the cloud was measured as a function of wavelength from 367 nm to 670 nm. A similar technique for particle production has been employed by Japanese investigators (Kamijo et al., 1975).

In a recent study (Tanabé et al., 1986 and this workshop) grains were produced by mixing several gases in a plasma jet. This technique, as that by Nuth et al. discussed later, requires using volatile compounds of refractory materials, e.g. SiH_4 to include silicon. The extinction of the cloud was again measured with a spectrophotometer to obtain wavelength dependence.

In the last three sets of experiments, the grains were collected and examined by electron microscopy to obtain shape and size distribution.

The bell-jar system of Nuth and Donn described in Section III (Nuth and Donn, 1982) and a similar apparatus (Day and Donn, 1978) was used to study the

condensation products from vaporized SiO. The cover of the furnace or the furnace wall
served to collect the particles which formed. In every instance the condensate was Si_2O_3.
This was unexpected as it requires disproportionation of the 8.3 eV SiO bond. Annealing
yielded amorphous quartz.

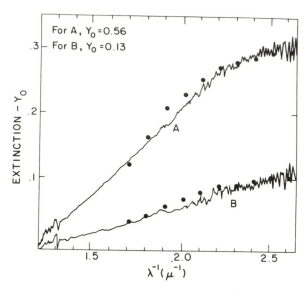

Figure 6: Extinction by iron grains. The lines are measured extinction, points are Mie
calculations for a size distribution giving best fit. Curve A is for steady state cloud of iron
particles, points are for size distribution from 65–100 nm. Curve B taken after particles
settled for 100 s, points are for distribution with four times the ratio of small to large
particles as for a. (Hecht and Nuth, 1982, Ap. J. **258**, 878).

Using an optical multi-channel analyzer which obtains the entire spectrum
at once on a narrow two-dimensional array detector, Hecht and Nuth (1982) obtained the
extinction of a cloud of small iron particles in the bell-jar apparatus of Nuth and Donn.
First they measured the extinction of the cloud in a steady state during nucleation. After
cutting off nucleation and waiting 100 s while larger grains preferentially settled, they
obtained a second extinction spectrum. These results appear in figure 6. Also shown
are the calculated curves for the size distribution givin the best fit to the measurements.
The distributionn matching curve B contains relativity two and a half times as many
small particle as that matching curve A.

In further experiments as described in the nucleation section (Day and Donn,
1978, Nuth and Donn, 1982) magnesium was added to the system. The product was
amorphous non-stoichiometric magnesium silicate grains. The infrared spectra and X-
ray diffraction patterns obtained during controlled annealing (Nuth and Donn, 1982) is
shown in figure 7. Note the shift in the absorption peaks in the spectra and the steadily
increasing development of structure in the 10 and 20 μm features. In every instance, the

completely annealed grains showed the characteristic spectra of olivene, Mg_2SiO_4.

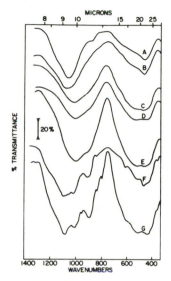

Figure 7a: Effect of annealing on infrared spectra of armorphous magnesium silicate smoke particles. A-G, annealing at 1000 K in vacuo for 0,1,2,4,8,16.5 and 30 hrs respectively.

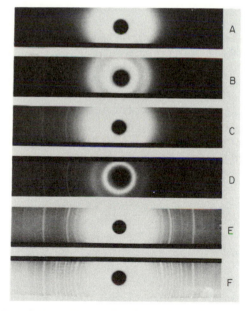

Figure 7b: Annealing effects on X-ray diffraction patterns of amorphous silicate smokes. A-E, annealed at 1000 K in vacuo for 0,1,16 1/2, 30, 167 hrs respectfully, F annealed 1250 K for 100 min. (Nuth and Donn, 1983, J. Geophys. Res. **88** Suppl. A 847).

The diffraction pattern in figure 7 for the unannealed grains gives no indication of crystallinity. However, a later, more refined analysis (Rietmeijer et al., 1986) showed a degree of crystallinity in X-ray and analytical electron microscopy examinations. It is not clear whether this occurs in the condensation or during an annealing process while the system is cooling before the grains can be removed.

In addition to the annealing experiments on magnesium silicate grains, at Goddard we carried out hydration experiments both in liquid water and water vapor. These are described in the Nuth at al. (this workshop).

A significant drawback of the bell-jar system is the occurrence of condensation before the vapor clouds get well mixed. This precludes the introduction of additional components into the grains. The plasma jet apparatus overcomes this deficiency. At Goddard a flow condensation system has been built for the same purpose. Here also, volatile compounds of refractory elements are used, silane SiH_4, iron carbonyl $Fe(CO)_5$, trimethyl aluminum $(CH_3)_3Al$, and titanium tricloride $TiCl_3$. These are mixed with oxygen and a hydrogen carrier in the high temperature mixing chamber before expanding into the condensation chamber where particles form and are collected. More details and early results are given in the paper by Nuth et al. at this workshop.

5   Optics of particles

The poster by Bliek and Lamy at the workshop describe another technique to measure scattering by a cloud of particles. A rather standard technique was used by Donn and Powell et al. (1967) to compare scattering by spheres with that by non-spheriral particles. In these investigations clouds of magnesium oxide cubes or zinc oxide fourlings (approximately, two v-shaped particles intersecting at 90°) were produced by vaporizing the metal in an argon-oxygen stream. Mie calculations for an equivalent size distribution of spheres gave a good fit to the cubic particles. However, with the fourlings, more large particles were needed but the Mie calculations could not give a satisfactory match to the angular or the wave length variation of scattering.

A number of experiments have studied scattering from isolated single particles. Such measurements yield the most definitive data as no clustering effects occur as discussed by Huffman at this workshop. Further, no size distribution washes out fine details of the scattering. On the other hand, astrophysical observations always include rather broad size distributions and integration of shape orientation for the expected non-spherical particles.

Single particle scattering experiments were discussed in some papers of these proceedings. Several techniques for isolating and trapping single particles have been reported. These include laser levitation (Ashkin, 1970; Ashkin and Dziedzic, 1971, 1979), quadrupole trap (Philip et al., 1983) and electrostatic levitation (Marx and Mulholland, 1983).

Much valuable information can be obtained by using cm microwaves and scaling particle size and wavelength from several tenths μm to the cm range. In this size range particles can be custom built to essentially any shape. It is necessary to use material that has the same index of refraction in the microwave region that the original material had at optical wavelengths. This is not always easily done.

There are two centers of microwave analogue research, the Space Astronomy Laboratory at the University of Florida (Schuerman, 1972; Greenberg and Gustufson, 1981) and Ruhr University in Bochum, FGR (Zerull, 1980). Same papers at this workshop include a description of the Bochum laboratory and recent results.

6 Experiments on clusters

Astrophysical settings generally have very different temperatures, pressures and time scales than do laboratory experiments. In order to apply the terrestrial data to astrophysics a knowledge of the mechanism of the process is needed to assure reliability or. make necessary modifications. In the case of grain formation from cosmic gas mixtures a very valuable step is to able to trace the growth of clusters which are the first stages of condensation. The proposal that classical nucleation theory or its modifications cannot handle multi-component systems (Donn, 1976) was based on the expected formation of metastable clusters that effect the ultimate grain composition. In recent years the studies of cluster has been an active field of research. This work is reviewed in a number of conference proceedings (e.g. Jena et al., 1986; Experiments on Clusters, 1984). These deal mainly with metal clusters. More recently, the reactivities of clusters have been studied (Guesir et al., 1985; Ricktmeier et al., 1985; Whetten et al., 1985; Martin, 1984a,b).

At Goddard we have been working on the cluster problem for several years. A number of cluster beam systems have been built and modified as problems were uncovered. A major difficulty has been the closing up of small nozzles or apertures by the refractory species being studies. The latest version seeks to avoid this problem by enabling larger openings to be used and by maintaining the opening at high temperatures. A schematic diagram of this system is given in the paper by Nuth et al. Two examples of the potential contribution of experiments on cluster to nucleation and condensation arose during the Goddard experiments. The vanishing of any contribution to silicate and metal nucleation by metal atoms at high temperatures can be readily investigated. The composition of clusters formed in a metal-SiO gas mixture will show if the hypothesis that Mg does not add to SiO cluster above 950 K is correct. Another observed result is that SiO vapor condenses to Si_2O_3 grains. In our apparatus we expect to be able to detect clusters up to about $(SiO)_{27}$. If the disproportionation to Si_2O_3 occurs before that size is reached we will be able to observe it. With this information it may be possible to delineate the molecular basis for the transition.

In addition to studying condensation from the monomer to the grain in the gas phase, this process may also investigated using matrix isolation techniques. These are reviewed by Moskovits and Ozin (1976) and examples are given in Gole and Stwalley (1982). Spectroscopy of low temperatures matrix isolated systems are given in Meyer (1971). Reviews emphasizing the study of molecules prepared at high temperatures are given by Weltner (1967, 1969).

The behavior of SiO in N_2 as the film was slowly warmed was studied by Khanna et al., (1981). Changes in the concentration of SiO, $(SiO)_2$ and $(SiO)_3$ were studied using the sharp vibrational bands of these species. No unambiguous evidence for larger SiO clusters was seen. When the N_2 vaporized at 50 K a non-volatile residue was left which had a infrared spectrum quite similar to that of the Si_2O_3 smoke. Similar results were obtained for films formed by the simultaneous deposition of Mg+SiO in a large excess of N_2, (Mg+SiO): $N_2 \approx 1:300$. Again the spectra of the residue was a good match to that for the amorphous Mg+SiO condensate (Donn et al., 1981). A comparison of the infrared spectra for SiO and Mg+SiO matrices, residues and smokes is displayed in figure 8.

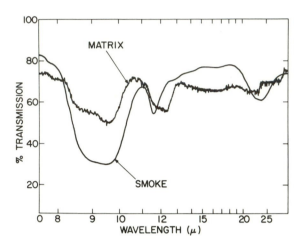

Figure 8a: Infrared spectrum of Si_2O_3 smoke and of annealed residue of SiO-N_2.

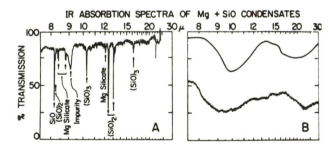

Figure 8b: A. Initial deposit of Mg+SiO in N_2 matrix at 12 K. B. Upper curve: infrared spectrum of amorphous magnesium silicate smoke; lower curve: spectrum of residue after sublimation of N_2 matrix (Donn et al., 1981, Surface Science **106**, 576).

7 Sputtering

A process that has become of increasing significance for understanding the evolution of cosmic grains in interplanetary (Wehner et al., 1983; Strazzulla, 1985, Rocard et al., 1986; Johnson and Lanzerotti, 1986) and interstellar space (deJong and Kamijo, 1973; Barlow, 1978; Seab and Shull, 1983, 1985) is sputtering. The primary experimental data has been sputtering yields, defined as atoms ejected per incident atom. There are two serious limitations to this data. Very little work has been done on targets of astrophysical interest and almost all has been on targets much thicker than dimension of cosmic grains which are essentially always under a micron. The lower limit to grain size would be about 10 nm. Possible effects of grain size have been included in some analyses, e.g., (deJong and Kamijo, 1973; Barlow, 1978; Draine and Salpeter, 1979a) but no detailed analyses were attempted. Experiments on sputtering of sub-micron grains are very desirable. Some theoretical calculations have been done by Onaka and Kamijo (1977, 1978) for small grains.

There appear to be several energy regimes of incident ions in space. For interplanetary particles bombarded by the solar wind, ion energies are 1 keV. Solar cosmic rays have energies in the range $1-10^3$ MeV (Brown et al., 1981; Chupp, 1984). Interstellar shocks produced by expanding supernovae ejecta can yield relative ion-grain energies up to or greater than 500 km^{-1} or 1.3 keV for protons. Low energy galactic cosmic rays with energies of 2 MeV have generally been adopted (deJong and Kamijo, 1973; Draine and Salpeter, 1979b) as the effective energy for interstellar effects.

The great majority of experiments have been on metal surfaces. A comprehensive summary of the present status of sputtering experiments and theory is given in Behrisch (1981, 1983). Extensive tables, diagrams and references are presented. Of considerable relevance for astrophysical problems is the chapter by Betz and Wehner (1983) on composite materials. A substantial number of oxides are included in their table 2.5 among which are Al_2O_3 and SiO_2.

Wehner et al. (1963) report measurements on a variety of rock samples but only quote the thickness loss relative to iron. This ranged from one to two times the iron loss. Rocks, which are composed of a variety of minerals, can have a considerable variation in composition. The dominant constituent is SiO_2 which ranged from 50% to 75% for the rock samples measured. Sputtering measurements on three meteorites, an iron and two chondrites, were carried out by Heymann and Fluit (1962). The rate in terms of mass per incident particle was about 50% greater for the iron. In terms of volume, the stones erode twice as fast, in general agreement with Wehner et al..

For multielement systems, sputtering generally causes an enrichment of one or more components in the surface, an SiO_2 surface being enriched in Si (Betz and Wehner, 1983).

Wehner et al. (1983) discuss the effect of surface roughness on sputtering yield and point out the difficulty of estimating this effect. They also show that courses irregularities tend to be smoothed out on a smaller scale, irregularities are introduced. Because of different sputtering rates from differently oriented crystal planes, a smooth surface develops microscope roughness. Experiments on amorphous silicates are required to determine what the surface of an interstellar grain would become under sputtering.

A question of considerable significance for sputtering of interstellar grains with dimensions of a few tenths μm is the effect of the small grain size, as already mentioned. For ^{3}He with energies from 20 to 60 keV impacting on carbon, Bottiger et al. (1978) found the probable range to vary from 30 to 70 μg cm^{-2}. These correspond to linear renges of 0.15 to 0.35 μm. Proton ranges would be a factor of ten greater (Bethe and Ashkin, 1953).

In a more detailed numerical simulation (Biersack and Haggmark, 1980) calculated the tracks of 10 keV ^{4}He$^+$ ions in niobium incident at a fixed point. The results, displayed in figure 9 show that they cover a volume approximately 70 nm deep and 100 nm wide. For protons impacting carbon or silicates, the ranges would again be about an order of magnitude greater according to the range-energy nomograph of Wilson (Bethe and Ashkin, 1953). The range-energy tables of Barkas and Berger (1964) show that in carbon and SiO$_2$ ranges are very similar. Here also, the paths cover an approximate radius up to an order of magnitude larger than the grain size. Hence, a high proportion of incident particles will escape from the grain with a large fraction of the initial energy. The energy loss will be enhanced by the escape of knock-on lattice particles, and by a distribution of incident particles over the surface rather than at one point.

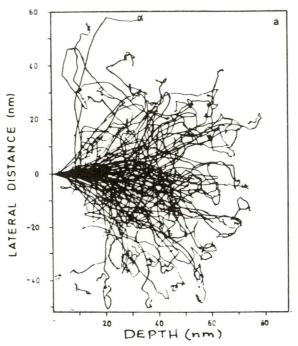

Figure 9: Numerical simulation of particle paths of keV ^{4}He$^+$−ions in Nb at normal incidence. Paths are projected on a plane parallel to direction of incidence. (Biersack and Haggmark, 1980 Nucl. Instr. Meth. **174**, 1194).

On the other hand, the escape of secondary particles increases the sputtering yield. The net result is not obvious and additional experiments or simulations for small grains are needed. An approach to such experiments would be sputtering from small islands which form on thin films during the first stage of condensation.

8   Temperature effects

Temperature appears to have only a small effect on the sputtering yield (Anderson and Bay, 1981; Nelson et al., 1962) although very few experiments have been done. Theoretically, the yield drops a factor of two from room temperature to the melting point (Nelson, 1962). Optical properties do have a significant dependence upon temperature although almost all measurements have been made at room temperature. For interstellar and circumstellar grains the temperature can vary from about 20 K to a maximum of at least 1000 K. Interplanetary and cometary grain temperatures fall within that interval.

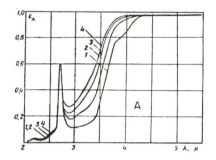

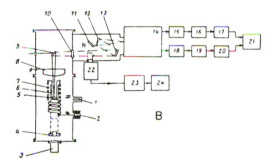

Figure 10: Temperature dependence of infrared emission by quartz. A. Diagram of experimental apparatus (3) pyrometer, (5) sample, (6) graphite oven and model blackbody, 10, CaF$_2$ window, (14) monochrometer, (15–21) detector-amplifiers-recorder.
B. Infrared emission of 6mm thick sample. (1) 677 K, (2) 1080 K, (3) 1465 K, (4) 1664 K (Dvurechenskii et al. High Temperatures, 16,641. [trans. Tep. Vysok.Temp. 16, 749]).

The emissivity of polished quart was measured by Dvurechenskii et al. (1979) from 673–1673 K. Figure 10 shows the apparatus and wave length dependence of emissivity both as a function of temperature and of sample thickness. Problems and methods of measurement of thermal properties from cryogenic temperatures to 3000 K are reported in NASA SP-31 (Richmond, 1962).

Two measurements of the temperature dependence of the infrared spectra of silicates have been carried. Donn et al. (1970) measured the 7–13 μm absorption of several silicate minerals. No significant difference was found between 300 K and 90 K for samples of 75–250 μm grains. Day (1976) examined the 8–25 μm spectra of 600 Å crystalline olivine (Mg_2SiO_4) spheres and larger, ground enstatite (Mg,Fe) SiO_3 grains. In each case, the three 20 μm absorption become much greater at 80 K compared to 300 K. However, in agreement with the earlier measurement, the 10 μm features were only slightly enhanced.

9 Particulate aggregates

A recent development of considerable significance for grain evolution has been experimental and theoretical investigations of the structure of particulate aggregates.

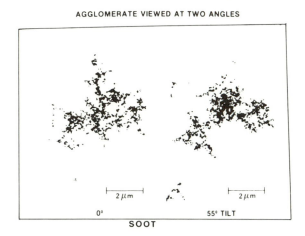

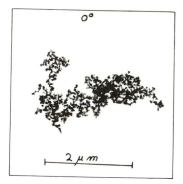 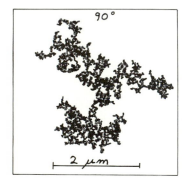

Figure 11: Fractal structure of particle aggregates. (Soot-courtesy of G.W. Mulholland; fractal simulation-courtesy of P. Meakin).

This work is continuing at a rapid pace. Examples of experimental investigations may be found in Forrest and Witten (1979), Weitz and Oliveria (1984), Martin et al., (1986) and Samson et al., (1987). Theoretical analyses based on numerical simulations of particle-cluster or cluster-cluster aggregation have been reported, among others, by: Meakin (1983; 1984a,b; 1985), Botet, Julian and Kolb (1984). Reviews of recent work on fractal aggregates are given by Hermann (1986) and Meakin (1987). A number of conference proceeding have appeared: Family and Laudau, (1984) and Pietronero and Tosatti (1986).

From the experimental investigations referred to one can conclude that with relative grain or cluster velocities less than several meters per second fractal aggregates will form. This is comparable to the velocity of a 10 nm radius aerosal particle in a air. The theoretical simulations support the fractal nature of the aggregates as is shown in figure 11 which compares the structure of a soot particle seen at two angles with perpendicular views of a computer simulation.

Experiments on the scattering and extinction of fractal aggregates using microwave techniques are being developed as a collaborative enterprise with P. Meakin, J. Stephens, B. Gustofson and R. Wang.

References

Abraham, F.F.: 1974, Homogeneous Nucleation Theory, Academic Press, N.Y.

Andersen, H.H., Bay, H.L.: 1981, in "Sputtering by Particle Bombardment" I. ed. R. Behrisch, Springer Verlag, Heidelberg, p. 145.

Ashkin, A.: 1970, Phys. Rev. Lett., **19**, 283.

Ashkin, A., Dziedzic, J.M.: 1971, Appl. Phys. Lett. **24**, 586.

Ashkin, A., Dziedzic, J.M.: 1979, in "Light Scattering by Irregular Particles", D.W. Schuerman, ed. Plenum Press. N.Y. p. 233.

Barkas, W.H., Berger, M.J.: 1964, "Tables of Energy Losses and Ranges of Heavy Charged Particles", NASA SP-3013.

Barlow, M.J.: 1978, Mon. Not. R.A.S. **183**, 357.

Behrisch, R.: 1981, 1983, ed., "Sputtering by Particle Bombardment" I (1981), II (1983).

Bethe, H., Ashkin, L.: 1953 in "Experimental Nuclear Physics" Vol. I, ed. E. Segre, ed. Wiley.

Betz, G., Wehner, G.K.: 1983, in "Sputtering by Particle Bombardment, II", R. Behrisch, ed. Sringer, Berlin p. 11.

Biersack, J.P., Haggmark, L.G.: 1980, Nucl. Instr. Meth **174**, 1194.

Botet, R., Julien, R., Kolb, M.: 1984, J. Phys, A **17**, L75.

Bottiger, J., Jensen, P.S., Littmark, V.: 1978, J. Appl. Phys. **49**, 965.

Brown, J.C., Smith, D.F., Spicer, D.S.: 1981 in "The Sun as a Star" ed. S. Jordan NASA SP-450, NASA, Washington, D.C. p. 181.

Cayrel, R., Schatzman, E.: 1954, Ann. d'Astr. Phys. **17**, 555.

Chupp, E.L.: 1984 in Ann. Rev. Astron. Astrophys.**22**, 359.

Crittenden, E.C., Jr: 1950 private communication.

Day, K.L.: 1976, Ap. J. Lett. **203**, L99.

Day, K.L., Donn, B.: 1978a Science **202**, 307.

Day, K.L., Donn, B.: 1978b Ap. J. **222**, L45.

deJong, T., Kamijo, F.: 1973, Astron. Astrophys. **25**, 363.

Donn, B., Hodge, R.C., Mentall, J.M.: 1968, Ap. J. **154**, 135.

Donn, B., Krishna Swamy, K.S., Hunter, C.: 1970 Ap. J. **160**, 353.

Draine, B., Salpeter, E.E.: 1979a, Ap. J. **231**, 77.

Draine, B.T., Salpeter, E.E.: 1979b Ap. J. **231**, 438.

Dvurechenskii, A.V., Petrov, V.A., Reznik, V.Yu.: 1978, High Temp. **16**, 641 (trans. Tep. Vysok. Temp. 16, 749).

Experiments on clusters, 1984, Ber. Bunsen-Gesellschaft Phys. Chemie. **88**.

Family, F., Landau, D.P.: 1984 "Kinetics of Aggregation and Gelation" Elsevier.

Forrest, S.R., Witten, T.A.: 1979, J. Phys. A. **172**, L109.

Frankl, D.R., Venable, J.A.: 1970 Adv. Phys. **19** 409.

Gole, J.L., Stwalley, W.C.: 1982 "Metal Bonding and Interactions in High Temperature Systems", Am Chem. Soc. Wash., D.C..

Guesir, M.E., Marse, M.D., Smalley, R.E.: 1985, J. Chem. Phys. **82**, 590.

Gunther, K.G.: 1957 Z. Phys. **149**, 538.

Hecht, J.R.: 1981, private communication.

Hecht, J.R., Nuth. J.A.: 1982, Ap. J. **258**, 878.

Herrmann, H.J.: 1986, Physics Reports, **136**, 153.

Heymann, D., Fluit, J.M.: 1962 J. Geophys. Res. **67**, 2921.

Hirth, J.P., Pound, C.M.: 1963 "Condensation and Evaporation", Pergamon Press.

Hudson, J.B.: 1970, Vac. Sci. Tech. **7**, 53.

Hudson, J.B., Sandejas, J.S.: 1967 J. Vac. Sci. Tech. **4**, 230.

Jena, P., Rao, B.K., Khanna, S.N.: 1986, "Physics and Chemistry of Small Clusters", Plenum Press, N.Y..

Johnson, R., Lanzerotti, L.J.: 1986 Icarus **66**, 619.

Kamijo, F., Nakada, Y., Igvchi, T., Fujimoto, M.K., Takada, M.: 1975 Icarus **26**, 102.

Khanna, R.K., Stranz, D.D., Donn, B.: 1981, J. Chem. Phys. **74**, 2108.

Lefevre, J.: 1970, Astron. Astroph. , 37.

Martin, J.E., Schaefer, D.W., Hurd, A.J.: 1986, Phys. Rev. A. **33**, 3540.

Martin, T.P.: 1984a, J. Chem. Phys. **80**, 170.

Martin, T.P.: 1984b, J. Chem. Phys. **81**, 4426.

Marx, E., Mulholland, G.W.: 1983, Nat. Bureau of Standards U.S., J. Res. **88**, 321.

Meakin, P.: 1983 Phys. Rev. a **27**, 604.

Meakin, P.: 1984a J. Coll. Interface Sci.**105**, 240.

Meakin, P.: 1984b Phys. Rev. B **29**, 997.

Meakin, P.: 1985 Phys. Rev. B **31**, 564.

Meakin, P.: 1987 CRC "Critical Reviews in Solid State and Material Science" **13**, 143.

Meyer, B.: 1971, "Low Temperature Chemistry", Elsevier, N.Y.

Moskovits, M., Ozin, G.A.: 1976, "Cryochemistry", John Wiley N.Y.

Nelson, R.S.: 1962, Phil. Mag. **7**, 515.

Nelson, r.s., Thompson, M.W., Montgomery, H.: 1962, Phil. Mag. **7**, 1385.

Nuth, J.A., Donn, B.: 1982 J. Chem. Phys. **77**, 2639.

Nuth, J.A., Donn, B.: 1983 J. Chem. Phys. **78**, 1618.

Onaka, T., Kajimo, F.: 1977, Jap. Jour. of Applied Phys. **16**, 359.

Onaka, T., Kajimo, F.: 1978, Astron.Astrophys. **64**, 53.

Phillip, D.Q., Wyatt, P.J., Bergman, R.M.: 1970, J. Coll. Interface Sci, **34**, 159.

Pietronero, L., Tosatti, E. 1986, Eds. "Fractals in Physics" North Holland, Amsterdam.

Pound, G.M.: 1972, Phys. Chem. Ref. Data **1**, 135.

Richmond, J.C.: 1962 "Measurement of Thermal Radiation Properties of Solids" Ed. J.C. Richmond NASA, Washington, D.C..

Ricktmeier, S.C., Parke, E.K., Liu, K., Pobo, L.G., Riley, S.J.: 1985, J. Chem. Phys. **82**, 3659.

Rietmeijer, F.J.M., Nuth, J.A., Mackinnon, I.A.: 1986, Icarus **66**, 211.

Rocard, F., Benit, J., Bibring, J-P., Ledu, D., Meunier, R.: 1986, Rad. Effects **99**, 97.

Rosenblatt, G.: 1975, J. Chem. Phys. **64**, 3942.

Rutner, E., Goldfinger, P., Hirth, J.P.: "Condensation and Evaporation of Solids", Gordon and Breach, N.Y. 1964.

Ruth, V., Moazed, K.L., Hirth, J.P.: 1966, J. Chem. Phys. **44**, 2093.

Samson, R.J., Mulholland, G.W., Centry, J.W.: 1987, Langmuir (Am. Chem. Soc.) **3**, 272.

Seab, C.G., Shull, J.M.: 1983, Ap. J. **275**, 652.

Schuerman, D.W.: 1979 in "Light Scattering by Irregularly Sloped Particles" D.W. Schuerman, ed. Plenum Press, N.Y. p. 227.

Somorjai, G.A.: 1969 in "Adv. High Temp. Chem" L. Eyring Acad. Press, N.Y. p. 203.

Stephens, J., Bauer, S.: 1981 Meteoritics **16**, 388.

Strazzulla, G.: 1985, Icarus **61**, 48.

Tanabé, T., Onaka, T., Kamijo, F., Sakata, A., Wada, S.: 1986 Jap. J. Appl. Phy. **25**, 1914.

Voorhoeve, R.J.H.: 1974, J. Cryst. Growth **23**, 177.

Wehner, G.K., Kennight, C., Rosenberg, D.L.: 1983, Planet. Sp. Sci. **11**, 885.

Weitz, D.A., Oliveria, M.: 1984, Phys. Rev. Lett. **52**, 1433.

Wiltner W.: 1967, Science **155**, 155.

Wiltner W.: 1969, Adv. in High Temp. Chem. **2**, 85.

Whitten, R.L., Cox, D.M., Trevar, D.J., Kaldor, A.: 1985, J. Phys. Chem. **89**, 566.

Wood, R.W.: 1915, Phil. Mag. **30**, 300.

Zerull, R.H., Giese, R.H., Schwill, S., Weiss, K.: 1980, in "Light scattering by irregularly shaped particles", D.W. Schuerman, ed., Plenum, press, NY, p. 273.

SOLID CARBON IN SPACE

E. Bussoletti[1], **L. Colangeli**[2], **V. Orofino**[3]

[1]*Istituto Universitario Navale, Napoli, Italy*
[2]*Space Department, ESA-ESTEC, Noordwijk, The Netherlands*
[3]*Dipartimento di Fisica, Università di Lecce, Lecce, Italy*

1 Introduction

Interstellar grains represent an important constituent of matter in space. They are necessary to the cooling and heating mechanisms and play an important role in the gas phase composition of molecular clouds. In addition, metallic components are important because they carry the charge inside dense clouds regulating the ion-molecule chemistry in these regions. Despite these facts and active research which is presently lasting since over 50 years, some major questions remain still controversial. According Mathis (1986a) it is crucial to note that the words "interstellar dust" refer to different materials as soon as we consider different space regions as the diffuse ISM, the outer edges of dense clouds, the dark central regions of these clouds.

Following Tielens and Allamandola (1986a) we can divide the dust components in two classes depending upon their formation history: a) star dust, b) dust formed in the ISM. In the first case silicates or different kinds of carbonaceous material are respectively produced in the oxigen-rich or carbon-rich outflow from late-type giants, planetary nebula, novae and supernovae. In the second case we face instead to icy mantles consisting of simple molecules (i.e. H_2O, NH_3, CH_3OH and CO) accreted on preexisting cores inside dense clouds, or to an organic refractory dust components, made of more complex molecules.

Good reviews on the subject are reported in many papers to which we refer (see, for instance, Willner 1984, Allamandola 1984, Bussoletti 1985, Mathis 1986a, Jura 1986, Draine 1986, Tielens & Allamandola 1986a,b). It is well known that about half of the volume of interstellar dust is made up of silicate grains while the other half is in different forms of carbon-bearing components. Aim of the paper is to discuss about this last subject showing particularly that graphite, a material usually invoked to account for many astronomical abservations, seems highly unlike to exist due to several physical and chemical reasons. On the contrary carbon is expected to exist in space under many different forms with some sort of continuously varying structure, from an ordered to a disordered form.

63

E. Bussoletti et al. (eds.), Experiments on Cosmic Dust Analogues, 63–73.
© 1988 by Kluwer Academic Publishers.

2   The extinction curve and related models

The extinction curve represents one of the main tools which may give indications about the actual nature of interstellar grains. Observationally the curve can be separated into four different portions:

a) far UV 1000–2000 Å
b) ultraviolet 2000–4000 Å
c) visible 4000–7000 Å
d) the hump region centered around 2175 Å.

Recent IUE results have definitely confirmed that the definition of an "average interstellar curve" as it was usually accepted, is no longer significant. Though the peak wavelength remain stable at 2175 Å the width and the strength of the bump may vary from star to star, as well as its general spectral trend. These variations are a first indirect indication of a multicomponent grain mixture in ISM because a different interaction with stellar ligth.

Greenberg and Chlewicki (1983) found some correlation between the different portions of the extinction curve by analizing IUE data:

a) a strong correlation is seen between the bump intensity and visible extiction
b) a weak correlation exists between the FUV extinction and the bump or visible intensity.

Carnochan (1986) has compared the UV extinction towards 154 "normal" stars (82/68 UV Sky Survey) studying three parameters: 1) E(2740 Å), 2) the UV continuum, 3) the hump strength. He found a linear correlation between E(2740 Å) and the UV slope.

On the basis of the observations and of the above mentioned correlation, different models have been produced: Mathis et al. (1977) and subsequent papers; Greenberg and Chlewicki (1983) and subsequent papers; Draine and Lee (1984), Hecht (1986). All these models suggest the existence of a mixture of graphite and silicates in addition, in some case, with other less important components. We note here that, though some sort of amorphous silicate is commonly accepted, less clear is the nature of carbon material. Usually graphite, as a major constituent, is considered unavoidable because, apparently, no other carbonaceous material seems to account for the observed extinction hump within cosmic abundance constraints of carbon. Only very recently Duley et al. (1987) have proposed a different model in which interstellar grains are composed by hydrogenated amorphous carbon-coated silicate particles. In this case the small silicate cores give rise to the 2200 Å extinction feature.

3   The existence of graphite in space

Among the various papers appeared in the literature, the review by Czyzak et al. (1982) represents, in our opinion, the most complete one about the different forms of carbon potentially present in space. According to these authors, to whom we refer, we point out that the usual arguments presented in favour of the existence of graphite are, at best, highly speculative and are sometimes lacking real scientific justification.

Actually we have to keep in mind that the kinds of carbon grains which we expect to find depend critically upon their formation conditions as well as their history once ejected into the ISM. In principle we could have single crystals, polycrystalline aggregates, carbon blacks and pyrolytic carbons of varying degrees. We recall here that graphite is made up of a single crystal with a stable exagonal lattice composed by a system of layers formed by close packed having a C-C bond distance of about 1.4 Å and an interplanar separation of about 3.35 Å.

Amorphous carbon is a material which presents a lack of order for distances of about 15 Å. Pyrolitic carbons are formed by a mixture of crystallites and amorphous material while carbon blancks are essentially amorphous. Finally, polycrystalline materials may be instead defined as a "mosaic" of single crystals oriented in different directions. Graphite is obtained following a two-step process: it starts from the condensation of disordered particles which attain an ordered lattice structure when treated by proper heat process, at temperatures of about 10^3–10^4 K and high pressures applied for long time lags. It is important to point out that both for natural and sinthesized graphite, due to their formation processes, the material occurs in form of flakes. This in contrast with the usual assumption, made in many astrophysical simulations, of spherical grains. This last form is however found only for graphitized carbon blacks (Czyzak & Santiago 1973) where, in any case, the long range order is no longer saved.

A second critical point against graphite formation in space is the use of classical nucleation theory. It actually does not take into account the quantum statistical contribution to the free energy of formation, that is the free energy of the translational, vibrational and rotational partition functions. In addition, the inaccuracy about the value of the free energy surface can be as large as two orders of magnitude. Actually, according to Tabak et al. (1975), who have taken into account the different contributions to the heterogeneous nucleation, we note that the surface energy of particles and the supersaturation ratio existing in the stellar atmophere are crucial for the process itself.

A third fundamental point is that the nucleation temperature and pressure are not the only significant parameters necessary to determine the character of the carbon material which is produced. Actually, the kinetics of grain formation plays, in addition, a not negligible role. When one takes into consideration the temperatures and the processes needed for nucleation and graphite formation in stellar atmospheres, the actual physical conditions appear quite unfavourable to this kind of particles. Single crystals do not have much chance to form due to the high sublimation rate expected in these regions of the sky. On the contrary, laboratory experiments indicate that polycrystalline grains must be formed (Czyzak et al., 1982 and references quoted in). In addition, we have to remind also that the effects of radiation damage on crystalline structure are expected to be so large that, whenever graphite is formed, it would suffer a massive damage which will render difficult its survival.

Finally, let us remember that the absorption peak for bulk graphite has been found at about 2600 Å and, as long as the size of grains remains within the bulk optical properties, the absorption peak does not change significantly. Spherical grain composed by a material having a dielectric tensor similar to that measured for crystalline graphite have been found to match acceptably the shape and the peak position of the interstellar hump (Mathis et al., 1977, Draine and Lee 1984). On the other hand, these results remain questionable for the following reasons:

a) there still exists uncertainty in the dielectric function of bulk graphite,

b) the peak position and band profile do vary according to the particle dimensions and size distributions,

c) spherical grains are considered despite experimental evidence,

d) Mie theory application remain doubtful for submicron grains (Bohren & Huffman 1983).

4   Laboratory carbons

The interpretation of observations or theoretical analyses of interstellar phenomena requires the knowledge of relevant data which, many times, can only be obtained by experiments carried out under appropriate conditions. At present, there exist many good measurements performed for other purposes which can however be used for astrophysical problems.

A good review on the subject has been recently published by Donn (1986), who, first, pointed out the need of caution and good judgement when applying data obtained for terrestrial purposes to an astrophysical situation. Actually space local conditions differ greatly from laboratory conditions and therefore it is crucial to evaluate the reliability of the data for the specific problems. The spectroscopic and optical properties of many materials have been summarized in the above mentioned paper to which we refer for further details. Table 1 reports an updated summary of some of the most relevant papers appeared in the literature concerning graphite, amorphous carbon and silicon carbide. This list is not intended as a comprehensive review of experimental results or techniques but rather it is intended to serve as a guide to some sources of data.

Despite their interest, most of the laboratory data present some negative characteristic:

a) very seldom clear information about the experimental parameters and conditions are given so that, usually, the experiments cannot be duplicated;

b) no one has extendend the analysis of a specific sample over the entire wavelength range 1200 Å-1 mm.

On the other hand, the theoreticians have usually utilized experimental data without an "a priori" evaluation of the internal consistency of the different data sets used in their simultations. To overcome the above mentioned problems the Cosmic Physics Group of the University of Lecce has performed in the last five years a systematic analysis of different carbonaceous materials such as amorphous carbon, PAHs and SiC.

The experimental details of such analysis are reported in the papers mentioned in table 1 and will not reported here.

Table 1: Experiments on carbonaceous and silicon carbide grains.

Graphite	**Reference**
Polarization by aligned grains	Cayrel and Schatzman (1954)
Normalized extinction coeff. 350–650 nm	Donn et al. (1968)
(a < 0.2 μm)	
Mass extinction coeff. 25–250 μm	Tanabé et al. (1983)

Amorphous carbon	**Reference**
Size and shape of condensate from vapor	Lefevre (1967)
Extinction	Lefevre (1970)
Extinction by (amorphous) carbon smokes,	Day and Huffman (1973)
120–600 nm (a = 30 nm)	
Size, structure, formation of particles	Kappler et al. (1979)
Extinction, 130–800 nm	Stephens (1980)
Extinction efficiences, 0.21–340 μm	Koike et al. (1980)
Mass extinction coeff., 25–250 μm	Tanabé et al. (1983)
Optical constants, 260–800 nm	Smith (1984)
Optical constants, 0.15–6.6 μm	Duley (1984)
Extinction efficiences, 0.11–300 μm	Borghesi et al. (1983a, 1985a)
	Colangeli et al. (1986)
	Bussoletti et al. (1987a)

Silicon Carbide	**Reference**
Mass absorption coeff. (a < 1.5 μm)	Dorschner et al. (1977)
Extinction by silicon carbide smokes,	Stephens et al. (1980)
130–800 nm	
Mass absorption coeff. (a < 4 μm)	Friedemann et al. (1981)
Mass extinction coeff., 25–250 μm	Tanabé et al. (1983)
Extiction efficiencies, 2.5–300 μm	Borghesi et al. (1983b, 1985b)
	Borghesi et al. (1986)

5 Experimental results

Table 2 summarizes the most relevant informations concerning our samples. In the following, however, we will focus our attention about amorphous carbon submicron grains.

Table 2: Summary of relevant parameters of the material studied in our laboratory.

Material	Aver. size	Wavel. range	Optical properties
Amorphous carbon by arc discharge (HAC)	40Å	0.1–300 μm	Extinction efficiences
Amorphous carbon by hydrocarbon burning (BE/XY)	150 Å	0.2–300 μm	" "
Silicon carbide (various types α/β)	0.01–1 μm	2.5–300 μm	" "
PAHs mixtures	—	2.5–40 μm	" "
(PAHs + HAC) mixtures	—	2.5–40 μm	Transmission
(HAC + SiC) mixtures	—	2.5–40 μm	Ext. cross sections

5.1   UV and visible range

Figure 1 reproduces the extinction properties of the grains:
a) a pronounced hump is visible between 2300–2400 Å
b) a shoulder emerges at about 1800 Å
c) a second peak is seen at about 1500 Å
d) a very steep increase is appearing below 1500 Å

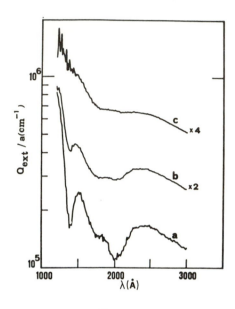

Figure 1

The hump is attributed to surface plasmon mode of surface electrons in the ground state while the steep increase is attributed to electron transitions to excited states. The other two features may instead be explained in terms of quantum size effects by the density of states, DOS, at the surface of the grains. Outside the band region, i.e. beyond 3000 Å, the extinction curve can be fitted by a $\lambda^{-\alpha}$ power law with $\alpha \simeq 1.1 \pm 0.1$. Further details are reported in the paper by Bussoletti et al. (1987a).

5.2   Infrared range

The extinction spectrum of the samples follows (see figure 2) a continuous trend $\lambda^{-\alpha}$ with α varying from 1.1 to 0.7 according to the type of raw materials and the wavelength range.

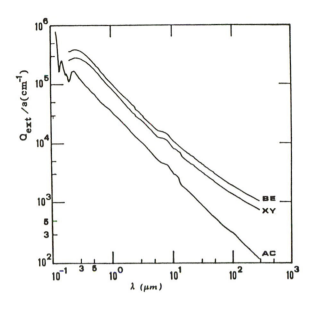

Figure 2

Table 3 summarized all these results, including those concerning par. 5.1, and shows a comparison with similar results produced by other authors. Weak bands are also superimposed on the continuum between 2 μm and 13 μm. They are produced mainly by CH and CH_3 groups bonded to active sites lying onto the surface of the grains. These results have been recently discussed by Bussoletti et al. (1987) and by Borghesi et al. (1987) in terms of possible weak bands which may account for the unidentified IR bands. A detailed discussion on these features will be given in this book to which we will refer (Blanco et al., 1988).

6 Astrophysical implications: the interstellar amorphous carbon

The amorphous carbon experimental results have been recently discussed by Bussoletti et al. (1987b) as possible candidates for cosmic dust. Particles obtained in the laboratory do not correctly reproduce the portion of the interstellar extinction curve commonly attributed to graphite.

Grains with 40 Å respect the carbon cosmic abundance constraints as they need only 38% of the available carbon to account for the interstellar hump. However, they show a peak at 2350 Å and give a peak to visual ratio $\Gamma = A(\text{peak})/A(\text{vis}) \simeq 2.5$ which is lower than the usual value observed for diffuse clouds, $\Gamma_{obs} = 3.1$.

Table 3: Spectroscopic analyses on amorphous carbon particles.

Authors	Production method	Spectroscopic measurements					
		UV peak position (Å)	Range (µm)	Spectral index α			
				UV-Vis	IR1	IR2	FIR
Bussoletti et	BE	2500	0.2–300	1.1±0.1	1.1±0.1	0.8±0.1	0.7±0.1
al. (1987a)	XY	2500	0.2–300	1.1±0.1	1.0±0.1	0.9±0.1	0.7±0.1
	AC-C	2350	0.1–300	1.1±0.1	1.1±0.1	1.1±0.1	0.9±0.1
Stephens	LS(1)	2500	0.12–1	(*)	—	—	—
(1980)	LS(2)	2350	0.12–1	(*)	—	—	—
Koike et al.	BE	2600	0.2–340	1.1±0.1	1.0±0.1	0.9±0.1	0.8±0.1
(1980)	XY	2600	0.2–340	1.0±0.1	1.0±0.1	0.9±0.1	0.7±0.1
	AC-G	2400	0.2–25	1.1±0.1	0.9±0.1	0.9±0.1	—
Tanabé et al.	AC-G(1)	—	25–250	—	—	—	0.6
(1983)	AC-G(2)	—	25–250	—	—	—	0.6

(*): not reported by the authors.

Notes:
LS(1):ambient pressure 760 Torr   LS(2): ambient pressure 7.6 Torr
AC-G(1): argon atmosphere   AC-G(2): hydrogen atmosphere
UV-VIS: 0.35–1.0 µm IR1: 1–5 µm   IR2: 10–30 µm FIR: >30 µm

On this basis, our experimental data have been used to construct an "interstellar amorphous carbon", IAC, whose properties are able to resolve the above-mentioned difficulties. Extrapolation of the results show that particles with mean radius of about 10 Å have the peak at the observed position, 2175 Å, and match satisfactorily the shape of the interstellar hump (see figure 3). Γ in this case is 3.9 and it justifies the existence of other dielectric materials which may account for the observed linear polarization. In addition, IAC requires only 20% of the available carbon to produce the interstellar hump.

7   Conclusions

In this paper we have discussed about the possible structure of solid carbon in space. Besides silicon carbide which represents a minor contributor, the astronomical observations seem to indicate that carbon exists under various forms. Graphite, at least with the well known structure observed on the earth, appears very unlikely due to several chemical and physical constraints. On the contrary, amorphous carbon seems heavily present in circumstellar envelopes of carbon stars as indicated by their spectra consistent with a λ^{-1} dependence of emissivity (see, for instance, Rowan-Robinson and Harris 1983, Sopka et al., 1985, Jura 1986 a,b). In addition Orofino et al. (1987), using experimental extinction data for AC and BE/XY amorphous carbon grains (see table 2) have found that the FIR spectral trend of 78 optically thin sources and the whole IR spectra of CIT6 and IRC+10216 are better explained by the presence of circumstellar grains whose state should be amorphous rather that crystalline.

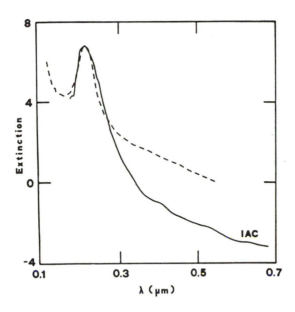

Figure 3

A clear indication in this sense comes from two direct examples the IR observations of the planetary nebula Abell 30 (Greenstein 1981) and the two stars RY Sgr and R CrB (Hecht et al., 1984). These two objects show an extinction peak at 2400–2500 Å which can be attributed to surface plasmon oscillations of particles with mean radii between 40 Å and 150 Å.

Indirect indications about the presence of disordered carbon are also obtained by Draine (1984) for IRC+10216 due to the absence of a "graphite" specific feature expected at 11.52 μm in its spectrum. Goebel (1987) shows that the 5–15 μm continuum spectrum of NGC due to particles at 200–300 K can be accounted for by hydrogenated amorphous carbon. On the other hand, FIR spectrophotometry of the object shows most of the dust is at 90 K and has a spectral index $\sim$ −2 (Mc Carthy et al., 1978) which can be attributed to a graphitic structure of the grains.

Finally, we have to remember the presence of PAHs which partly account for the UIR bands.

In conclusion, though the picture is still far to be completely clear, we may conclude that solid carbon in space may appear in various forms. Purely amorphous carbon represents one end of a scale while a continuum of various degrees of disorder and crystallinity exists. According to Mathis (1986 b), probably some annealing may occur in the stellar atmosphere or while the grains are temporarily heated by passage through a shock in the interstellar medium.

Allamandola et al. (1985) argue that carbon grains are made subunits consisting of PAHs while Duley and Williams (1986) argue that PAHs are made from hydrogenated amorphous carbon. Actually probably both are right as the two kinds of materials are strongly interconnected in their birds places and it seems presently quite impossible to ascertain which of the two pictures may be the most correct.

References

Allamandola, L.J.: 1984, in "Galactic and Extragalactic Infrared Spectroscopy", XVIth ESLAB Symposium, eds. M.F. Kessler, J.P. Phillips and T.D. Guyenne (Dordrecht: Reidel), p. 5.

Allamandola, L.J., Tielens, A.G.G.M., Barker, J.R.: 1985, Astrophys. J. (Letters) **290**, L25.

Blanco, A., Borghesi, A., Fonti, S., Orofino, V., Bussoletti, E., Colangeli, L.: 1988, "Experimental evidences for HAC and PAHs soup in space", present book.

Bohren, C.F., Huffmann, D.R.: 1983, "Absorption and scattering of light by small particles", Wiley (New York).

Borghesi, A., Bussoletti, E., Colangeli, L., Minafra, A., Rubini, F.,: 1983a, Infrared Phys. **23**, 85.

Borghesi, A., Bussoletti, E., Colangeli, L., De Biasi, C.: 1983b, Infrared Phys. **23**, 321.

Borghesi, A., Bussoletti, E., Colangeli, L.: 1985a, Astron. Astrophys. **142**, 225.

Borghesi, A., Bussoletti, E., Colangeli, L., De Biasi, C.: 1985b, Astron. Astrophys. **153**, 1.

Borghesi, A., Bussoletti, E., Colangeli, L., Orofino, V., Guido, M., Nunziante-Cesaro, S.: 1986, Infrared Phys. **26**, 37.

Borghesi, A., Bussoletti, E., Colangeli, L.: 1987, Astrophys. J., **314**, 422.

Bussoletti, E.: 1985, Rivista Nuovo Cimento **8**, 1.

Bussoletti, E., Colangeli, L., Borghesi, A.: 1986, in "Polycyclic Aromatic Hydrocarbons and Astrophysics", NATO Advanced Research Workshop, eds. A. Leger, L. d' Hendecourt and N. Boccara (Dordrecht: Reidel), p. 63.

Bussoletti, E., Colangeli, L., Borghesi, A., Orofino, V.: 1987a, Astron. Astrophys. Suppl. Ser. **70**, 257.

Bussoletti, E., Colangeli, L., Orofino, V.: 1987b, Astrophys. J. (Letters) **321**, L87.

Carnochan, D.J.: 1986, Mon. Not. R. Astr. Soc. **219**, 903.

Cayrel, R., Schatzman, E.: 1984, Ann. Astrophys. **17**, 555.

Colangeli, L., Capozzi, V., Bussoletti, E., Minafra, A.: 1986, Astron. Astrophys. **168**, 349.

Czyzak, S.J., Santiago, J.J.: 1973, Astrophys. Space Sci. **23**, 443.

Czyzak, S.J., Hirth, J.P., Tabak, R.G.: 1982, Vistas in Astr. **25**, 337.

Day, K.L., Huffman, D.R.: 1973, Nature Phys. Sci. **243**, 50.

Donn, B., Modge, R.C., Mentall, J.E.: 1968, Astrophys. J. **154**, 135.

Donn, B.: 1986 in " Interrelationships among Circumstellar, Interstellar and Interplanetary Dust", NASA CP2403, eds. J.A. Nuth III and R.E. Stencel, p. 109.

Dorschner, J., Friedman, C., Gurtler, J.: 1977, Astron. Nach. **298**, 279.

Draine, B.T.: 1984, Astrophys. J. (Letters) **277**, L71.

Draine, B.T., Lee, H.M.: 1984, Astrophys. J. **285**, 89.

Draine, B.T.: 1986 in "Interstellar among Circumstellar, Interstellar and Interplanetary Dust", NASA CP2403, eds, J.A. Nuth III and R.E. Stencel, p. 19.

Duley, W.W.: 1984, Astrophys. J. **287**, 694.

Duley, W.W., Williams, D.A.: 1986, Mon. Not. R. Astr. Soc. **219**, 859.

Duley, W.W., Jones, A.P., Willliams, D.A.: 1987, "Hydrogenated amorphous carbo-coated silicates particles as a source of interstellar extinction" Astrophys. J. submitted.

Friedemann, C., Gurtler, J., Schmidt, R., Dorschner, J.: 1981, Astrophys. Space Sci. **79**, 405.

Goebel, J.M.: 1986 in "Polycyclic Aromatic Hydrocarbons and Astrophysics", NATO Advanced Research Workshop, eds. A. Leger, L. d' Hendecourt and N. Boccara (Dordrecht: Reidel), p. 329.

Greenberg, J.M., Chlewicki, G.: 1983, Astrophys. J. **272**, 563.

Greenstein, J.L.: 1981, Astrophys. J. **245**, 124.

Hecht, J.H., Holm, A.V., Donn, B., Wu, C.C.: 1984, Astrophys. J. **280**, 228.

Hecht, J.H.: 1986, Astrophys. J. **305**, 817.

Jura, M.: 1986a in "Interrelationships among Circumstellar, Interstellar and Interplanetary Dust", NASA CP2403, eds. J.A. Nuth III and R.E. Stencel, p. 3.

Jura, M.: 1986b Astrophys. J. **303**, 327.

Kappler, P., Ehrburger, P., Lahaye, J., Donnet, J.B.: 1979, J. Appl. Phys. **50**, 308.

Koike, C., Hasegawa, H., Manabe, A.: 1980, Astrophys. Space Sci. **67**, 495.

Lefevre, J.: 1967, Ann. Astrophys. **30**, 731.

Lefevre, J.: 1970, Astron. Astrophys. **5**, 37.

Mathis, J.S., Rumpl, W., Nordsiek, K.H.: 1977, Astrophys. J. **217**, 425.

Mathis, J.S.: 1986a, in "Interrelationships among Circumstellar, Interstellar and Interplanetary Dust", NASA CP2403, eds. J.A. Nuth III and R.E. Stencel, p. 29.

Mathis, J.S.: 1986b, in "Light on Dark Matter", IRAS Symposium, ed. F.P. Israel (Dordrecht: Reidel), p. 171.

Mc Carthy, J.F., Forrest, W.J., Houck, J.R.: 1978, Astrophys. J. **224**, 109.

Orofino, V., Colangeli, L., Bussoletti, E., Strafella, F.: 1987, "Amorphous carbon around stars", Astrophys. Space Sci., in press.

Rowan-Robinson, M., Harris, S.: 1983, Mon. Not. R. Astr. Soc. **202**, 797.

Smith, F.W.: 1984, J. Appl. Phys. **55**, 764.

Sopka, R.J., Hildebrand, R., Jaffe, D.T., Gatley, I., Roelling, T., Werner, M., Jura, M., Zuckerman, B.: 1985, Astrophys. J. **294**, 242.

Stephens, J.R.: 1980, Astrophys. J. **237**, 450.

Tabak, R.G., Hirth, J.P., Meyrich, G.G., Roark, T.: 1975, Astrophys. J. **196**, 457, 19D11.

Tanabé, T., Nakada, Y., Kamijo, F., Sakata, A.: 1983, Publ. Astron. Soc. Japan **35**, 397.

Tielens, A.G.G.M., Allamandola, L.J.: 1986a, in "Physical Processes in Interstellar Clouds", eds. G. Morfill, M. Scholer.

Tielens, A.G.G.M., Allamandola, L.J.: 1986b, in "Composition, Structure and Chemistry of Interstellar Dust" , NASA TM-88350.

Willner, S.P.: 1984, in "Galactic and Extragalactic Infrared Spectroscopy", XVIth ESLAB Symposium, eds. M.F. Kessler, J.P., Phillips and T.D. Guyenne (Dordrecht: Reidel).

PAHs, THE STATE OF ART

A. Léger
*Groupe de Physique des Solides de L'E.N.S., Université Paris VII,
Paris, France*

1 Tools for studying the interstellar matter

The Interstellar Matter (IM) is made of 99% in mass by H and He gas and 1% by heavy element compouns. Although the latter are in a minority, they are essential for the optical properties of the medium because they can condense and make solids or large molecules which have optical wide bands whereas atom ions or small molecules (H, H^+, He, H_2...) only have narrow lines and cannot give a continuous absorption in a large range of wavelengths.

Basically, we have two ways of getting information on the IM: the absorption of the star light when it travels across and the emission of photons by the IM itself when it is heated. Let us briefly review the features which are observed and the information which is deduced.

1.1 Absorption

Star spectra exhibit very specific features that allow the identification of the star type even if its light has been somewhat modified when travelling. We can then compare the spectra from two stars of the same type, one behind a large column density of IM, the other close to us and deduce the specific extinction (absorption+scattering) by the IM.

Let us summarize the extinction curve main features in the UV-Visible and IR ranges:

– A general up slope from long to short wavelengths. Comparing it with the scattering theory, a size is inferred for the solid particles present in the IM (a distribution from r< 50 Å to 0.2, 1 μm, according to the regions);

– A hudge and broad absorption feature at 2200 Å, which has been tentatively attributed to *graphitic material*;

– More than 40 very well defined absorption bands in the Visible, called "Diffuse Interstellar Bands". Their origin is completely unknown and the corresponding information disregarded;

– Features in the Visible due to small molecules as H_2, CH, CH^+, CN;

75

E. Bussoletti et al. (eds.), Experiments on Cosmic Dust Analogues, 75–94.
© 1988 by Kluwer Academic Publishers.

– Several bands in the IR observed only in dense interstellar regions (3.1 – 4.62 – 4.67 –
4.9 – 6.0 – 6.8 μm...) which are attributed to ices (H_2O, CO...);
– Two bands at 9.6 and 18 μm ubiquitously observed in both dense and diffuse regions.
They are attributed to silicates.

1.2   Emission

The IM is heated and emits resolved bands in the IR mainly in two cases:
– when in dust shells close to stars. The main observed feature is at 9.7 μm and is
also attributed to silicate vibrational modes. This material is therefore observed both in
absorption and in emission.
– when dust is irradiated by star light, even if the distance is large. Several bands are
observed. The main ones are at 3.28 – 6.2 – 7.7 – 8.6 – 11.3 μm. They are observed in a
wide variety of astronomical objects: reflection nebulae (an interstellar cloud close to a
star), bipolar nebulae (matter probably ejected from a star), planetary nebulae (matter
around a very hot star) and whole regions of active galaxies. Although the luminosity of
these objects varies by 7 orders of magnitude (10^4 to 10^{11} solar luminosities) the positions
of the features are always the same, see figure 1.

These last bands have been discovered since 1973 and were one of the longest-
standing puzzle in IR astronomy (see Allamandola 1984, Willner 1984). They were called
"Unidentified IR Emission Bands".

The aim of this paper is to show that Polyclyclic Aromatic Hydrocarbons
(PAHs) are likely their carriers. If this interpretation was correct, we would have significantly progressed in our understanding of the spectroscopic message from the IM.

2   Origin of the emission bands

Any explanation have to account for the following points: a) the five main bands always
appear together; b) the regions where they are observed have Interstellar Matter and
strong UV-Visible irradiation; c) they emit a substancial fraction of the whole energy
radiated by these regions.

2.1   Equilibrium temperature of interstellar dust

Let us consider the case of a reflection nebula: a star of luminosity L is illuminating
Interstellar matter located at distance d. Typical values are $L_1 = 3 \cdot 10^3$ L_0, where L_0 is
the sun luminosity, $d_1 = 0.6$ ly $\simeq 6 \cdot 10^{17}$ cm. An interstellar dust grain mainly exchanges
energy with its surroundings by absorption and emission of photons. If it is large enough,
its temperature fluctuations are negligible and its mean temperature is determined by
balancing its absorption and emission power: $P_{abs} = P_{em}$. For a spherical grain with
radius a:

$$P_{abs} = \pi a^2 Q(\lambda_{UV}, a)\Phi_{in}; \quad P_{em} = 4\pi a^2 Q(\lambda_I R, a)\sigma T_{eq}^4$$

where $Q(\lambda, a)$ is the emissivity or absorptivity of the sphere at wavelength λ, Φ_{in} the star flux, σ the Stephan constant. For typical interstellar grain material, one finds:

$$T_{eq} = 65\ K.\left(\frac{L}{L_1}\right)^{0.22}\left(\frac{d}{d_1}\right)^{-0.44}\left(\frac{a}{0.1\mu}\right)^{-0.22}.\qquad(2.1)$$

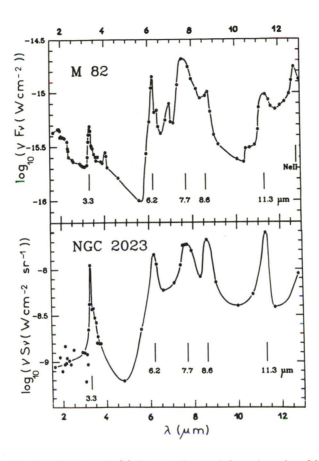

Figure 1: Mid-infrared spectra of: (1) the central part of the active galaxy M82, adapted from Willner et al. (1977) and Gillett et al. (1975). The Emission Bands at 3.3 – 6.2 – 7.7 – 8.6 – 11.3 μm are dominating the emission of this region which luminosity is a large fraction of the whole galaxy luminosity. The additional emissions at 4.05 – 7.0 – 12.8 – μm are identified as ionized gas lines, H^+, Ar^+, Ne^+ respectively. (2) The reflection nebula NGC 2023 adapted from Sellgren et al. (1986). For spectra (1) and (2), the continuous line between measured points is only for clarity. The luminosity of the first object is over 10^6 times larger than that of the second, The Emission Bands are at the same position, pointing to an universal process.

It is important to notice that *such a grain emits in the Far-IR ($\lambda_{max} \simeq 50\ \mu m$) and definitely not in the range [3–12 μm]*. For instance, there is no way to obtain realistic parameters that would give a significant emission at 3.3 μm. We conclude that dust at equilibrium temperature cannot explain the observed Emission Bands in reflection nebulae.

2.2   Tentative explanations of the emission bands in the past

Before 1984, several propositions have ben made for explaining the Bands:

- Allamandola et al. (1979) suggested that simple molecules or radicals (CH_4, NH_3, H_2O,...) frozen in grain mantles could be excited and emit with their vibrational modes. But the ubiquity of the observed spectra would imply an unlikely uniformity in the composition of the mantles.

  In addition, the observed yield of conversion of UV photon to IR photon would have implied a quantum yield about unity. Now, for an excited radical in a solid, the non-radiative transfer of the energy to the matrix is many orders of magnitude faster than the radiative decay because vibrational modes have a long radiative life time. Most of the UV energy is then transferred to the bulk of the grain that will re-radiate it in the far IR because of its low equilibrium temperature, not at the wavelengths of the observed bands.

- Duley and Williams (1981) showed that radicals bound at the periphery or graphitic grains would produce very interesting bands. Specifically, CH groups would have modes at 3.3 and 11.3 μm, fitting nicely two of the observed bands. Unfortunately, to produce the others, they had to invoke groups such as NH_2 that would also produce unobserved bands (2.9 μm for instance). Moreover, the problem of the excitation was still unsolved because of the competition between radiative and non-radiative decays for a vibrationally excited radical bound to a solid.

2.3   The idea of emission during temperature fluctuations

Sellgren et al. (1983) measuring the IR spectrum of reflection nebula noticed that in addition to the prominent emission bands, there was a continuum that could not be explained by scattered light from the illuminating star. This continuum could be fitted by a diluted black body emission whose temperature (T∼ 1500 K) was far too high to be an equilibrium temperature of dust in the nebula (see equation 2.1). In 1984, K. Sellgren suggested that such temperatures could be achieved during the thermal impulse reached after the absorption of a UV photon by a grain if it is small enough. The temperature increase, ΔT, is given by:

$$h\nu_{UV} \sim 3Nk_B\Delta T \tag{2.2}$$

if one approximates the N atoms solid specific heat its high temperature limit, k_B being the Boltzmann constant. With $h\nu_{UV} = 10$ eV, one get $\Delta T = 10^3$ K for $N= 50$ atoms.

The very small grain emits most of its energy when cooling from the high temperature peaks. The color temperature of its emission is then high whereas its mean temperature is low as dominated by the long periods between two absorptions. This last temperature is similar to that of bigger grains where the fluctuations are negligible (Equation 2.1). If they had a smooth emittivity, such very small grains could explain the nebula emission background but the Bands would still be unexplained.

2.4 Nature of these very small grains

In 1984, Léger and Puget pointed out that in order to resist to sublimation, these grains must be refractory. They computed the erosion rate for typical interstellar materials taking into account single and double photon processes. They found that ices and silicates are too volatile whereas graphite can stand such high temperature excursions. Considering the large cosmic abundance of carbon, they concluded that *graphite-like very small grains are good candidates for the quantum heating by UV photon.* Hydrogen is very abundant in the astronomical environment. In addition, the carbon atoms at the periphery of the graphitic planes have unsatisfied bonds. It is therefore very likely that hydrogen atoms are bound to the peripheric carbon atoms as in organic molecules. This gives a new argument to the suggestion of Duley and Williams (1981) to account for the 3.3 and 11.3 μm bands with CH modes with the possibility of explaining the excitation mechanism because the whole system is heated. However, a key feature to explain is the simultenous occurrence of the five Bands. With that aim, let us consider the whole IF spectra of a hydrogenated graphitic cluster.

2.5 Optical properties of hydrogenated graphitic clusters

To calculate the absorption of a sphere of graphite surrounded by CH groups, the first idea is to use bulk graphite optical constants and CH oscillators. Such an electromagnetic calculation (Mie calculation) gives a disappointing results: besides the 3.28 and 11.3 μm modes, the spectrum is dominated by a strong continuum due to the graphite; the other spectral features are minute and cannot account for the observed Bands. However, the use of bulk optical constants for a small cluster may be uncorrect. The optical absorption of a system in the 3–15 μm range is governed by its electronic and lattice excitations, so let us compare them for bulk graphite and for a graphitic cluster. The electron band structure of a graphitic plane is reported in figure 2a. It exhibits a full band (π) and an empty one ($\pi*$) which are $\sim$ 10 eV broad and touch at the Brillouin zone points (Dresselhaus and Dresselhaus, 1981). Therefore, there is a continuum of allowed transitions from 0 to 20 eV. The presence of these interband electronic transitions in the IR is responsible for the absorption continuum and the high IR dielectric constant of graphite that screens the vibrational modes.

The electron band structure of a finite piece of graphite can be crudely approximated by applying boundary conditions to the infinite solid (see Ashcrof and Mermin, 1976). The electronic states are still on the same curves but they are in a finite number and equally spaced in $\vec{k}$ space (Figure 2b). The interband transitions do not form a continuum anymore, they are discrete and their lowest energy is of a few eV for a 50 atoms cluster. The optical constants of a graphitic cluster are then *completely different from those of bulk graphite* in the range of interest (3–15 μm, 0.4–0.08 eV) because the cluster has no electronic transition at these energies (Figure 2c). *The strong electronic absorption continuum vanishes and the lattice modes appear as they are no more screened by electrons.* So, we expect more spectroscopic features in the absorption curve than indicated by the first Mie calculation. But before going further, we need a more precise model for our hydrogenated graphitic clusters.

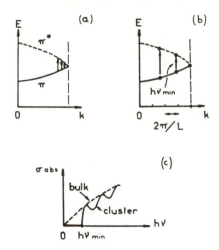

Figure 2: Band structure sketch of: (a) bulk graphite, all the values of k are allowed; (b) a graphitic cluster, only discrete values are allowed. L is the linear size of the cluster; (c) coresponding optical absorption cross sections. For photon energies less than $h\nu - min$, the cluster exhibits no electronic absorption whereas the bulk sample does absorb.

2.6   Small grains or large molecules?

A 50 carbon atoms graphite "sphere" would have a radius of ~ 5 Å and could contain only two layers of aromatic planes because the interplane distance is 3.35 Å (Figure 3).

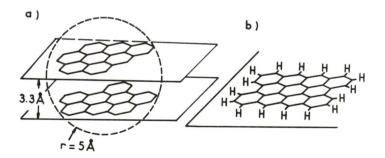

Figure 3: a) What a 50 atom graphite "sphere" would look like. b) Proposed structure of hydrogenated graphitic cluster.

As the grains are subjected to high temperature heatings, the planes would split because the interplane binding of carbon atoms is much weaker than the intraplane binding (0.06 eV and 7.5 eV per atom respectively). Such hydrogenated pieces of graphitic plane are known in organic chemistry as large Polycyclic Aromatic Hydrocarbon (PAH) molecules (Clar, 1964). *We suggest that the Interstellar Very Small Grains proposed by K. Sellgren are large PAH molecules.* These species exhibit a very good stability against heating and UV photolysis (Turro, 1978) and can stand the expected hot events and direct photolysis in the Interstellar Medium. With this molecular model we can now estimate the emission spectrum of a cluster when it has absorbed an UV photon.

2.7   Emission spectrum of a PAH molecule: Coronene

The energy diagram of a closed shell isolated molecule having absorbed an UV photon is well known in the one electron approximation. The absorbed photon makes a transition between the ground singlet state S_o to an upper singlet state S_i. The molecule being isolated cannot exchange energy with its surroundings but by radiation. When there is no fluorescence de-excitation nor intersystem crossing ($\equiv$ transfer to a triplet state T_i), the fast ($t = 10^{-12} - 10^{-8}$ s) non radiative processes transfer the energy to the vibrationally excited electronic ground state S_o (Birks, 1970). The energy exchange between vibrational modes (Internal Vibrational Randomization, Parmenter, 1982) is fast enough to reach a regime where their population can be described by a vibrational temperature after $t = 10^{-11} - 10^{-8}$ s. The validity of this approximation is due to the long time ($t \sim 10^1$ s before the first IR photon is emitted. If the energy is transferred to a vibrationally excited triplet state as suggested by Allamandola et al. (1985), there is also a thermalization of the vibrational modes in the triplet state. If the molecules are ionized, as it is likely in many astronomical objects, the states are doublets and quartets and a similar thermalization of the vibrational excitations is expected.

The IR emission of the molecule can be described as that of a system at an equilibrium vibrational temperature because *it has enough time between two photon emissions to redistribute energy between its vibrational levels.* The hypothesis of thermal physics is fulfilled. Besides fluorescence or phosphorescence, this IR photon emission is the only energy release process for an isolated molecule. This point is crucial as *it overcomes the problem of non-radiative de-excitation* that Allamandola et al. (1979) were faced with. The emission intensity I_λ of a physical object is, at wavelength and equilibrium temperature T:

$$I_\lambda = B_\lambda(T)\epsilon_\lambda^{emis}(T) \tag{2.3}$$

where $B_\lambda(T)$ is the black-body emission intensity and ϵ^{emis} the emissivity of the object at that temperature. The second thermodynamic law implies that the sample emissivity is equal to its absortivity at the same temperature, $A_\lambda(T)$, (Kirschoff law, see Reif, 1965). This quantity can be measured in an absorption experiment at temperature T, where an incident intensity J_o is trasmitted as J_λ:

$$A_\lambda = \frac{J_o - J_\lambda}{J_o} = 1 - e^{N\sigma_\lambda} \tag{2.4}$$

where N (mol cm^{-2}) is the column density of the molecule assembly and σ_λ (cm^2 mol^{-1})

the cross section per molecule. In the optically thin limit, $N\sigma_\lambda \ll 1$, one reads:

$$I_\lambda = B_\lambda(T)N\sigma_\lambda \tag{2.5}$$

Unfortunately, IR absorption spectra of large PAHs are not available at high temperature. Room temperature of molecules as pyrene ($C_{16}H_{10}$), perylene ($C_{20}H_{12}$) and coronene ($C_{24}H_{12}$) have been published and their comparison can give an idea of what can be expected for larger molecules (Sadtler spectra, 1959). The emission of heated coronene has been calculated (Léger and Puget, 1984) from relation (2.5), approximating the absorption at high temperature by the spectrum at room temperature published by Sadtler (1959). *Although the interstellar PAHs are likely a complex mixture of molecules,* *the main emission bands of coronene already give a suggestive fit to the observed features.* The 3.3 μm (3030 cm^{-1}) band has already been discussed and corresponds to an aromatic C-H stretch. The 5.2 μm (1920 cm^{-1}) peak of coronene has not yet been observed in astronomical spectra. The 6.2 μm (1610 cm^{-1}) absorption is "highly characteristic" of C=C stretch in an aromatic ring (Bellamy, 1966, p.69); its presence in most astronomical spectra and the absence of 6.7 μm (1500 cm^{-1}) absorption indicates that most aromatic rings are in a compact arrangement. The 7.65 μm (1310 cm^{-1}) feature of coronene is present in astronomical spectra. It is not found in small aromatic compounds and may be specific to larger species. The 8.85 μm (1130 cm^{-1}) band is an in-plane aromatic C-H bend whereas the 11.9 μm (840 cm^{-1}) is an out-of-plane aromatic C-H bend. The exact position of this last mode depends upon the number of adjacent H in which are on a ring: solo (11.0–11.6 μm , 910–860 cm^{-1}), duo (11.6–12.5 μm, 860–800 cm^{-1}) or trio (12.4–13.3 μm, 810–750 cm^{-1}) (Clar et al., 1981). The observed spectra have a peak at 11.3 μm (885 cm^{-1}) and a broad structure up to 13.5 μm. This indicates the presence of different hydrogen sites with possibly a predominance of solo.

In conclusion, from the point of view of the analytic IR spectroscopy, the observation of the 3.3, 6.2 and 11.3 μm bands *is highly characteristic of PAH compounds.*

3   New IR laboratory data

3.1   More IR Absorption Spectra of large PAHs

The absorption IR spectra of several large PAHs have being recently measured with emphasis on the intensity of the bands (Léger, d'Hendecourt, Schmidt, 1988). The integrated intensities of the main bands are given in table 1.
It can be noticed that:

• The integrated cross sections of the different bands are not too scattered from one compact species to another. The average value can then be used for calculations in Astrophysics on typical large compact PAHs.

• The cross sections per solo hydrogen (11.3 μm) are quite larger than per duo hydrogen (11.9 μm) in the two measured species that exhibit both modes. If this is confirmed with more molecules, it may be important to explain that the 11.3 μm band is dominating in the astronomical spectra.

3.2 Calculated emission spectra: effect of temperature

The emission spectrum (I_λ) of a molecule, at a temperature T, depends on its absorption spectrum (σ_λ) and on the temperature through the Planck function $B_\lambda(T)$ (relation 2.5).

Table 1: Integrated *absorption cross sections* $(A_i = \sigma_i \Delta\lambda_i)$ for different compact PAHs as measured in the laboratory. The average values can be used for astrophysical calculations on "typical" compact PAHs. The resulting oscillator strengths, $F_i = 1.13 \cdot 10^{20}\,(A_i/1\,cm^3)(\lambda/1\,\mu m)^{-2}$, are also given.

λ_i (µm)	3.3	6.2	7.7 (±.3)	8.8	11.3 ± 11.9	
Units for A_i $10^{-25}\,cm^3$ ×	H^{-1}	C^{-1}	C^{-1}	H^{-1}	H^{-1}	
Coronene $C_{24}H_{12}$	1.2	.54	2.25	.90	47	
Dicoronene $C_{48}H_{20}$	1.7	1.04	2.32	1.0	50	$\begin{cases} 110/H\ solo \\ 34/H\ duo \end{cases}$
Methylcoronene $C_{25}H_{14}$	1.4	.90	2.32	1.6	38	
Ovalene $C_{32}H_{14}$	1.5	.45	1.13	.89	43	$\begin{cases} 160/H\ solo \\ 23/H\ duo \end{cases}$
Circobiphenyl $C_{38}H_{20}$	1.1	.59	2.0	1.6	33	
Typical compact PAH	1.4	.70	2.0	1.2	42	
$f_i/10^{-6}$	1.45 (H^{-1})	.21 (C^{-1})	.38 (C^{-1})	.18 (H^{-1})	3.4 (H^{-1})	

3.3 Calculated emission for compact PAHs

The calculated emissions of several large compact PAHs are reported in figure 4 and are compared with an astronomical observed spectrum. The fit has only one adjustable parameter: the emission temperature. *The similarity between the spectra indicates that the "unidentified" IR Emission Bands can be explained by a mixture of compact PAHs.* Coronene is therefore not the only "miracle" molecule that can account for the observations. The agreement for the 3.3 and 6.2 µm bands is quite good with any of the PAH spectra. This confirm that these bands are characteristic of large PAHs (see also figure 5) and supports their identification in the Interstellar Medium. A band is observed in all the calculated spectra at 5.2 µm. This region of the spectrum has not yet been spectroscopically well observed. *We strongly recommend in the future to look for this feature in astronomical objects.*

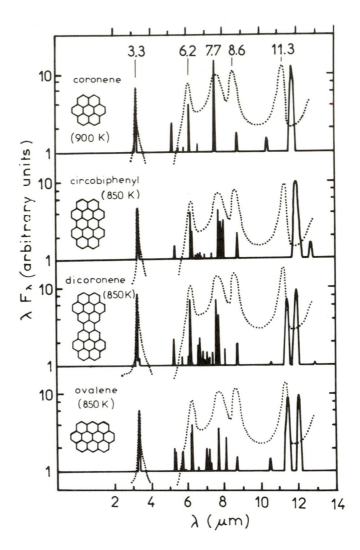

Figure 4: Emission spectra of several compact PAHs calculated from their absorption spectra measured in the laboratory at room temperature (Léger et al., 1988) using relation (2.5). The observed spectrum of the reflection nebula NGC 2023 is reported for comparison (dotted line). The only free parameter in the calculation is the mean emission temperature (in brackets) which is adjusted to reproduce the observed ratio of the two CH bands at 3.3 and 11.3 μm. For clarity, the bands attributed to C-C modes are filled whereas only the contour of the C-H bands is drawn.

In the 6.5–8.5 μm range, the main bands of the compact PAHs appear within the nebula 7.7 μm broad feature. This region is important as it permits to discriminate between different aromatic compounds. As already mentioned, it seems possible to explain the observed broad band at 7.7 μm by a mixture of compact PAHs.

The main spectroscopic disagreement still concerns the 11–13 μm region. It probably indicates a larger abundance of solo H atoms (11.3 μm) in the molecules present in space than in those which have been studied here and where most H atoms are duo (11.9 μm).

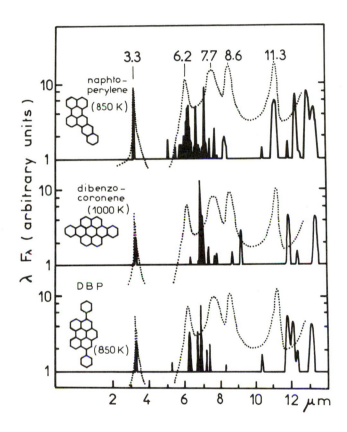

Figure 5: Emission spectra of non-compact PAHs with same conditions as in figure 4. DBP is diphenylbenzoperylene. The fit with astronomical band is poor.

3.4 Calculated emission for non-compact PAHs

Similar emission spectra are reported in figure 5 for non-compact PAHs. A mode is present in these spectra, about 6.7 μm (1500 cm^{-1}) that is characteristic of *protuberant* aromatic *rings*, as opposed to rings in compact arrangements. The latter have only a mode at 6.2 μm (1600 cm^{-1}) (Figure 4; Russell, 1984). On the contrary, the astronomical spectra exhibit a dip in the range 6.5–7.2 μm when they have not the line at 7.0 μm which has a different origin Ar II, Nϱ II. This lack of fit indicates that non-compact PAHs, if present, are minute constituents of the Interstellar Medium. We conclude that *we can make the identification of PAHs in the astronomical IR spectra more accurate: the dominating species are large compact molecules as opposed to molecules with protuberant cycles.* This conclusion agrees with the higher thermodynamic stability of the former species (Clar, 1964). The comparison with astronomical bands indicates that the presence of alkane groups is possible and can probably explain the observed bands about 3.4 μm. On the other hand, large amount of C=O groups, for instance, would give structures at 5.9 μm (1700 cm^{-1}) which are not present in the observed spectra.

3.5 Emission temperature

Equation 2.5 relates the integrated absorption in a band, $A_i = \sigma_i \Delta \lambda_i$, to its integrated emission intensity, $E_i = I_{\lambda i} \Delta \lambda_i$, using the Planck function, $B_\lambda(T)$. The mean emission temperature T_{em} can be deduced from the ratio of two bands, of a same group, that occurs at different wavelenghts. Considering the νC-H (3.3 μm) and γC-H (11.3 μm) modes gives:

$$\frac{E_{11.3}}{E_{3.3}} = \frac{B_{11.3}(T_{em})}{B_{3.3}(T_{em})} \cdot \frac{A_{11.3}}{A_{3.3}} \tag{3.1}$$

Table 2 gives the corresponding emission temperatures for different astronomical objects using the absorption of a typical compact PAH as reported in table 1. The dependence of T_{em} upon the absorption cross whether the value for solo H is larger than for duo H.

3.6 Dehydrogenation of PAHs in the interstellar medium

A quick look at spectra in figure 4 indicates that C=C modes are more intense, relatively to C=H modes, in the spectrum of the nebula than in the PAH spectra. This favors the idea of partial dehydrogenation of PAHs in Space.

Quantitatively, the H coverage, $x_H = (H\ present)/(sites\ for\ H)$, can be deduced from the intensities of H and C bands if one assumes that dehydrogenation does not upset the oscillator strength of the remaining modes:

$$\frac{E_{\lambda H}}{E_{\lambda C}} = x_H \frac{B_{\lambda H}(T) A_{\lambda H}}{B_{\lambda C}(T) A_{\lambda C}} \tag{3.2}$$

where T has already been determined from the ratio of two C-H modes. This hydrogen coverage is reported in table 2 for three astronomical objects. *This estimate points to very strong dehydrogenation of PAHs in Space ($x_H \sim 10\%$).*

Theoretical studies of PAH dehydrogenation by photons favors either no dehydrogenation (large molecules) or complete dehydrogenation (small ones). We suggest that this value of hydrogen coverage mainly reflects the fraction of hydrogenated species versus total number of species rather than partial coverage of a given species.

Table 2: Mean emission temperature (T_{em}) and hydrogen coverage (x_H) deduced from comparison between astronomical observations and laboratory absorption of PAHs. $A_{\lambda i}$ are the integrated absorption cross sections for a typican PAh (Table 1). The value in brackets are deduced from data on solo H in dicoronene and ovalene. $E_{\lambda i}$ are the observed integrated band intensities. (a)(c): Values are from Léger, d'Hendecourt and Schmidt (1988), Cohen et al. (1986); (b)(d): Values are deduced from spectra by Sellgren et al. (1986), Willner et al. (1977).

λ_i (μm)	3.3	6.2	7.7	11.3	
$A_{\lambda i}/A_{3.3\mu m}$	1	1.0	2.9	30 (97)	a
$E_{\lambda i}/E_{3.3\mu m}$ (NGC 2023)	1	4.6	9.8	4.2	b,c
—— (Red Rect.)	1	4.3	8.1	2.0	c
—— (M 82)	1	4.1	17.6	1.8	d
x_H (NGC 2023)		12%	11%		
—— (Red Rect.)		9%	8%		
—— (M 82)		9%	3%		
T_{em} (NGC 2023)	780 K (1190 K)				
—— (Red Rect.)	980 K				
—— (M 82)	1020 K				

4 Implications of this identification

Let us first determine some properties of isolated PAHs that are useful for their study. We shall then derive different implications of their identification as their size, their abundance in Space and the different fields in Astrophysics where they can play a role.

4.1 Specific heat of PAHs

The specific heat of a PAH depends on: (a) N_t, its total number of atoms, (b) N_H/N_C, its relative number of hydrogen and carbon atoms and (c) the specific molecule considered. Taking into account points (a) and (b), the dependence on point (c) is probably weak within a category of compounds (e.g. compact PAHs). The relative number of H versus C atoms is rather constant in the serie: coronene (.50), ovalene (.44), circobiphenyl (.53). So, we derive the specific heat for hydrogenated compact PAHs from data on coronene. For dehydrogenated molecules, we shall use the values for graphitic planes, an infinite dehydrogenated PAH. A free molecule with N_t atoms has $4N_t-6$ vibrational modes. If the frequency V_i of each mode is known, its vibrational specific heat is given by an Einstein multi-frequencies model:

$$C = k \sum_{i}^{3N_t-6} \left(\frac{\theta_i}{T}\right)^2 \frac{e^{\theta_i/T}}{(e^{\theta_i/T} - 1)^2} \qquad (4.1)$$

where $k\theta_i = hV_i$. This approximation is in fact quite good as it depends only on the validity of the harmonic one. Fortunately, the frequencies of all the modes in coronene have been calculated by Cyvin (1982) and Cyvin et al. (1984), using a force field model for condensed aromatics. The quality of such a semi-empirical model is indicated by the agreement with the measured active modes (IR and Raman) which are much more numerous than the free parameters of the model. Their resulting energy distribution and the specific heat has been published (Léger and d'Hendecourt, 1987, p.242). We consider *this curve as representative for the specific heat of hydrogenated PAHs in Space*. A crude approximation for it is:

$$C(T) = 3(N_t - 2)k(T/1300 \ K), \ for \ T < 1300 \ K$$

$$(4.2)$$

$$C(T) = 3(N_t - 2)k, \quad for \ T > 1300 \ K$$

For completely dehydrogenated PAHs, the curve for graphite should be used. It is derived from Krumhansl and Brooks (1953).

4.2   Cooling time for a PAH in space

In the Interstellat Medium, the main cooling process for a hot PAH is IR radiation besides fluorescence or phosphorescence in the Visible. Let us evaluate this cooling time.
       One can show, using equation (2.5), that the radiative power, P, emitted by a molecule at temperature T, is: $P = \int P_\lambda \, d\lambda$, with:

$$P_\lambda = 4\pi B_\lambda(T)\sigma_\lambda$$

This temperature decreases according to: $P \, dt = -C(T) \, dt$,
And a cooling time τ can be defined as:

$$\tau = \frac{C(T)T}{P}.$$

The emitted power is:

$$P = 4\pi \sum_i B(\lambda_i, T) A_i^c N_c$$

where $B(\lambda, T)$ is the Planck function, $A_i^c = \sigma_i^c \Delta\lambda_i$ the integrated cross section per carbon of the ith band and N_c the number of carbon atoms in the molecule. Using values of table 1 for A_i and $C(T)$, with $N_H/N_C = 0.5$, one finds the *molecule cooling time* τ:

$$T = 1000 \ K \rightarrow \tau = 3.8 \ s$$

$$(4.3)$$

$$T = 1500 \ K \rightarrow \tau = 2.4 \ s$$

It is independent of the molecule size as both the emitted power and the specific heat are proportional to the number of atoms. The cooling time should be compared to the radiative life time, t_o, of an excited vibrational mode of the molecule:

$$t_o = \bar{A}^{-1} = \frac{mc}{8\pi^2 \bar{\nu}^{-2} e^2 f} = \frac{1.50}{\bar{\nu}^{-2} f} \quad in \ CGS \ units,$$

where $\bar{A}$ is the Einstein emission coefficient and $\bar{\nu}$ is in cm^{-1}. The oscillator strength, f, is related to the integrated cross section A_λ as: $f = 1.13 \cdot 10^{20}(A_\lambda/1 \text{ cm}^3)(\lambda/1 \text{ } \mu m)^{-2}$ From table 1, a typical value for an aromatic C-H stretch (3.3 μm) is $f = 1.45 \cdot 10^{-6}$, leading to a radiative life time:

$$t_o = 0.12 \text{ } s \qquad (4.4)$$

The radiative life time of this mode is therefore significantly shorter than the cooling time of the whole molecule. This can be understood as followed. The emission from the νC-H (3.3 μm) modes of the molecules is efficient (45 % of the total power when T=1500 K) but these modes are $1/9^{th}$ of the total number of modes and, at T=1500 K, they are excited only in a fraction: $n_{3.3}^1/n_{3.3}^0 \simeq e^{-h\nu/kT} = 5.6 \cdot 10^{-2}$. Each radiative emission from these modes takes out $h\nu$=3000 cm^{-1} from the molecule. They occur at rate t_o^{-1} per excited mode. One finds:

$$\frac{t_o}{\tau} = \frac{h\nu}{kT} \frac{l}{9\alpha\beta} e^{-h\nu/kT},$$

where $\alpha = C(T)/C(\infty)$ and $\beta(T)$ is the fraction of the total emitted power that goes through the emission of the considered modes. For νC-H in PAHs: $t_o/\tau = 4.6 \cdot 10^{-2}$ which is small and is in agreement with (4.3) and (4.4). *The mean time between two emissions of IR photon* for a molecule at $\sim$ 1000 K or between the UV absorption and the first IR emission is $t_i \simeq < h\nu_{IR} > /h\nu_{UV}$, or: $t_i \sim 0.1s$.

4.3 Size of interstellar PAH molecules

4.3.1 Evaluation from the mean emission color temperature

Considering the situation in a reflection nebula, the illuminating star has a temperature $T_* \sim 20000$ K and a spectrum with a maximum at 1500 Å. PAH having a strong absorption around 2000–2200 Å, we assume that the mean energy of absorbed photon is:

$$< h\nu_{abs} > = 6\text{eV} (\lambda = 2100 \text{ Å}).$$

The comparison between laboratory data and the spectrum of NGC 2023 indicates a mean emission temperature $T_{em} \simeq$800 K (Table 2). This value is an average. The actual process is a heating up to T_{peak} by the absorption of the UV photon and subsequent emissions, at variable temperatures, during the cooling. Assuming that the average T_{em} is close to the temperature at which the molecule has lost half of its energy, one reads:

$$\int_0^{T_{em}} C(T)\,dT = \frac{1}{2} \int_0^{T_{peak}} C(T)\,dT,$$

and the approximation (4.2) for $T_{peak} <$1300 K gives:

$$T_{em} \simeq 0.71 \text{ } T_{peak}. \qquad (4.5)$$

The relation between the energy of absorbed UV photon and the peak temperature,

$$h\nu = \int_0^{T_{peak}} C(T)\,dT,$$

gives, using (4.2):

$$h\nu_{abs} \simeq (N_t - 2)kT_{peak}^2/867 \text{ K}. \tag{4.6}$$

For the reflection nebula environment, one deduces a mean number of atoms:

$$< N_t >= 50. \tag{4.7}$$

This derivation is an approximation for a more complete calculation (Puget, Léger and Boulanger, 1985) that includes summations over: (i) the incident photon spectrum, (ii) the molecule temperature during cooling and (iii) the molecule size distribution.

It should be noted that the PAHs reported in figure 4 are relevant for molecules in reflection nebulae as far as the number of atoms is concerned.

4.3.2 Evaluation from stability against photo-thermodissociation

When the energy $h\nu$ of an UV photon has been absorbed by the molecule, the probability for ejecting one atom can be calculated by the theory of unimolecular processes. A preliminary result for reflection nebulae is that for:

$$N_t > 25 \text{ atoms}. \tag{4.8}$$

The molecule is destroyed slower than it can be reconstructed by the available mechanisms. Conditions (4.7) and (4.8) indicate a size distribution for PAHs, in the reflection nebula environment, that starts above 25 atoms and has a mean value of 50 atoms.

4.4 The most abundant known organic molecules in gas phase

The evaluation of the abundance of PAHs in the Interstellar Medium results directly from the interpretation of its Near IR and Far IR emissions. Grains and molecules absorb the incoming stellar flux according to their respective abundance N_i and specific cross section σ_i. As they re-emit in separate domains of wavelength, their emission fluxes Φ_i can be measured separately. One has:

$$\frac{\Phi_{NIR}}{\Phi_{FIR}} = \frac{< \sigma_{PAH}^{abs} > \cdot N_{PAH}}{< \sigma_{gr}^{abs} > \cdot N_{gr}}$$

The average absorption cross section $< \sigma^{abs} >$ can be estimated from molecular and solid state physics in different radiation fields (UV-Visible). Then, the relative abundance of PAHs results from the ratio Φ_{NIR}/Φ_{FIR}.

The IRAS mission gives an opportunity for estimating this ratio in many astronomical objects. Typical values are reported in table 3. Inspection of table 3 indicates that:

1. Involving several percent of the cosmic carbon, *PAHs are the most abundant organic molecules known at that date in the IM*. The abundance of other detected interstellar molecules is reported in table 4 (Duley and Williams, 1984). PAHs are in third position, far above molecules which are easily observed in radioastronomy. It is remarkable that such abundant species have remained undetected for a long time. This

is due to their *lack of simple signature in the radiowavelength domain* which used to be the royal, if not unique, way of detecting interstellar molecules.

Table 3: Ratios of Near IR to Far IR emissions of different objects and deduced fraction of the cosmic carbon involved in PAHs. The relatjve geometrical surface (S) of these molecules to that of graphite grains in current model emphasizes their possible rôle in the Interstellar Medium. Their expected contribution to the 2200 Å feature of the Extinction Curve is also reported. A_x^{2200} is the integrated cross section of the species x around 2200 Å.

	$\frac{\Phi_{NIR}}{\Phi_{FIR}}$	$\frac{M_{PAH}^c}{M_{cosmic}^c}$	$\frac{S_{PAH}}{S_{graph}}$	$\frac{A_{PAH}^{2200}}{A_{graph}^{2200}}$
Reflection Nebula	15%	3%	1.9	16%
Diff. Interst. Medium	30%	6%	3.8	33%
R. Cor. Austr.*	100%	20%	13.0	110%

* A specific molecular cloud, see Leene (1986).

Table 4: Abundance of different interstellar molecules expressed in number of atoms included versus total number of hydrogen nuclei.

Molecule	n_{at}/n_H
H_2	0.5
CO	10^{-4}
PAHs	10^{-5}
HCN, H_2CO	$< 10^{-7}$
$HC_{11}N$	10^{-9}

2. PAH geometrical surface is comparable or larger than that of classical grains. This points to their possible rôle in interstellar chemistry as catalyst (Omont, 1986).

3. The contribution of PAHs to the 2200 Å bump of the Extinction Curve is important. However, in most regions, these molecules alone seem not to be able to explain the feature intensity.

4.5 Are PAHs observable in absorption?

If PAHs are as abundant as we claim, a question immediately rises: why do not we observe them in absorption? Typical absorption cross sections are:

$$\nu \text{ C-H}, \quad 3.3 \ \mu\text{m}, \quad \sigma = 2.4 \cdot 10^{-20} \text{ cm}^2$$
$$\gamma \text{ C-H}, \quad 11.9 \ \mu\text{m}, \quad \sigma = 1.2 \cdot 10^{-19} \text{ cm}^2$$
$$\nu \text{Si-O}, \quad 9.7 \ \mu\text{m}, \quad \sigma = 1.4 \cdot 10^{-18} \text{ cm}^2$$

If a cloud, with optical depth in the Visual $\tau_v \sim A_v = 2.5$, is in front of a source its optical depth in the PAH bands will be:

$$\tau = N_{CH}\sigma, \text{ with } N_{CH} \sim 10^{-2} N_c,$$
$$N_c = 4 \cdot 10^{-4} N_H \text{ and } N_H = 2 \cdot 10^{21} A_v. \text{ Then :}$$
$$\tau_{3.3\mu m} = 10^{-2}$$
$$\tau_{11.3\mu m} = 5 \cdot 10^{-2}$$

These value are small and are quite difficult to observe. It is our hope that better signal/noise in the future will permit their detection. However, there is no conflict between the presence of PAHs at the level we have inferred and the present IR observations in absorption.

It should be noticed that the easily detected silicate absorption at 9.7 μm corresponds to the Si-O group that has an exceptionally high specific cross section.

A feature around 3.4 μm is observed towards the Galactic Center (Butchard et al., 1986). It is attributed to alkane Ch stretching modes (Bellamy, 1966). *The observation of a 3.4 μm band in absorption and a 3.3 μm band in emission* is curious at first glance. However, it can be explained if the saturated compounds are more abundant but *incorporated into grains*, whereas the aromatic ones are less abundatn but are *free flyer molecules*. Such a situation would lead to a strong selection effect for the emission: as already explained, the latter are the only ones that can emit at short wavelengths.

5 Open questions

The identification of PAHs in astronomical IR spectra as presented in this paper was not obtained by looking for the best spectrum in a big IR Atlas. The presence of large PAHs was inferred from discussing the possible composition of interstellar Very Small Grains and the spectroscopic agreement was obtained *afterwards*, as a check.

The possibily of *explaining at once the five main bands* by a single family of molecules is also an argument in favour of this explanation because the bands are always observed to occur together.

However, it is *not* a one to one spectroscopic identification with a given molecule but an identification with a family of molecules which have characteristic bands. Some points are still not clear such as the precise interpretation of the 11.3 μm band.

The content of this molecular family has to be made more precise in the future. Advances in that direction have been presented in the present paper: *compact PAHs give a better spectroscopical agreement than non-compact ones.* Coronene is no more the only molecule that fits the observations.

In this explanation, *tha fact that the molecules are free flyers is essential* in order to give the high conversion yield from UV to Near IR. It is needed to explain that 30% of the interstellar radiation is emitted by this process.

This identification of an important component of the Interstellar Medium, which was ignored before, has numerous implications on the physics of the medium (IR emission, Extinction Curve) and on its chemistry. It also gives new ideas for explaining the mystery of the Diffuse Interstellar Bands in the Visual.

Several problems remain however, some of them are:

5.1 Ionization of PAHs

In regions as Reflection Nebulae most of the PAHs should be ionized (Omont, 1986). *What is the IR spectrum of ionized PAH* is therefore an important question. Would it give such good a fit to the observed bands as the neutral PAHs?

The extracted electron of a PAH$^+$ is one delocalized π electron out of tens. One expects then only a small change in the vibration force constants and in the frequencies. But *is it as small as 1% which is the accuracy of the agreement on the 3.28 and 6.20 µm bands?* Laboratory data are needed to answer that question.

5.2 Dehydrogenation of PAHs in space

This point has to be investigated seriously. The initial arguments were: (1) the observation of solo H modes whereas in the laboratory species many H are in duo position; (2) the comparison of C-H and C-C mode intensities in the observed and in the calculated spectra points to a strong deficiency in H; (3) the binding energies of H and C atoms to the molecule are quite different (4.8 eV and > 7.5 eV) and allow dehydrogenation.

The study of photo-thermodissociation (Désert and Léger, 1987) tends to invalidate argument argument (1) as dehydrogenation seems to be either weak (large molecule) or almost completed (small ones). The range of size for partially dehydrogenated species is rather narrow and it seems difficult to explain the dominating presence of the solo H band this way.

A dehydrogenated PAH is a small fragment of graphite and there is a severe need for laboratory data on IR spectra of carbon clusters specially with the new suggestions of spheroidal particles.

Obtaining *laboratory spectra of partially and totally dehydrogenated PAH* would be a significative advance in the field.

Acknowledgements

The authors want to thank W. Schmidt, E. Clar, J.C. Roussell and Setton for important discussions.

References

Allamandola, L.J., Greenberg, J.M., Norman C.A.: 1979, Astron. Astrophys. **77**, 66.

Allamandola, L.J.: 1984 In "Galactic and Extragalactic IR Spectroscopy", eds. Kessler M.F., Phillips J.P., Reidel, Holland.

Allamandola, L.J., Tielens, A.G., Barker, J.R.: 1985, Astroph. J. Lett. **290**, L25.

Ashcroft, N.W., Mermin, N.D.: 1976, "Solid State Physics", ed. Holt, Rinehart and Winston.

Bellamy, L.J.: 1966, "IR Spectra of Complex Molecules", Wiley.

Birks, J.B.: 1970, "Photophysics of Aromatic Molecules", Wiley-Interscience.

Butchard, I., McFadzean, A.D., Whittet, D.C.B., Geballe, T.R., Greenberg, J.M.: 1986, Astron. Astrophys. Lett. **154**, L5.

Clar E.: 1964, "Polycyclic Hydrocarbons", Academic Press.

Cohen, M., et. al.: 1986, Astrophys. J., in press.

Combes M., et. al.: 1986, Nature **321**, 266.

Cyvin, S.J.: 1982, J. of Mol. Struct. **79**, 423.

Cyvin, B.N., Brunvoll, J., Cyvin, S.J.: 1984, Spectrosc. Let. **17** (9), 559.

Désert, F.X., Boulanger, F., Léger, A., Puget, J.L., Sellgren, K.: 1986, Astron. Astrophys., **159**, 328.

Dresselhaus, M.S., Dresselhaus, G.: 1981, Adv. in Phys. **30**, 139.

Duley, W.W., Williams, D.A.: 1981, M.N.R.A.S. **196**, 269.

Duley, W.W., Williams, D.A.: 1984, Interstellar Chemistry, Acad. Press.

Gillett, F.C., Kleinmann, D.E., Wright, E.L., Capps, R.W.: 1975, Astrophys. J. Lett. **198**, L65.

Huffman, D.R.: 1977, Adv. in Phys. **26**, 129.

Krumhansl, J., Brooks, H.: 1953, J. Chem. Phys. **21**, 1663.

Leene, A.: 1986, Astr. Astrophys. **154**, 295.

Léger, A., Puget, J.L.: 1984, Astron. Astrophys. Let. **137**, L5.

Léger, A., d'Hendecourt, L.: 1987, in "Polycyclic Aromatic Hydrocarbons and Astrophysics", ed. A. Léger, L. d'Hendecourt and N. Boccara, Nato ASI Serie, Reidel, 1987.

Léger, A., d'Hendecourt, L., Schmidt, W.: 1988, in preparation.

Omont, A.: 1986, Astron. Astrophys., **164**, 159.

Puget, J.L., Léger, A., Boulanger F.: 1985, Astron. Astrophys. Let. **142**, L19.

Parmenter, C.S.: 1982, J. Phys. Chem. **86**, 1735.

Reif, F.: 1965, "Fundamental of Statistical and Thermal Physics, McGraw Hill.

Roussell, J.C.: 1984, private communication.

Russell, R.W., Soifer, B.T., Willner, S.P.: 1978, Astrophys. J. **220**, 568.

"Sadtler Standard Spectra": 1959, Midget edition.

Sellgren, K., Werner, M.W., Dinerstein, H.L.: 1983, Astrophys. J. Lett. **271**, L13.

Sellgren, K.: 1984, Astrophys. J. **277**, 623.

Sellgren, K., Allamandola, L.J., Bregman, J.D., Werner, M.W., Wooden, D.H.: 1985, Ap. J. **299**, 416.

Turro, N.J.: 1978, "Modern Molecular Photochemistry", Menlo Park, Ca.: Benjamin/Cummings.

Willner, S.P., Soifer, B.T., Russell, R.W., Joyce, R.R., Gillett, F.C.: 1977, Astrophys. J. Lett. **217**, L121.

Willner, S.P.: 1984, Same book as Allamandola, 1984.

LABORATORY PRODUCTION OF AMORPHOUS SILICATES

W. Krätschmer
Max-Planck-Institut für Kernphysik, Heidelberg, F.R.G.

1 Introduction

This is a report on a technical subject, namely how to get a silicate into an amorphous state. Usually, the reverse process, i.e. to crystallize an amorphous material, takes place by itself due to the combined action of temperature and time and thus does not present experimental problems. Because of its spontaneous decay, the amorphous state is metastable and the conversion to the crystalline phase at a given temperature may take seconds or billions of years depending on the activation energy involved. For the silicates which are of interest here, the transition takes place within a reasonable amount of time at temperature in the order of 500 to 1000°C.

It should be noted that the term "amorphous" does not define a specific state (as e.g. "crystalline" does); it names a variety of states with may differ from one another by the degree of small (i.e. molecular) scale order. Thus the optical properties of amorphous silicates of a given chemical composition usually show variations depending on the degree of the structural disorder of the material.

Since a perfect crystal (in the sense of solid state physics) has to be infinitely large, any grain boundary introduces distortions. Thus small grains, for which the number of molecules in surface sites is not negligible compared to the number of molecules in the interior may, in a more general sense, be called amorphous as well.

The report will therefore start with very small grains, i.e. molecular clusters. The further topics are chosen according to the techniques applied to make the silicate amorphous. The list of references is not intended to be complete; only a few works are picked out to characterize each topic. Additional references can be found in these papers.

2 Silicate molecular clusters

Spectroscopic data on the molecular clusters $(SiO)_n$ have been obtained by trapping SiO vapour into a cryogenic matrix. Upon warming up, the trapped SiO molecules diffuse through the matrix and can stick together to form larger clusters. In a recent paper by Khanna et al. (1981), the structure of clusters with n=2 and 3 were investigated. The dimer is unsymmetrical while the trimer seems to be cyclic, i.e. the Si and O atoms are

95

E. Bussoletti et al. (eds.), Experiments on Cosmic Dust Analogues, 95–102.
© 1988 by Kluwer Academic Publishers.

located at the corners of equal-sided triangles. When the matrix annealing is enhanced up to the sublimation point of the matrix and beyond, the molecular features disappear and the bulk silicate absorptions become pronounced.

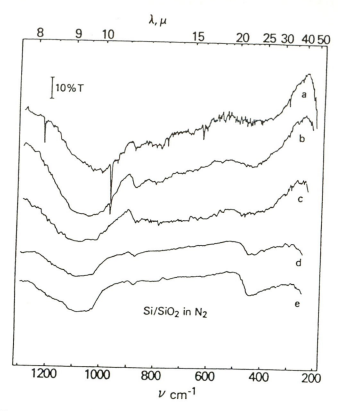

Figure 1: Transition from molecular to bulk material monitored using infrared spectroscopy: (a) after 8 h at 30 K plus an additional 2 h at 32 K; (b) after 8 h at 30 K plus 4 h at 32 K; (c) after warming the sample to ∼ 50 K and subliming the N_2 host matrix; (d) after 3 days at room temperature in the evacuated cell; (e) after annealing the residue at 500 K for 6 h.

This is illustrated in figure 1, where, according to the authors, the bulk silicate has the composition Si_2O_3. In the initial spectrum, narrow lines by molecular SiO (at 8.2 μm) and the trimer $(SiO)_3$ (at 10.2 μm) can be recognized. In the later states, two broad absorptions can be discerned in the bulk material; the feature near 10 μm originates from the Si-O stretching and the band near 20 μm from the Si-O bending vibration in the solid. These two bands are prominent in all kinds of silicates: they may be called the fingerprints of silicates. The details in band shapes and positions depend on the structure and chemical composition of the silicate. The authors observed that the absorption features tend to become narrower after progressive annealing. This indicates a gradual transition from the amorphous to a more ordered structure. One notices in figure 1 that the bulk material does not exhibit distinct sub-structures within the absorption features. The presence of sub-bands is usual an indicator for the well crystallized state, in which the optical properties depend on the orientation of the polarisation of the radiation with respect to the crystallographic directions.

It would be interesting to study molecular clusters (or very small partices) of more interstellar-like silicates i.e. compounds containing Fe and Mg and the tetrahedal SiO_4 ion rather than SiO.

3   Evaporation and quenching

The classical method to reach the amorphous or glassy state is by rapid quenching of the liquid. This has in fact been applied by Dorschner et al. (1986) to produce amorphous bronzite $FeSiO_3$. The molten silicate was quenched in a mercury bath. In the technique which is more commonly applied, the silicate to be studied is vapourized in the presence of a (usually inert) quenching gas. The gas cools the vapour molecules by collisions, leading to supersaturation of the vapour. Finally, condensation takes place into small particles which are, if the conditions are chosen properly, amorphous in structure. The chemical composition of the grains has to be checked to make sure that no modification of the silicate took place during condensation.

As an example for silicate evaporation by a laser beam I want to mention the work of Stephens and Russell (1979), and for evaporation by a burning carbon arc that of Day (1975) and Krätschmer and Huffman (1979). The latter authors used a hollowed out carbon electrode, in which the silicate —in this case crystalline olivine Mg_2SiO_4— was deposited. The arc burned in oxygen or air of a few 100 Torr. Day (1975) reported that he obtained small grains of crystalline olivine. However, in later studies, particles of amorphous olivine were produced as well. These difficulties indicate that the conditions of arc burning and gas quenching are crucial if one wants to produce an amorphous olivine condensate. The laser evaporation of olivine Mg_2SiO_4 and enstatite $MgSiO_3$ when performed in the presence of a suitable quenching gas leads to amorphous small particles. Stephens and Russell (1979) used oxygen for quenching since they observed that the silicate tends to decompose otherwise. The measured grain absorptivities and emissivities for crystalline and amorphous enstatite and olivine are shown in figure 2. Notice the substructure in the crystalline dust absorption, which is absent in the optical isotropic glassy material. The most pronounced effects can be recognized in the shifts and changes in the widths of the features of the amorphous as compared to the crystalline material. Within the spectra of the condensates, the wavelength position of the peak absorption depends on the chemical composition of the substance. The peak of the silicate with the lower molecular oxygen concentration (i.e. $MgSiO_3$) is shifted to shorter wavelength. Effects of this kind certainly help to gain information on both, the composition and structure of the interstellar silicates.

In the techniques so far described, the chemical composition of the condensate is largely determined by the initial material. To overcome this inflexibility, the different ingredients required to synthesize a silicate (e.g. MgO and SiO) can be vaporised from two (or more if necessary) different furnaces in such a way, that the vapours of the constituents are well mixed at the onset of gas quenching and condensation.

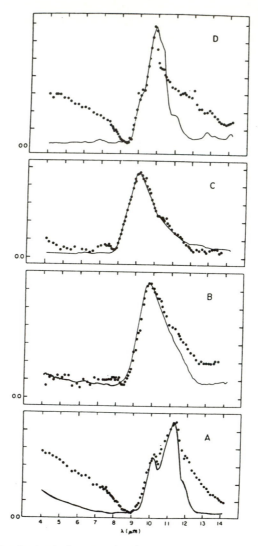

Figure 2: Absorption (extinction) spactra (*dots*) of A: ground olivine; B: olivine condensate; C: enstatite condensate; and D: ground enstatite, superposed on the corresponding
emissivity data (*solid line*).

These kind of experiments were carried out by Day and Donn (1978) and Nuth and Donn
(1982), especially to study the nucleation process.

A shortcoming of methods which merely produce small particles lies in the
fact that data are produced from which no reliable optical constants (i.e. the refraction
and absorption index as function of wavelength) can be obtained. The reason is that the

small particle extinction spectra not only depend on the silicate composition but also on the size and shape of the grains. These parameters are rather difficult to control. Even in the case that all the individual grains are spherical, the particles tend to clump together, and in this geometrical arrangement the small particle extinction is not the same as for a system of isolated spheres. Therefore it is advantageous to obtain the silicate samples in such a form (e.g. as coated films) that the optical constants can be derived by trasmission and/or reflection measurements.

4   Sputtering

The non-thermal evaporation of silicates by ion sputtering and the subsequent deposition of the silicate molecules on a substance leads to amorphous coatings. The sputtering of olivine by argon ions of the keV energies and the measurement of the optical constants of the obtained coating of amorphous olivine were performed by Krätschmer (1980). In a series of papers, Day (1979 and 1981) has extended the study to Fe and Mg silicates of the type -SiO$_3$ and -SiO$_4$ by a modification of the sputtering technique which he called reactive sputtering. He used electrodes of Si-Mg and Si-Fe alloys of appropriate composition and sputtered these in an oxygen containing atmosphere. The small particle absorption calculated from the optical constants for the variuos silicates are shown in figures 3a and 3b. He finds a remarkably small peak absorption in the case of Fe$_2$SiO$_4$. This is interesting since IR astronomical data suggest that the interstellar silicates do not absorb very strongly at 10 μm. (Capps and Knacke, 1976). According to the results of Day, additional information on the concentration of Fe and Mg in the interstellar silicates may therefore be obtained.

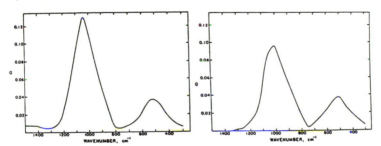

Figure 3a: Extinction efficiency Q plotted versus wavenumber for amorphous MgSiO$_3$ sphere with $a = 0.1$ μm and for amorphous MgSiO$_4$ sphere with $a = 0.1$ μm.

5   Ion bombardment

It has long been known that high doses of ionizing radiation, especially those produced by Heavy ions, very efficiently disrupt crystalline structure (see e.g. Fleischer et al., 1975). For heavy ions, the damage is large close to the end of range, where the electronic stopping power reaches a maximum. The saturation dosage, at which the damage becomes stationary, is usually reached when the surface of the solid is hit by about one ion per square Ångström.

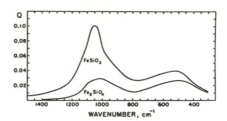

Figure 3b: Calculated extinction efficiency for spheres ($a = 0.1$ μm) of FeSiO$_3$ and FeSiO$_4$ in the mid-infrared.

Krätschmer and Huffman (1979) have exposed crystalline olivine to such a dose of 1.5 MeV Ne ions and have measured the optical constants of the radiation damaged silicate. The changes in the optical properties are striking, and the small particle extinction is similar to that of amorphous olivine grains obtained by the evaporation and quenching technique described above. Figure 4 compares both kinds of data. Even though the doses appear to be rather high, intense stellar winds may in fact contribute to the amorphous state of the interstellar grains.

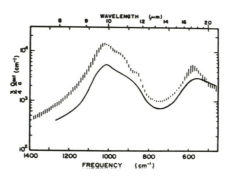

Figure 4: The upper curve shows the calculated extinction spectrum for Rayleigh particles of the disordered olivine. The lower, solid curve shows measurements of extinction taken on amorphous olivine smoke produced in a carbon arc in air. The extinction is normalized per unit volume of the solid which, in the case of spheres, is $3Q_{abs}/4a$. Both kinds of structurally disordered olivine show similar extinction spectra.

6 Precipitation from solution

Silicates of olivine and enstatite-like chemical composition were produced by Day (1976) in a wet-chemical manner, namely from the water soluble sodium silicates and iron or magnesium chlorides, which, when mixed together, give a precipitation of an unsoluble Fe or Mg silicate containing water and, under specific conditions, OH groups as well. The IR spectroscopic evidence and the appearence of the silicate under the electron microscope suggests, that the precipitate contains layer silicates similar to e.g. talc or serpentine, however with a very distorted structure. This synthetic material appears to be related to the silicates found in a class of primitive meteorites called carbonaceous chondrites. The spectra of both kinds of silicates are shown in figure 5. It is not yet clear whether the carbonaceous chondrites are directly related to comets or to some other form of interstellar material. Layer silicates of highly distorted structure were proposed as carriers of the interstellar 10 μm feature (Zaikowsky et al. 1975), however the spectral fits are less good as compared to those amorphous silicates which are water and OH-group free.

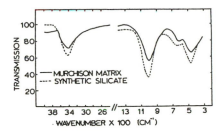

Figure 5: Infrared absorption spectrum (via KBr pellet technique) of fast-grown silicate, compared to Murchison matrix material. These spectra are qualitative; nothing should be inferred from the relative strengths of the absorption features in the two spectra.

7 Conclusions

As already mentioned, amorphous states may differ from one another by the degree of order. It may be of interest to known which of the methods yields the most disordered silicate. To investigate this problem, the empirical rule can be applied, according to which the width of an absorption feature increases and the peak absorption decreases with the degree of disorder. Figure 6 compares the small particle extinctions of amorphous olivine Mg_2SiO_4 obtained by the methods 1 to 4. The figure includes extinctions which were calculated from optical constants and data which were measured on small particles. Since the experimental data may be distorted by particle clumping effects, which also tend to systematically reduce the absolute values of extinction, all data were normalized to unity in the maximum. The sputtered silicate of Day (1979) exhibits the largest, the radiation damaged olivine of Krätschmer and Huffman (1979) the smallest width. However, the differences in width and peak positions of all the amorphous silicate are relatively small. Thus the most disordered silicates apparently are produced by sputtering. Following the empirical rule, the sputtered olivine shows the lowest peak absorption as well.

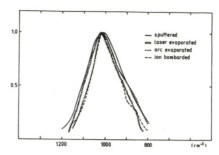

Figure 6: Normalized small particle extinction of amorphous MgSiO$_4$, produced by the various techniques described in the text.

It is of considerable advantage if the method allows the reliable measurement of the optical constants. At least in theory, the extinctions of any kind of particles can be calculated from these. Since interstellar grains probably represent a system of very well isolated small particles, a state which can hardly be simulated experimentally, the calculated extinction may allow a more realistic comparison to the interstellar data than the extinctions measured in the laboratory. This does not mean that the dust extinction measurements are worthless. They provide valuable data to cross-check the calculated extinctions. For example they allow a test of whether the bulk optical constant are, as in most cases is tacitly assumed, in fact valid for the small particles as well. Having at hand both sets of data, namely measured optical constants and measured small particle extinctions, the interstellar silicates can be interpreted in a safer way in terms of their laboratory analogous materials.

References

Capps, R.W., Knacke, R.F.: 1976, Ap. J. **210**, 76.

Day, K.L.: 1976, Icarus **27**, 561.

Day, K.L., Donn, B.: 1978, Ap. J. **222**, L45.

Day, K.L.: 1975, Ap. J. **199**, 660.

Day, K.L.: 1979, Ap. J. **234**, 158.

Day, K.L.: 1981, Ap. J. **246**, 110.

Dorschner, J., Friedemann, C., Gürtler, J., Henning, Th., Wagner, H.: 1986, Mon. Not. R. Astr. Soc. **218**, 37.

Fleischer, R.L., Price, P.B., Walker, R.M.: 1975, Nuclear Tracks in Solids, Berkeley, University of California Press.

Khanna, R.K., Stranz, D.D., Donn, B.: 1981, J. Chem. Phys. **74**, (4), 2108.

Krätschmer, W., Huffman, D.R.: 1979, Astrophys. Space Sci. **61**, 195.

Krätschmer, W., Halliday, I., MacIntosh, B.A. (eds.): 1980, Solid Particles in the Solar System, IAU 1980, 351.

Nuth, J.A., Donn, B.: 1982, Ap. J. **257**, L103.

Stephens, J.R., Russell, R.W.: 1979, Ap. J., **228**, 780.

Zaikowski, A., Knacke, R.F., Porco, C.C.: 1975, Astrophys. Space Sci. **35**, 97.

ION BOMBARDMENT: TECHNIQUES, MATERIALS AND APPLICATIONS

G. Strazzulla

Istituto di Astronomia, Università di Catania, Catania, Italy

1 Introduction

The interaction between fast colliding ions and solid targets produces several effects many of which have been studied in recent years with a view to their astrophysical relevance. The penetrating ion loses energy both as a consequence of elastic collisions with target nuclei and of excitations and ionizations of target atoms and molecules. The bombarded sample can be deeply modified both because material can be removed from it (sputtering) and because of possible alteration of its bulk chemical composition. The study of some of these modifications is the object of experimental research in some laboratories through the world, among which in Catania (Italy). To observe the induced modifications one has to use techniques that permit to characterize directly the sample before, during and after the bombardment or to look to the products of the irradiation as e.g. to emitted atoms, ions, molecules, electrons or photons (from visible to X wavelengths). The purpose of this paper is to review some work made in our laboratory in the last 5 years (with a quotation to the results obtained in other laboratories), by describing irradiated materials, used techniques and astrophysical applications.

2 Materials

Since the measurements, in 1978 (Brown et al. 1978), of very large sputtering yields for water ice bombarded by keV-MeV ions, molecular solids have been intensitively studied. In particular they have been measured:

– sputtering yields (number of target atoms or molecules released per impinging ion)
– type and amount of (mainly neutral) molecules released during the bombardment or during the warm-up (after irradiation) phase of the experiment
– alteration in the stoichiometry of the bombarded target.

The bombarded targets (e.g. H_2O, CO_2, NH_3, SO_2, S) have been chosen because of their relevance in astrophysics. Being them important and even dominant constituents of external planets and satellites, comets, interstellar and interplanetary grains. In space these targets are continuously bombarded by energetic ions from solar

E. Bussoletti et al. (eds.), Experiments on Cosmic Dust Analogues, 103–113.
© 1988 by Kluwer Academic Publishers.

wind and flares, planetary magnetospheres, stellar winds and galactic cosmic rays. Another class of materials that have been extensively studied are frozen hydrocarbons and various organic materials. We have studied in particular:

- Frozen gases: $C_6 H_6$, $C_6 D_6$, CH_4 and mixtures with H_2O. These are rapidly converted to new organic materials stable at above the room temperature.
- Aliphatic and aromatic polymers: polyethylene, polypropylene, polystyrene, polyimide. These target are astrophysically relevant both because they could get insight into the ion-induced modification processes and because of their possible use in space technology for which radiation-resistant materials are requested.
- Complex molecules: 1,4 diaminobutane. The existence in space of complex molecules is well supported by lots of evidences. 1,4 diaminobutane could be, in particular, representative of a fraction of the cometary organic inventory (Krueger and Kissel, private communication).
- Biological compounds: amino acids, proteins, biological heteropolymers (melanin), bacillus subtilis. These target are relevant both to test the reliability of the panspermia hypothesis (i.e. Hoyle and Wickramasinghe 1981) and to get insight into problems connected with risks from space flights and of radio-protection in general.

3   Techniques

The above described targets have been bombarded by ion beams (H, He, heavy ions) at energies between 10 keV and 2 MeV. The beams were obtained at the ion implanter and VdG accelerator of the Catania University. Ion current between few nA/cm^2 and 100 $\mu A/cm^2$ can be obtained with fluences between 10^{10} and 10^{19} ions/cm^2. Vacuum in the scattering chamber is better than 10^{-7} mbar and the target temperature can be changed from 10 to 400 K. In table 1 some of the effects we have studied are reported together with the methods chosen for their quantitative measurements and used techniques. Some of these techniques are discussed in some detail in the next.

3.1   Ion backscattering

For sake of simplicity I will describe this technique in a particular experiment: the bombardment of frozen methane by 1.5 MeV H^+ beams and the measurement of the cross section for the conversion of the target to a new organic residue stable at room temperature and above (Foti et al. 1984). Frost methane layers were accreted on silicon substrates placed on a cold finger into the scattering chamber, by admitting methane gas as a broad stream. During film accretion the gas inlet has been monitored by a quadruple mass spectrometer. The proton backscattering technique (PBS) has been used to measure "in situ" the thickness of the accreted layer. The backscattered protons were detected with a solid-state detector, placed at $\theta = 165°$ with respect to the incoming beam, and the energy spectrum was recorded in a 4096 multichannel analyser. For protons colliding with surface atoms, the energy detected will be $E = kE_o$, where E_o is the energy (1.5 Me)V of the beam and k is a tabulated quantity depending on the target elemental composition (Mayer and Rimini, 1977).

Table 1: Summary of some of the effects induced on targets bombarded by fast fast ions together with methods and techniques used in our laboratory to measure them.

Effect	Method	Technique
sputtering	thickness variation after a given dose	Ion Backscattering Ion induced X-ray emission Interferometry
released molecules	analysis of gas-phase products	Quadrupole mass spectrometry
production of organic residue (o.r.)	amount of carbon left over after warm-up	Ion backscattering Ion induced X-ray emission
Emitted photons	detection of emitted photons	Photometry/spectroscopy
characterization of newly produced o.r.	Remote analysis Spectroscopy	IR UV-Vis SIMS ESCA RAMAN

The experimental resolution in the proton spectrum is about 25 keV and the signals which come from carbon or silicon are well separated in energy. The depth perception of PBS depends on the energy loss of protons in the given target; when a proton penetrates the deposited layer, the initial energy is reduced of a amount:

$$\Delta E = |\epsilon|\, T \tag{1}$$

where T is the film thickness in atoms cm^{-2} and $|\epsilon|$ a constant (eV/atoms/cm^2) which depends on the mass and on the specific energy loss of protrons in the given target. For frozen CH$_4$ $|\epsilon| = 13.8$ (keV/10^{18} atoms/cm^2). The film thickness can be converted in μm if the target mass density (g/cm^3) is known. The measured thickness of deposited methane layers ranged between 10^{18} and $4 \cdot 10^{19}$ carbon atoms/cm^2 which correspond to 0.4 and 15 μm respectively. This is better described in figure 1, where the spectra obtained from three different films of CH$_4$ (10 K) frozen, deposited on a bulk silicon crystal are shown. Whichever is its thickness, CH$_4$ being the first material that protons see, the carbon edge in the spectrum remains unchanged as shown, for three thicknesses, in figure 1. The back edge of the carbon peak shifts to lower values of energy according to equation 1 giving a peak area proportional to the thickness of the layer. Again the proton looses energy before colliding with silicon substrate, so that the Si edge exhibits a shift to lower energies as the layer of the material deposited on it becomes thicker (see figure 1). The measurement of the Si shift is once again a measurement of the thickness of the deposited layer.

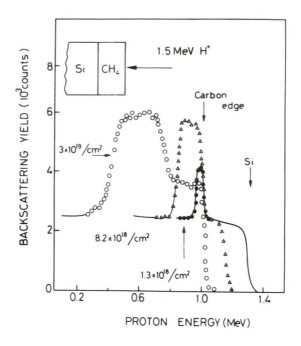

Figure 1: Energy spectra of 1.5 MeV protons backscattered from three different films of CH_4 frozen at 10 K on a bulk silicon cristal.

After deposition at 10 K, the CH_4 films have been bombarded at different doses by the same 1.5 MeV proton beam. During bombardment the thickness has been monitored and found not to change appreciably implying that sputtering is negligible. After bombardment one expects that by warming up the sample from 10 K to room temperature it evaporates and vapour is expelled by the pumping system. This has been, in fact, observed, for example, after the bombardment of water ice. But, when hydrocarbon containing samples are warmed up after the bombardment, this is not the case. Under the action of bombarding beam, CH_4 film has been trasformed into a stable organic material that at eyes appears as a yellow-brown residue on the silicon substrate. The thickness of these residues can be measured in situ (in C atoms/cm^2) by the same tecnique. In figure 2 the energy spectra obtained at $T = 300$ K from three residues, whose measured thickness is given, are shown. From figure 2 we also see that the carbon peak does not exhibit sharp edges in the backscattering spectra, this points out that the solid residue is not uniform. To test the dependence of the polymerization process on the thickness of the initial CH_4 layer, we have bombarded films as thick as some 10^{19} C-atoms/cm^2 that for a methane frost density of 0.55 g cm^{-3} corresponds to tens of micrometers.

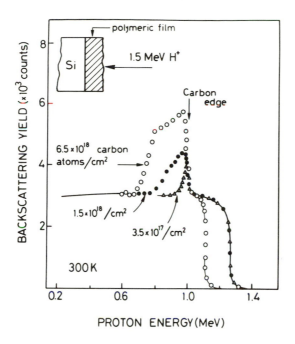

Figure 2: Energy spectra of 1.5 MeV protons backscattered from three organic residues at room temperature.

However, the thickness of the layers has been always maintained lower than the penetration depth (range) of 1.5 MeV proton in CH_4 ($1.3 \cdot 10^{20}$ C-atoms cm^{-2}).

It has been shown (Foti et al. 1984) that the thickness of the residue is correlated to the initial thickness of the methane layer by:

$$T = T_0(1 - e^{-\sigma\phi}) \tag{2}$$

where

T in C-atoms cm^{-2}	is the thickness of the residue,
T_0 in C-atoms cm^{-2}	is the deposited CH_4 thickness,
σ in cm^2	is the cross-section of the process,
ϕ in protons cm^{-2}	is the total implanted dose.

The σ value measured for 1.5 MeV H^+ is $0.4 \cdot 10^{-16}$ cm^2. The exponential dependence of the thickness of the residue points out that the investigated phenomenon is a volume process interesting all of the ion path into the film. Unfortunately, the used technique does not give information on the hydrogen content, however, quadrupole mass spectrometry, the nuclear-reaction technique performed on thin samples and remote IR analys show that the initial CH_4 loses preferentially hydrogen and the residues contain long-chain molecules probably mixed together with some carbon precipitates; the ratio C:H changes from 1:4 to 1:2.5. This as a consequence of the high-energy deposition rate $(100 \text{ J s}^{-1} \text{ g}^{-1})$ from the bombarding proton which induces in a small region (50 Å) around the ion path the formation of H and CH_m radicals. The recombination of H radicals gives rise to the observed high yield H_2 production and the recombination of CH_m radicals produces the stable residing molecules in the target. The production of the new materials depends on the specific energy deposition by incoming ion and on the mechanism of its redistribution among the atoms of the original material.

3.2 Ion induced X-ray emission

Also in this case I will describe a concrete example: the measurements of the sputtering yield of sulphur bombarded by 1 MeV He^+ beams. This material has been studied because of its relevance on Io, whose sulphur rich surface is bombarded by intense fluxes of ions from the Jovian magnetosphere. 1.0 MeV He beams were used to bombard thin $(1-20 \cdot 10^{17} \text{ S atoms/cm}^2)$ accreted, by depositing sulphur vapours, onto a cold bulk finger (Figure 3). The temperature of the finger was changed between 77 and 300 K. The thickness of accreted targets has been measured "in situ" by using the ion induced X-ray technique.

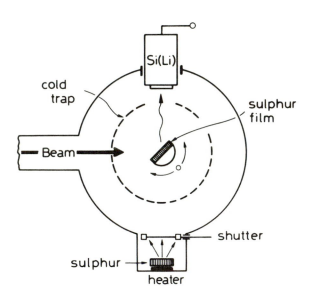

Figure 3: Schematic view of the experimental apparatus.

To this end a Si (Li) X-ray detector (FWHM=200 eV at 6 keV) has been interfaced with the scattering chamber through a mylar window. The solid angle subtended by the detector was 13 msterad. Targets were positioned at 45° into respect to the beam and detector axes, as shown in figure 3. By bombarding the sample we detect the characteristic X-lines emitted from sulphur (figure 4, upper section, dashed area) at energy $K_\alpha = K_\beta = 2.3$ keV. In the same figure 4 (lower section) the characteristic X-ray lines ($H_\alpha = 7.5$ keV and $K_\beta = 8.3$ keV) emitted from a thin Ni film ($1.4 \cdot 10^{17}$ Ni atoms/cm^2) used as a marker, are reported.

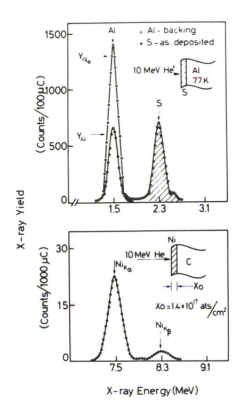

Figure 4: X-ray spectrum obtained by 1 MeV He bombardment of aluminum bulk (Y-AL$_o$) and sulphur-covered aluminum (Y-S and Y-A1) (upper section) and of a standard (calibrating) Ni target (lower section).

By comparing the total yields (counts) from the sulphur peak with those from the Ni marker, we can determine the thickness of the sulphur film deposited on the aluminum substrate (for details see Torrisi et al. 1986). The X-ray energy spectrum recorded with the Si(Li) detector contains also the lines ($K_\alpha = K_\beta = 1.5$ keV) emitted from the aluminum substrate (see figure 4, upper section). This experimental datum, allow us to determine the sulphur thickness by following another procedure. The Al yield, before sulphur deposition, is shown in figure 4 (open circles) and after sulphur deposition it decreases of an amount (a factor two in the case in figure 4) depending on the thickness of sulphur layers deposited on it. The reduction of Al yield is related to the strong energy dependence of X-rays production, because of the energy loss of incoming beam in sulphur. The reduction, measured in terms of normalized yield YAl (after sulphur deposition)/YAlo (before sulphur deposition) can be related to the S–thickness. Figure 5 shows an example of erosion experiment and the comparison between sulphur thickness measurements executed by Al yield variations and by S–yield variations (using Ni standard reference).

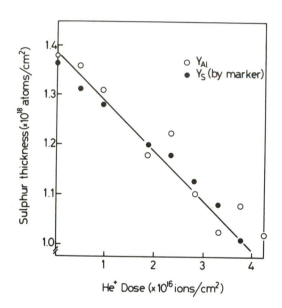

Figure 5: Sulphur thickness variation of a deposited film plotted vs. He dose for two different tecknique of thickness measurements.

The thickness of the as deposited film is estimated, with both methods, about $1.36 \cdot 10^{18}$ atoms/cm^2 and erosion with 1.0 MeV helium beam (300 nA) is followed point by point, up to a dose of $4 \cdot 10^{16}$ ions/cm^2. We get an erosion yield Y=9.5 S atoms/ion.

4   Applications

In "simulation" experiments of this kind the important parameters to reproduce are: mass and energy of incoming ion, composition and physical state (pressure, temperature, density, etc.) of the target, dose rate and total implanted dose. Let me briefly examinate these points:

(i) Ions mass and energy. In astrophysics we have virtually to do with all masses (generally H$^+$:He$^+$: (C,N,O)$^+$: (other ions) = 1:10^{-1} : 10^{-3} : 10^{-4}) and energies between few eV (thermal energy of interstellar gas to GeV and more (energetic cosmic rays). However the solid state effects we are talking about are attributable to ions between 1 keV and few tens of MeV. In the laboratory these energies are easily obtained by using low-energy electrostatic accelerators.

(ii) Target composition and state. In astrophysics we can roughly divide the solid state components in three species: a refractory component (mainly amorphous silicates and carbonaceous materials) as, in interstellar and circumstellar grains or rocky planets, satellites and asteroids; a frozen volatile component (mainly ices of H_2, NH_3, CH_4 etc.) as in mantle of interstellar dust or ice caps and surfaces on external planets satellites and comets; and exotic organic solid materials.

Although requiring an accurate technology (e.g. for the preparation of cold surfaces) the simulation of the target is not impossible "in principio" in the laboratory. The problems arise from the fact that often one do not know the initial composition of the target i.e. that at time t=0 before the beginning of the irradiation often occuring together with the formation process of the solid. Moreover, some astrophysical targets are surrounded with an atmosphere (e.g. a planet or a comet near the Sun) that changes the spectrum and the effects of incoming particles. However these atmosphere have been also simulated in the laboratory by maintaining controlled pressure in the vacuum chamber. Experiments with solid refractory materials, molecular solids and exotic compounds (as polymers) also surrounded by atmospheres have been performed in a temperature range between 10 and 500 K and pressure between 10^{-4} and 10^{-8} mbar.

(iii) Dose rate. Although in astrophysics the flux of energetic particles varies of many orders of magnitude in the various scenarii (see table 2) however it is many order of magnitude lower than attainable in the laboratory (10^{10} ions cm^{-2} sec^{-1}) and all of the effects (erosion, molecular production, clustering of fluffy particles which depend on the dose rate should be then regarded as meaningless for astrophysical applications.

(iv) Total dose. A rough evaluation of the total ion dose in astrophysics (column 5 in table 2) is obtainable by multiplying the dose rate for the duration of the exposure (column 4 in table 2). Likely the total doses are quite higher than usable in the laboratory. However resurfacing agents (e.g. vulcanism or meteorite impact or ice mantle accretion) often change the considered target whose layers suffer an effective irradiation much lower than reported e.g. in table 2.

With this in mind it is however important to note that the application of the experimental results to astrophysical scenarios ranging from interstellar dust to most of the objects in the Solar System has been and is a very fruitful research field. The best examples of succesful applications come, in my opinion, from the findings of Voyager encounter with the Uranian system and of Giotto and Vega encounter with Halley's comet. Voyager discovered in fact that the surfaces of rings, shepherds and minor satellites could be an organic polymer probably formed by the intense ion fluxes discovered by the same spacecraft. Those fluxes are capable of turning the methane that is likely incorporated in water ice at Uranus into a black amorphus polymer (Kerr 1986, Smith et al. 1986). The presence of residual carbon left over by ion irradiation of Uranian ring was predicted by Cheng and Lanzerotti (1978) and the build-up of organic polymers on/in the Uranian satellites by our group (Calcagno et al. 1985).

Table 2: Short summary of typical astrophysical sources of energetic protons.

Proton source	Energy (MeV)	Flux	Irradiation time(yrs)(cm^{-2})	Total dose
Solar wind (at 1AU)	10^{-3}	$3 \cdot 10^8$cm^{-2}sec^{-1}	$4.6 \cdot 10^9$	$4.4 \cdot 10^{25}$
Solar Flares (1 large event per yr at 1AU)	≥ 1	10^{10}cm^{-2}yr^{-1}	"	$4.6 \cdot 10^{19}$
Modulated Galactic Cosmic Rays (E in MeV)	≥ 10	$6 \cdot 10^{-3}E^{-1.75}$cm^{-2}yr^{-1}	"	$3 \cdot 10^{22}E^{-1.75}$
Jovian Magnetosphere (R$_j$ = Jovian radius)				
Io(6R$_j$)	1	$2 \cdot 10^6$ cm^{-2}sec^{-1}	"	$3 \cdot 10^{23}$
Europa (9.5R$_j$)	1	$3 \cdot 10^7$ " "	"	$4.5 \cdot 10^{24}$
Ganimede (15R$_j$)	1	$3 \cdot 10^6$ " "	"	$4.5 \cdot 10^{23}$
Callisto (26.6R$_j$)	1	$6 \cdot 10^4$ " "	"	$9 \cdot 10^{21}$
Young Stars (T Tauri at 100/AU)	1	10^6cm^{-2}sec^{-1}	10^6	$3.2 \cdot 10^{19}$
Low Energy Galactic Cosmic Rays	1	10 cm^{-2}sec^{-1}	$3 \cdot 10^8$	10^{17}

Table 3: Summary of some astrophysical applications of ion irradiation experimental results.

Object	Source of Ion irradiation	Result	References
Grains around T-Tauri stars	Stellar flares	Destruction of Ice grains	Strazzulla et al. 1983a
Comets	Solar wind	Sputtering dominates sublimation at r $\geq$ 5–6 AU	Strazzulla et al. 1983b
Interstellar grains	Low energy Cosmic rays	Production of organic mantles	Strazzulla et al. 1983c
Pluto, Triton	Solar cosmic rays	Production of carbonaceous material	Strazzulla et al. 1984
Interplanetary grains	Flares during T-Tauri phase of the Sun	Essential to Sweep out material; chemical evolution	Strazzulla 1985
Saturnian and Uranian moons	Flares during T-Tauri phase of the Sun	Production of organic and carbon like materials	Calcagno et al. 1985, Strazzulla 1986a
Comets	Galactic cosmic rays	Production of dark organic materials down to 100 m	Strazzulla 1986b
Io	Jovian magnetospheric particles	Sputtering of S important but not sole supply of S for the Io torus	Strazzulla et al. 1987
Asteroids	Solar ions	Differential darkening Vs. Solar distance	Andronico et al. 1987

The "organic" nature of the cometary surfaces or, at least, the existence, on Halley's comet, of radiation-processed materials is also well supported by results from dust particle analyzer experiments (PUMA, PIA) on board Vega and Giotto (Kissel et al. 1986a,b; Kissel and Krueger, 1987). The table 3 it is presented a summary of some astrophysical applications of ion irradiation experiments made in recent years by our group in Catania. Informations on the contributions by other groups working in this field can be found in : Johnson et al. (1984, 1986), Moore et al. (1983), Eviatar et al. (1985), de Vries et al. (1984), and references therein.

References

Andronico, G., Baratta, G.A., Spinella, F., Strazzulla, G.: 1987, Astron. Astrophys., in press.

Brown, W.L., Lanzerotti, L.J., Poate, J.M., Augustiniak, W.M.: 1978, Phys. Rev. Lett. **40**, 1027.

Calcagno L., Foti, G., Torrisi, L., Strazzulla, G.: 1985, Icarus **63**, 31.

Cheng, A.F., Lanzerotti, L.J.: 1978, J. Geophys. Res. **83**, 2597.

de Vries, A.E., Pedrys, R., Haring, R.A.,Haring A., Saries, F.W.: 1984, Nature **311**, 39.

Eviatar A., Bar-Nun, A., Podolak, M.: 1985, Icarus **61**, 185.

Foti, G., Calcagno, L., Sheng, K.L., Strazzulla, G.: 1984, Nature **310**, 126.

Garcia, J.D., Forter, R.J., Kavanagh, Y.M.: 1973, Rev. of Mod. Physics **45** (2), 111.

Hoyle, F., Wickramasinghe, N.C.: 1981, New Scient. (31 Aug), 412.

Johnson, R.E., Lanzerotti, L.J., Brown, W.L.: 1984, Adv. Space Res. **4** n.9, 41.

Johnson, R.E., Cooper, J.F., Lanzerotti, L.J.: 1986, ESA SP-250, 269.

Kerr, R.A.: 1986, Science **231**, 793.

Kissel, J. et al.: 1986a, Nature, **321**, 280.

Kissel, J., et al.: 1986b, Nature, **321**, 336.

Kissel, J., Krueger, F.R.: 1987, Nature, **326**, 755.

Mayer, J.M., Rimini, E.: 1977, Ion Beam Handbook for Material Analysis (Academic Press, New York).

Moore, M.H., Donn, B., Khanna, R., A'Hearn, M.F.: 1983, Icarus **54**, 388.

Smith, B.A. et al.: 1986, Science, **233**, 43.

Strazzulla, G.: 1985, Icarus, **61**, 48.

Strazzulla, G.: 1986a, Icarus, **66**, 397.

Strazzulla, G.: 1986b, Icarus, **67**, 63.

Strazzulla, G., Pirronello, V., Foti, G.: 1983a, Astrophys. J., **271**, 255.

Strazzulla, G., Pirronello, V., Foti, G.: 1983b, Astron. Astrophys. **123**, 93.

Strazzulla, G., Calcagno, L., Foti, G.: 1983c, Mon. Not. Royal Astron. Soc. **204**, 59p.

Strazzulla, G., Calcagno, L., Foti, G.: 1984, Astron. Astrophys. **140**, 441.

Strazzulla G., Torrisi, L., Coffa, S., Foti, G.: 1987, Icarus, **70**, 379.

Torrisi, L., Coffa, S., Foti, G., Strazzulla, G.: 1986, Rad. Effects, **100**, 61.

THE STRUCTURE AND DYNAMICS OF AMORPHOUS WATER ICE AND TRAPPING OF GASES IN IT

A. Bar-Nun
Department of Geophysics and Planetary Sciences, Tel Aviv University, Tel Aviv, Israel

1 Introduction

The realization that water ice is the major constituent of comets, the satellites of the outer planets and their rings particles, and of the icy grain mantles in dense interstellar clouds, prompted us to study in detail the properties of water ice at very low temperatures, with an emphasis on its ability to trap various gases (Bar-Nun et al., 1985; Laufer et al., 1987). The experimental setup and procedure will be described here only briefly: The test chamber and its pump consisted of two 10-in cryogenic pumps, connected head-on by a 6-in gate valve. After rough pumping the chamber to 10^{-4} Torr by a sorption pump, the pressure in it was lowered to better than 10^{-8} Torr by the cryogenic pump. A thick 5 by 2.5 cm gold-coated copper plate was cooled cryogenically (by the chamber's cold finger) to ~ 19 K, and its temperature could be controlled by heating, between 19 and 250 K, within ± 1 K. A stream of water vapor, from a reservoir of throughly degassed triple-distilled water, was directed at the plate, through a capillary tubing with a diffuser on its tip. An ice layer, consisting of $10^{19} - 10^{20}$ water molecules was deposited on the plate during ~ 45 min. A stream of gas (Ar, CO, N_2, CH_4, H_2, D_2 or Ne) was then directed at the ice, through the same capillary tubing, at a pressure between $8 \cdot 10^{-8}$ and 0.1 Torr, for several minutes, keeping the gate valve fully open. Alternatively, a premixed gas-water vapor mixture was flowed on the cold plate and was codeposited on it. When the deposition was termined, the chamber was pumped for ~ 10 min, unit a constant pressure of $\sim 10^{-8}$ Torr was reached. The plate was then uniformly warmed, at a costant rate of 0.1–3 K min^{-1}. The evolution of gas and water vapor from the ice was monitored by a precalibrated quadrupole mass filter and the amounts of gas and water vapor emerging at each temperature range were obtained by integrating their fluxes over the time of their evolution from the ice.

The picture of the structure and dynamics of the amorphous ice, which emerges from this study is rather complex: water vapor deposition results in the growth of smooth ice needles, ~ 1 μm long and ~ 0.2 μm wide, which comprise at least 80% of the $10^{18} - 10^{19}$ water molecules cm^{-2} deposited. The very fluffy structure (Figure 1)

E. Bussoletti et al. (eds.), Experiments on Cosmic Dust Analogues, 115–119.
© *1988 by Kluwer Academic Publishers.*

accounts for the large surface area ($86 \text{ m}^2 \text{ g}^{-1}$) of the ice, as measured by Ar adsorption.

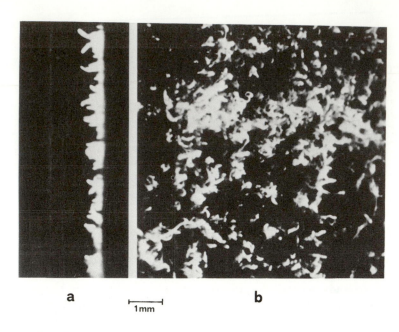

a |————————| b
                          1mm

Figure 1: Two views of an amorphous ice layer, seen edge-on (a) and at an oblique angle (b). Needles about 0.1 mm long are seen, with an ice layer containing $\sim 10^{20}$ water molecules cm^{-2}.

Altogether, gas emerges from the ice in eight temperature ranges (Figure 2), starting at: (a) 23 K, where the frozen excess gas, which was not trapped internally, evaporates; (b) 35 K, where the slow, stepwise and irreversible annealing of the ice releases some of the trapped gas; (c) 44 K, where the monolayer of adsorbed gas evaporates from the $86 \text{ m}^2 \text{ G}^{-1}$ surface of the ice; (d) 85 K, where the slow, stepwise and irreversible annealing of the ice, up to 120 K, releases some additional gas; (e) 122 K, which might be associated with a glass transition in the amorphous ice; (f) 136.8 K, during the transformation of the amorphous ice into cubic ice, when the ice matrix softens temporarily, letting out some of the trapped gas; (g) 160.0 K, during the trasformation of the cubic ice into hexagonal ice, and finally, (h) at $\sim$ 180 K, where the ice itself evaporates, releasing the gas which was trapped in the clathrate-hydrate cages. From this description, it is clear that all the processes of gas release, except the first one, are a result of the dynamics of the ice itself, and the release of the gas is merely a manifestation of the changes in the ice. Thus, the dynamic range of gas detection by a quadrupole mass-filter, which spans

10 orders of magnitude, enables the detection of very subtle changes in the ice.

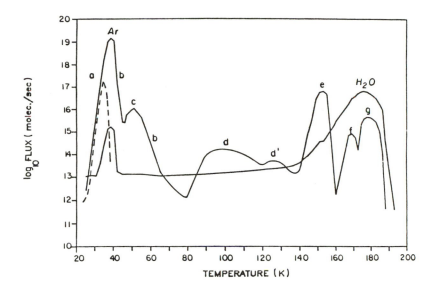

Figure 2: A plot of the fluxes of evolved argon and water vs. temperature, representing the **eight** ranges of gas evolution. The fluxes vary by up to 8 orders of magnitude as the ice temperature varies. The rise in the water flux at 30 K is due to ice-grain ejection. An evaporation curve of frozen argon from an ice-free plate (− − −) is added, for comparison.

Very large fluxes of gas jets, each containing about $5 \cdot 10^{10}$ gas molecules, and ice needles, each containing about 10^{10} water molecules, are emitted from the ice, whenever Ar, CO, CH_4, N_2 and Ne, but not H_2 or D_2, are emitted from gas-rich ice (Figure 3). Apparently, the ice needles are broken from their base and are propelled by the large gas jets. In addition to the large gas jets, all gas emission in the various ranges is not quiescent, but consists of numerous mini-jets, ~ 100 times smaller than the large ones, which result in a "noisy" gas signal at a frequency of $\sim 10^3$ sec^{-1}. Thus, the large gas jets are superimposed on a continuous stream of smaller gas jets. The relative amount of gas which is emitted in the large jets, as compared with the small ones, is proportional to the total amount of gas stored in the ice.

As to the internal structure of the ice, the channels which are formed by the water hexagons are wide enough, even in cubic ice, to allow the penetration of up to 63% of H_2 and D_2, but only in amorphous ice are the channels wide enough for Ar, CO, CH_4 and N_2 to squeeze in. The small Ne is an intermediate case. Amorphous ice which was annealed at 120 K, behaves already like cubic ice. The gas-filled channels in the amorphous ice close by the annealing ice or by additional ice layers. Further annealing or phase transition of the ice connect gas filled regions, through dynamic percolation, up to the surface and result in gas emission in mini-jets. Channels which open into internal holes, build up a pressure large enough to rupture the overlying ice and propel the $\sim 1 \cdot 0.2$ μm ice needles by the large gas jets, each containing $\sim 5 \cdot 10^{10}$ gas atoms.

Dynamic percolation behavior explains also the collapse of the amorphous ice matrix under a gas pressure of 2.6 dyn cm^{-2}, below 29 K, and the burial of enormous amounts of gas in it, up to a gas/ice ratio of 3.3.

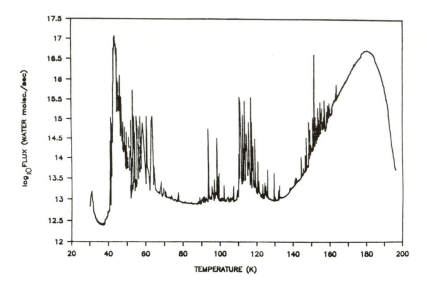

Figure 3: A plot of the flux of water vs. temperature, from a codeposition of a 1:1 Ar-H$_2$O mixture at 25 K, showing the ranges of ice-grain ejection. The time scale in this measurement did not allow the separation of single grains, which was shown earlier.

2   Implication to icy bodies

The experimental findings are readily applicable to icy bodies in the solar system and in dense interstellar clouds. Thus, the agglomeration of the fluffy, needle-like, ice to form comets, could have resulted in cometary nuclei of very low density, in accordance with the density of 0.17 g cm^{-3}, which was recently calculated for comet P/Halley by Rickman (1986).

The experimentally observed accumulation of large quantities of gas in pockets and their occasional explosion, with the propulsion of the ice fragments by gas jets, is also directly applicable to comets: upon warming by solar radiation, the outer, $\sim$ 50 m thick, layer of a pristine cometary nucleus, made of gas-rich amorphous ice, would reach 137 K and transform into crystalline ice (Prialnik and Bar-Nun, 1987a,b,c). A fraction of the gases released from this layer would be trapped in pockets, which would explode occasionally, giving rise to the frequent small bursts, which were observed on P/Halley (Feldman et al., 1986; Festou et al., 1986). These explosions would also blow off the accumulated layer of dark cometary dust, which keeps the nucleus well insulated and inactive (Prialnik and Bar-Nun, 1987c; Lamarre et al., 1986). The exposed ice would sublimate violently and widen the hole, until a large crater is formed, like the craters observed by the Giotto and Vega spacecrafts on P/Halley.

Yet another application of the experimental results is the trapping of up to 63% of hydrogen in ice below 20 K (Laufer et al., 1987). It is expected that the fluffy ice mantles of grains, in dense and cold ($\sim$ 10 K) interstellar clouds, would be hydrogen rich. Only about 2/3 of the hydrogen would be released upon warming to 16–35 K, 1/9 at 35–85 K and the remaining 2/9 will be released only between 85 and 150 K. Thus, if comets were formed by the agglomeration of unaltered, ice-coated interstellar grains, (e.g. Greenberg, 1986), the molecular hydrogen released from their nuclei, upon solar heating, would form a hydrogen coma. The rate of hydrogen release should exceed the rate of water sublimation at $\sim$ 3 AU preperihelion and $\sim$ 4 AU postperihelion (Bar-Nun and Prialnik, 1987). Thus, the detection of a hydrogen coma around comets at large heliocentric distances could indicate that comets were indeed formed by the agglomeration of unaltered ice coated interstellar grains.

The heating of icy bodies by radionuclides was studied as well (Prialnik and Bar-Nun, 1987b). It was found that, if the cometary ice were to remain amorphous, as suggested by the observations, the initial ^{26}Al abundance in comets should have been 100 times lower than that of the inclusion in the Allende meteorite. Thus, about $5 \cdot 10^6$ years should have passed from the formation of the Allende chondrules to the time when comets reached their final size, to allow the decay of ^{26}Al by two orders of magnitude.

Moreover, bodies larger than 100 km cannot release the heat generated by ^{40}K, ^{232}Th, ^{235}U and ^{238}U fast enough, so that their interiors all underwent transformation from amorphous to crystalline ice. The trasformation proceeds from the center outward and the released gases are accumulated at the trasformation front. Perhaps the chaotic terrain on Miranda is a result of the collapse of the overlying ice, when these huge quantities of gas were released from within (Prialnik and Bar-Nun, 1987b).

Altogether, it seems that the deeper insight into the structure and dynamics of amorphous water ice and into the interaction of various gases with it, could lead to a better understanding of the properties and behavior of icy bodies in the solar system and in dense interstellar clouds.

References

Bar-Nun, A., Herman, G., Lauref, D., Rappaport, M.L.: 1985, Icarus, **63**, 317, 332.

Bar-Nun, A., Dror, J., Kochavi, E., Lauref, D.: 1987, Phys. Rev. B. **35**, 2427-2435.

Bar-Nun, A., Prialnik, D.: 1987, submitted to Astrophys. J. Lett..

Feldman, P.D., Weaver, H.A., Wood, N.T., A'Hearn, M.F., McFadden, L.A., Festou, M.C.: 1986, Abstract #20.10, DPS meeting, Paris., Bull. Am. Astron. Soc., **18**, 795.

Festou, M.C., et al.: 1986, Nature, **321**, 361.

Greenberg, M.J.: 1986, in The Comet Nuleus Sample Return Mission, ESA **SP-249**, p. 47.

Lamarre, J.M. et al.: 1986, Abstract #20.02, DPS Meeting, Paris., Bull. Am. Astron. Soc. **18**, 794.

Laufer, D., Kochavi, E., Bar-Nun, A.: 1987, Phys. Rev. B. in press.

Prialnik, D., Bar-Nun, A.: 1987a, Astrophys, J., **313**, 893–905.

Prialnik, D., Bar-Nun, A., Podolak, M.: 1987b, Astrophys. J., **319**, 993.

Prialnik, D., Bar-Nun, A.: 1987c, submitted to Icarus.

Rickman, H.: 1986, in The Comet Nucleus Sample Return Mission. ESA SP-249, pp. 195-205.

A NEW IDENTIFICATION OF THE 6.2, 7.7 AND 8.5 μm CIRCUMSTELLAR EMISSION FEATURES WITH OXIDIZED QCC

A. Sakata[1], S. Wada[2], T. Onaka[3], A.T. Tokunaga[4]
[1]Lab. of Applied Physics, U. of Electro-communications, Tokyo, Japan
[2]Lab. of Chemistry, U. of Electro-communications, Tokyo, Japan
[3]Dept. of Astronomy, Faculty of Science, U. of Tokyo, Tokyo, Japan
[4]Inst. for Astronomy, U. of Hawaii, Honolulu, U.S.A.

1 Introduction

The unidentified infrared (UIR) emission bands constitute a family of emission features at 3.3, 3.4, 6.2, 7.7, 8.6, and 11.3 μm (Aitken, 1981; Barlow, 1983; Willner, 1984). Recently new features at 5.6 and 6.9 μm (Bregman et al., 1983; Cohen et al., 1986) and those at 3.46 and 3.52 μm (de Muizon et al., 1986; Nagata et al., 1987) as well as a plateau between 11.3 and 13.0 μm (Cohen et al., 1985) have been added to the family. A broad feature seen between 5.5 and 9.6 μm must also be related to the UIR emission bands (Sellgren et al., 1985; Cohen et al., 1986). No fine structure have been detected in the three μm features, indicating that they are solid state in origin (Tokunaga and Young, 1980). Latest observations indicate the absence of the 3.4 μm feature in an object emitting the 3.3 μm feature together with the variation of the width and the position of the 3.3 μm feature in some objects (Nagata et al., 1987; Tokunaga et al., 1987). There has been accumulating evidence for the carbonaceous origin of these features and attempts have been made to interpret all the features in terms of only carbon and hydrogen bonds of polycyclic aromatic hydrocarbons (PAHs) (Léger and Puget, 1984; Allamandola et al., 1985).

Sakata et al. (1984, hereafter Paper I) have shown that quenched carbonaceous composite (QCC) synthesized from a hydrocarbon plasma can reproduce most of the UIR features except for the 7.7 and 8.6 μm bands. The 7.7 μm band is the strongest member of the *generic spectrum* of the UIR emission bands (Cohen et al., 1986) and the identification of this feature is very significant. Duley and Williams (1981) have attributed the 7.7 μm feature to NH or OH of the surface functional groups on carbon grains. They did not give any assignment of the 8.6 μm feature. In the interpretation that the UIR emission bands arise fron PAHs the 8.6 μm feature is assigned to the in-plane C-H bending mode (Léger and Puget, 1984), but the assignment of the 7.7 μm feature is not so straightforward as the other UIR emission bands. Allamandola et al. (1985) point out the similarity to the Raman spectrum of auto exhast and attribute it to the combination of carbon skeltal vibration modes. Recently, Léger and d'Hendecourt (1987) have suggested that compact PAHs, such as coronene, circobiphenyl, and dicoronene, have a feature at 7.7 μm but the mode assignment has not been attempted.

121

L. Dussolotti ot al. (ads.), Experiments on Cosmic Dust Analogues. 121–127.
© 1988 by Kluwer Academic Publishers.

Recently, Sakata et al., (1987, hereafter Paper II), have shown that the 7.7 and 8.6 μm features can be attributed to a "cross-conjugate ketone" (CCK) structure. In this report, spectra of molecules with/without the CCK structure are shown to confirm the identification in addition to the results of Paper II.

2   Experimental results

Details of the experimental procedure to synthesize QCC have been given in Sakata et al. (1983). The plasmic gas of 4 torr methane is injected into a vacuum chamber and quenched, producing QCC. The products can be classified into at least three groups (Paper II): (i) gaseous products, involving CH_4, C_2H_4, C_2H_2, and linear and ringed molecules (Sakata, 1980), (ii) film QCC *(f-QCC)*, collected on the wall of the vacuum chamber, and (iii) granular QCC *(g-QCC)*, composed of grains of 450-500 nm in diameter. The 220 nm absorption of QCC, which has been proposed as a candidate for the interstellar extinction hump (Sakata et al., 1983; Onaka et al., 1986), can be attributed to g-QCC component. Spectra of f-QCC exposed to the air ("oxidized") and that without being exposed to the air ("unoxidized") were taken with a JASCO A-102 spectrometer and shown in figure 1.

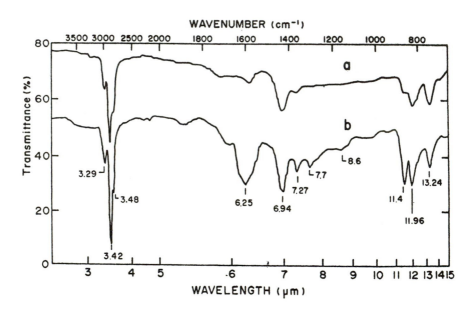

Figure 1: (a) Infrared spectrum of unoxidized f-QCC and (b) that of oxidized f-QCC (shifted by −20% in the ordinate).

We have confirmed that the spectrum of "raw" QCC reported in Paper I is the same of f-QCC except for the relative strengths of the band features to the continuum. The infrared features of raw QCC can be ascribed to f-QCC component within it. Figure 1 shows that oxidation changes the spectrum of QCC in the following ways: (i) new features at 7.7 and 8.6 μm clearly appeared, (ii) the strengths of the features at 6.2, 7.3, and 11.4 μm are increased, (iii) a broad feature between 5.5 and 9.6 μm appeared, and (iv) faint features are also seen at 5.3, 9.7, and 10.5 μm.

3   Interpretation

The experimental results indicate that the features at 7.7 and 8.6 μm in the f-QCC spectrum result from the addition of oxygen. Spectra of the molecules, such as 1.4-naphthoquinone, benzophenone, and anthraquinone, clearly show the band features at 6.2, 7.7, and 8.6 μm (Pouchert, 1981). These molecules have a special ketone structure in common:

This is named "cross-conjugated ketone" (CCK). The infrared spectra of molecules with/without the CCK structure were taken by a JASCO IR-810 spectrometer. Examples are shown in figure 2. The comparison of these spectra clearly confirms the assignment of the 6.2, 7.7, and 8.6 μm feature to the CCK structure. The attachment of oxygen atoms produces large dipole moments, raising strong IR active vibrations. The ketone structure C=O usually shows strong band features between 5.8 and 6.0 μm (Bellamy, 1958). The CCK structure shifts the band to longer wavelengths around 6.2 μm and also produces two new features at 7.7 and 8.6 μm. The latter two are unique features to the CCK structure, since molecules which have C=O bonds different from the CCK structure do not have either 7.7 or 8.6 μm features. Analogously to other vibration modes, the 7.7 and 8.6 μm features can probably be assigned to the asymmetric stretching vibrations of the CCK structure, respectively. Further experiments are needed to confirm the mode assignments. The 6.2 μm feature in celestial objects has been attributed to the C=C skeletal vibration (cf. Paper I). However the dipole moment of C=O is much larger than that of C=C and IR bands due to C=O should be stronger. The observed 6.2 μm feature can also be attributed to the CCK structure if the 7.7 and 8.6 μm feature are observed together. The electron spin resonance measurement suggests the presence of a large amount of dangling bonds in f-QCC (Paper II). These sites are quite reactive and efficiently attacked by oxygen. It is also well known that oxygen atoms are mainly incorporated in carbon molecules as ketones in the oxidation of activated carbons. Thus, we attribute the 7.7 and 8.6 μm features of oxidized f-QCC to the CCK structure. The oxidation process also prompts the polymerization. This increases the number of isolated "solo" hydrogen atoms on a ring, strengthens the 11.3

μm feature as observed. The new interpretation is summarized in table 1.

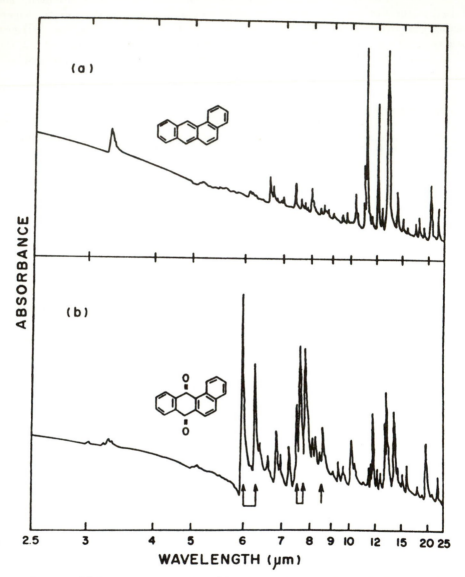

Figure 2: (a) Infrared spectrum of benz[a]anthracene (b) that of benz[a]anthracene-7,12-dione.

Table 1: Infrared features detected in oxidized f-QCC and their interpretations.

Observed Emission Features (μm)	Features detected in the Spectrum of Oxidized f-QCC (μm)	Interpretation
3.28	3.29	CH stretching of =C-H[a]
3.40	3.42	Asymmetric stretching of C-H
3.46	3.48	Symmetric stretching of C-H
5.63	–	
6.20	6.25	C=O stretching and skeltal in plane vibration of C-C
6.95	6.94	Asymmetric deformation of C-H
–	7.27	Symmetric deformation of C-H
7.70	7.80	Asymmetric stretching of cross-conjugated ketone[b]
8.60	8.60	Symmetric stretching of cross-conjugated ketone[b]
11.30	11.40	Aromatic CH out-of-plane deformation of one adjacent hydrogen atoms on a ring
plateau between	11.96	Aromatic CH out-of-plane deformation of two adjacent hydrogen atom on a ring
11.3 and 13.0 μm	13.24	Aromatic CH out-of-plane deformation of three, four, or five adjacent hydrogen atoms on a ring

[a] The symbol = means aromatic or olefinic.
[b] The assignments to the structure are secure but those to the modes are preliminary.

4   Discussion

Oxygen is expected to be in the atomic form around the regions where the UIR bands have been mostly observed (Paper II). Thus, it is reasonable to expect that the carbonaceous material emitting the UIR bands has oxygen incorporated in it and if so, it is probable that oxygen atoms are part of the CCK structure as discussed above.

In figure 3 the spectrum of oxidized f-QCC is compared with the observed spectrum of HD44179 (Russell et al., 1978) in the 6–14 μm region. All of the observed features in the celestial objects are seen in the f-QCC spectrum except for the weak 5.6 μm feature. The broad feature between 5.5 and 9.6 μm is also noticed in several objects emitting the UIR bands (Sellgren et al., 1985; Cohen et al., 1986). The features of f-QCC are all broad and show no fine structures, which is different from molecular spectra. They are in good agreement with the observations. Except for the fainter ones, the only "extra" feature of f-QCC, which has not been reported in celestial observations, is the 7.3 μm feature. The relative appearance of the band features must depend on the emission mechanism (Allamandola et al., 1985) and spectra of celestial objects could be different from absorption spectra taken at room temperature in relative appearance. However, we surmise that f-QCC has more hydrogen than the interstellar dust material, causing 3.4, 3.5, 7.3, 11.9, and 13.2 μm features. Qualitatively, deficiency of hydrogen in celestial objects could explain the discrepancy in band strengths and the absence of some features.

In this report, we present the infrared spectrum of oxidized f-QCC and it shows a remarkable agreement with the UIR emission bands. Nine of eleven UIR bands (3.3, 3.4, 3.5, 6.2, 6.9, 7.7, 8.6, and 11.3 μm) as well as the broad feature between 5.5 and 9.6 μm are seen in the oxidized f-QCC spectrum. A new identification of the 6.2, 7.7, and 8.6 μm feature with the cross-conjugated ketone structure is proposed. The 6.2 μm feature has been attributed to the C=C skeltal vibration. However, the present results indicate that it can also be ascribed to the CCK structure for objects where the 7.7 and 8.6 μm are observed together. The experimental results suggest that oxygen in addition to carbon and hydrogen plays an important role in the structures of the UIR emitting material.

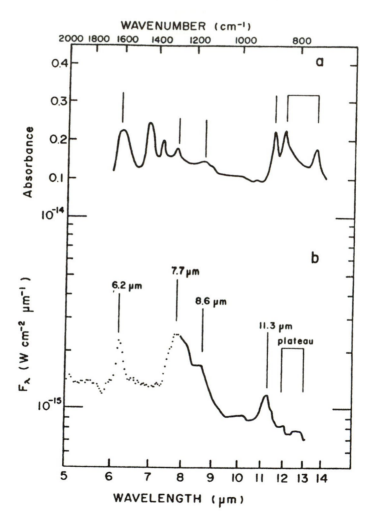

Figure 3: Comparison of the spectrum of unoxidized f-QCC (a: in the arbitrary unit of absorbance) with the observed spectrum of HD44179 (b: Russell et al., 1978). Position of the UIR emission bands are marked by vertical lines as well as plateau around 12 μm.

References

Aitken, D.K.: 1981, In IAU Symposium 96, Infrared Astronomy. eds. C.G. Wynn-Williams, D.P. Cruikshank, Reidel, Dordrecht, p.207.

Allamandola, L.J., Tielens, A.G.G.M., Barker, J.R.: 1985, Ap. J. Lett. **290**, L25.

Barlow, M.J.: 1983, in IAU Symposium 103, Planetary Nebulae, ed. D.R. Flower, reidel, Reidel, Dordrecht, p.105.

Bellamy, A.F.: 1958, The Infrared Spectra of Complex Molecules, Methuen, London.

Bregman, J.D., Dinerstein, H.L., Goebel, J.H., Lester, D.F., Witteborn, F.C., Rank, D.M.: 1983, Ap. J. **274**, 666.

Cohen, M., Tielens, A.G.G.M., Allamandola, L.J.: 1985, Ap. J. Lett. **299**, L93.

Cohen, M., Allamandola, L.J., Tielens, A.G.G.M., Bregman, J., Simpson, J.P., Witteborn, F.C., Wooden, D., Rank, D.: 1986, Ap. J. **302**, 737.

de Muizon, M., Geballe, T.R., d'Hendecourt, L.B., Baas, F.: 1986, Ap. J. Lett. **306**, L105.

Duley, W.W., Williams, D.A.: 1981, M.N.R.A.S. **196**, 269.

Léger, A., d'Hendecourt, L.: 1987, in Polycyclic Aromatic Hydrocarbons ans Astrophysics, eds. A. Léger, L. d'Hendecourt, N. Boccara, Reidel, Dordrecht, p.223.

Léger, A., Puget, J.L.: 1984, Astr. Ap. **137**, L5.

Nagata, T., Tokunaga, A.T., Sellgren, K., Smith, R.G., Onaka, T., Nakada, Y., Sakata, A.: 1987, Ap. J. in press.

Onaka, T., Nakada, Y., Tanabe, T., Sakata, A., Wada, S.: 1986, Ap. Sp. Sci. **118**, 411.

Pouchert, C.J.: 1981, The Aldrich Library of Infrared Spectra Edition III, Aldrich Chemical Company, Milwaukee.

Russell, R.W., Soifer, B.T., Willner, S.P.: 1978, Ap. J. **220**, 568.

Sakata, A.: 1980, in IAU Symposium 87, Interstellar Molecules, ed. B.H. Andrew, Reidel, Dordrecht, p.325.

Sakata, A., Wada, S., Okutsu, Y., Shintani, H., Nakada, Y.: 1983, Nature, **301**, 493.

Sakata, A., Wada, S., Tanabe, T., Onaka, T.: 1984, Ap. J. Lett. **287**, L51, Paper I.

Sakata, A., Wada, S., Onaka, T., Tokunaga, A.T.: 1987, Ap. J. Lett. **320**, L63 (Paper II).

Sellgren, K., Allamandola, L.J., Bregman, J.D., Werner, M.W., Wooden, D.H.: 1985, Ap. J. **299**, 416.

Tokunaga, A.T., Young, E.T.: 1980, Ap. J. Lett. **237**, L93.

Tokunaga, A.T., Nagata, T., Sellgren, K., Smith, R., Onaka, T., Nakada, Y., Sakata, A., Wada, S.: 1987, submitted to Ap. J. Lett.

Willner, S.P.: 1984, in Galactic and Extragalactic Infrared Spectroscopy, eds. M.F. Kessler and J.P. Phillips, Reidel, Dordrecht, p.37.

References

ON PAHs AS INTERSTELLAR GRAINS: INFRARED ABSORPTION COEFFICIENTS

D.W. Salisbury[1,2], **J.E. Allen, Jr**[2], **B. Donn**[2], **R.K. Khanna**[1],
W.J. Moore[3]
[1]*Chemistry Dept., U. of Maryland, College Park, U.S.A.*
[2]*Code 691, NASA/GSFC, Greenbelt, U.S.A.*
[3]*Code 6875, Naval Research Laboratory, Washington, U.S.A.*

1 Introduction

There are several inconsistencies with the interesting proposal that PAHs are the source
of infrared continuum and emission features and the visible diffuse bands. We report
absolute infrared cross-sections for eight PAHs and discuss problems with the proposal
that follows from these measurements. (1) A thermal continuum, the basis for the original
proposal, is not consistent with these spectra. (2) An array of normal molecules shows
a more complex spectrum than observed. Spectra for partially dissociated or ionized
molecules are still lacking. (3) The ultraviolet spectra has cross-sections 2–3 orders
of magnitude larger than for the infrared. It does not seem possible to simultaneously
account for all of the proposed observations without producing structure in the ultraviolet
extinction curve.

2 Experimental

To acquire additional information which is complemetary to the absorption data, a high
temperature cell has been designed and constructed. This cell can be attached to the
emission part of an FT-IR spectrometer and is being used to obtain thermal emission
spectra for temperatures as high as 825 K. Preliminary experiments on coronene exhibit
emission peaks. The emission data is currently being analyzed to identify the origin
of these features and to quantitatively measure the band intensities. In addition, the
cell is being modified to enhance control of the apparatus. Using the improved cell,
thermal emission experiments will be performed on other PAHs. Following this set of
experiments, a series of ultraviolet induced infrared fluorescence studies is planned which
will provide corresponding information concerning infrared emission from electronically
excited states. The purpose of these experiments is to attain spectra which more realis-
tically represent the proposed interstellar emission mechanism. Comparison of this data
with the UIR bands will permit a critical test of the PAH hypothesis.

E. Bussoletti et al. (eds.), Experiments on Cosmic Dust Analogues, 129–135.
© 1988 by Kluwer Academic Publishers.

3   Discussion

There are several problems with the interesting proposal that polycyclic aromatic hydrocarbons (PAHs) are the source of both infrared absorption/emission in various astronomical sources and also produce the diffuse interstellar bands.

1. The original suggestion for PAHs arose to explain the 1.5–5 μm continuum of reflection nebulae. Figures 1–3 show the absorption spectra of eight pericondensed PAHs obtained by us.

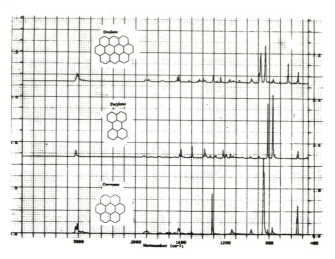

Figure 1

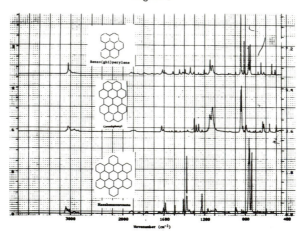

Figure 2

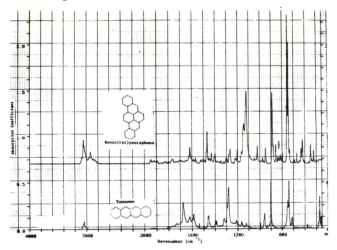

Figure 3

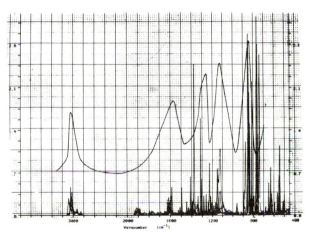

Figure 4

They are combined in figure 4. Unlike the observations of reflection nebulae obtained by Sellgren et al. (1983, 1984) there is no detectable continuum absorption in that interval and therefore there will be negligible emission. If a continuum is explained as overlapping features of one or several species, it will not represent thermal radiation at some temperature, T.

2. Figure 4 also contains a comparison of the room temperature absorption of the array of PAHs with the emission features of the reflection nebula NGC 2023 (from Léger and d'Hendecourt, 1986, attributed to Sellgren and colleagues). Although the comparison is suggestive, there are significant differences in wavelengths and relative intensities of features. Unequivacal identification of atoms or molecules requires better agreement. The comparison does suggest a carbonaceous material, rich in hydrogen.

3. Weighting the room temperature absorption with the Planck function is not generally a valid procedure to obtain the emission spectra of excited molecules, as hot bands will not be included. It is especially true for this case where the molecules are excited by absorption of stellar ultraviolet radiation extending from ~ 100 nm to the long-wave molecular absorption limit (350–700 nm). This is approximately 2–10 eV. For large molecules in a highly excited state, rapid intramolecular vibrational redistribution (IVR) populates an array of states. Emission can occur by radiative transitions from such states and this emission cannot be identical to features in the low temperature absorption spectrum arising from the ground state only.

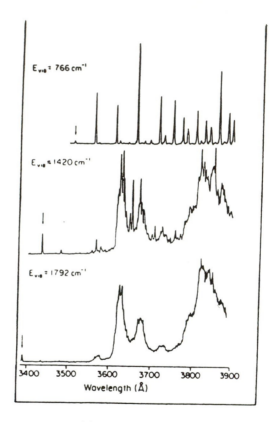

Figure 5: Fluorescence spectra of jet-cooled anthracene for three values of excess vibrational energy from laser excitation in S_1 electronic manifold. Note dependence of vibrational excitation as indicated by band width on excess vibrational energy. (From Felker and Zewail, 1984, Chem. Phys. Lett. **108**, 303) Arrows show position of exciting laser radiation.

The new features will run from the near ultraviolet through the infrared. Figure 5 shows the fluorescence spectrum of the isolated, vapor phase PAH anthracene resulting from excitation to three different single levels in an excited electronic manifold.

The broad emission indicates an array of vibrational levels were excited by IVR. Transitions between these will produce infrared fluorescence by PAHs. Such experiments are planned.

Measurements of the thermal infrared emission of the eight PAHs shown in figures 1–3 are underway. These thermal emission studies utilize the cell diagramed in figure 6.

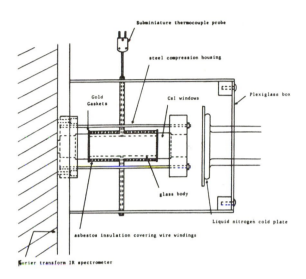

Figure 6: Schematic diagram of the cell used in the thermal emission studies.

This cell has been designed to be resistively heated to temperatures as high as 825 K. By mounting the cell directly against an emission port of a Fourier transform infrared spectrophotometer, the thermal emission can easily be detected. Thermal cycling of the cell prevents it from being kept vacuum tight during the course of an experiment. To overcome this problem the cell has been enclosed within a plexiglass box which is purged with an inert gas. This enables us to ensure that the cell is free of normal atmospheric species which may be infrared active or which may oxidize the thermally excited PAHs. These results will allow a comparison between the experimental emission spectra with those previously calculated and will be necessary for future ultraviolet induced infrared fluorescence studies in vapor phase.

4. The recent attempt (d'Hendecourt et al., 1986) to explain the visible spectrum of the Red Rectangle (HD 44179) by PAH fluorescence is not satisfactory. In figure 3 from that paper only the *phosphorescence* spectrum of hexabenzacoronene in a matrix using near ultraviolet radiation gives an encouraging match. The fluorescence from an array of PAHs excited by 100–700 nm radiation would produce even more strong features in the near ultraviolet-visible than figure 3 in the above reference shows.

5. PAHs have been proposed to explain the diffuse interstellar bands using ionized molecules which absorb in the visible. Figure 7, taken from Crawford et al. (1985) shows that only the naphthalene ion, the smallest and least stable PAH, provides a reasonable comparison with the diffuse bands. Also, all ions have intense, broad, visible-near ultraviolet, absorptions which would dominate the interstellar extinction spectrum if PAH ions caused the narrow diffuse bands.

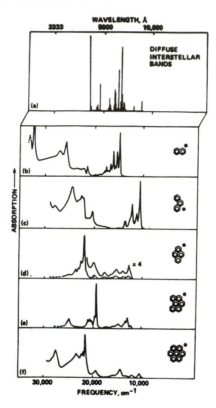

Figure 7: (a) Schematic representative of diffuse interstellar band spectrum. (b)-(f) Visible absorption spectra of five PAH cations isolated in solid matrices. See text for discussion. (From Crawford et al., 1985, Ap. J. Lett. **293**, L45).

6. One may try to avoid the strong ultraviolet visible absorption of PAHs by overlapping such features from many molecules. Donn and Krishna-Swamy (Physics 41, 133, 1969) pointed out that a large array of PAHs and their derivatives (heterpolar molecules with N, O, S atoms substituted for carbon in the rings, as radical side chains, e.g., CH, OH, NH_2, etc. replacing H) or ions smooth out the extinction curve and can be

made to yield a fair comparison with the observed extinction (see figure 8).

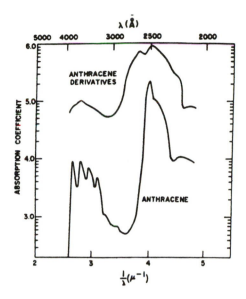

Figure 8: A comparison of the absorption of anthracene with anthracene plus thirteen derivatives or ions.

However, such an array would also show a very complex infrared spectrum, not the simple observed pattern.

4 Conclusions

There does not seem to be any substantial evidence that PAHs cause the infrared continuum or emission features, the Red Rectangle fluorescence or the diffuse bands. On the contrary, there are good reasons against such molecules being present in interstellar space. The general agreement of astronomical infrared features and PAH spectra lends support to a significant role for hydrogenated carbonaceous material. The room temperature condensate from a discharge in methane (Sakata et al., 1984) produced features at 3.29, 3.42, 3.48, 6.25, 6.94, 7.27, 11.40, 11.96 and 13.24 μm. This simple experiment thus yielded suggestive agreement with observations. There is much room for further experimentation with this hydrogenated amorphous carbon type material. However, the constancy of the infrared wavelengths severely constrains the hypothesis of carbonaceous material.

Amorphous grains with continuous absorption and emission, small enough to be heated to ~ 1000 K would be poor emitters in the 1–5 μm region where $a/\lambda \sim 3 \cdot 10^{-3}$. Thus, the observed infrared continuum is not readily explainable as thermal emission from PAHs or small carbonaceous grains. In view of this difficulty we conclude that the observed continuum is unlikely to be thermal emission.

EXPERIMENTAL EVIDENCES FOR HAC AND PAHs SOUP IN SPACE

A. Blanco[1], E. Bussoletti[2], L. Colangeli[3], S. Fonti[1], V. Orofino[1]
[1]*Physics Department, University of Lecce, Lecce, Italy*
[2]*Istituto Universitario Navale, Napoli, Italy*
[3]*ESA Space Science Department, ESTEC, Noordwijk,*
  The Netherlands

1 Introduction

Since its discovery in the spectrum of NGC 7027, the family of weak IR bands, the so called unidentified IR bands (hereinafter UIR), has been found in emission in a wide variety of galactic and extragalactic sources (Aitken, 1981; Bregman et al., 1983; Allamandola, 1984; Willner, 1984; Cohen et al., 1986). As the number of observations has increased, new features have been revealed (de Muizon et al., 1986) so that we have at present a group of eleven bands. Their relative intensity may vary a little (Simpson et al., 1984) while their widths seem to be source independent. Due to the ubiquity of the bands, a great effort has been put in identifying both the nature of the emitting material and the emission mechanism.

Recently Borghesi et al. (1987) have suggested that UIR bands can be better explained by a mixture of hydrogenated amorphous carbon, HAC, and a combination of polycyclic aromatic hydrocarbons molecules, PAHs. A similar idea has been put forward by Allamandola et al. (1987), who suggested that spectroscopic data indicate at least two components which contribute to the interstellar emission spectrum: free molecule sized PAHs producing narrow features and amorphous carbon particles contributing to the broad underlying components. Also Goebel (1987), analyzing the spectrum of the planetary nebula NGC 7027, reached a similar conclusion.

2 Laboratory measurements

Mixtures of HAC particles and PAHs have been prepared in laboratory. Details on the production and characterization of HAC particles have already been extensively provided by Borghesi et al. (1983, 1985). Collections of PAHs have been produced by means of vacuum pyrolysis of cellulose (Morterra and Low 1983). The char produced at different temperatures give rise to PAH combinations with prevalent aliphatic character at 330 C (PAH330) or aromatic character at 460 C (PAH460). The IR transmission spectra of different samples have been obtained in laboratory. Figures 1 and 2 report the portions of the transmission spectra of some samples. In table 1 are listed the peak wavelengths for some representative samples together with those of UIR bands.

E. Bussoletti et al. (eds.), Experiments on Cosmic Dust Analogues, 137–143.
© *1988 by Kluwer Academic Publishers.*

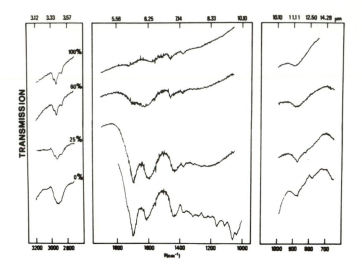

Figure 1: Ordinate-displaced segments of transmission spectra of HAC+PAH330 samples.
The percentage indicate the amount, by weight, of HAC in each sample.

Table 1: Infrared bands of the different carbonaceous materials taken into consideration in this work (see text) and the UIR bands. The last two columns refer to mixtures both with 25 % of HAC by weight.

UIR Bands	HAC	PAH330		PAH460	HAC+PAH330	HAC+PAH460	
				$\lambda(\mu m)$			
3.28	—	3.26	(*)	3.29	—	3.28	(*)
3.39	3.39	3.38	(+)	—	3.38	3.38	
3.42	3.42	3.44		3.44	3.44	3.42	
3.51	3.51	3.48	(+)	—	3.48	3.50	
5.62	5.78	5.88		5.90	5.86	5.92	
6.29	6.29	6.19		6.29	6.25	6.27	
6.90	6.85	6.94		6.99	6.96	6.97	
7.27	7.30	7.27		7.25	7.27	7.29	
7.70	—	7.60		—	7.60	—	
8.60	—	8.62		—	—	—	
		9.01					
		9.43					
		9.66					
11.30	11.30	11.40		11.40	11.40	11.40	
		12.80		11.90			
				13.00			

(*) weak; (+) shoulder

3   Discussion and conclusions

An inspection of table 1 show that single HAC or PAH samples are not able to reproduce completely the peak wavelengths of astronomical features. While HAC sample lacks some bands, PAH collections show, in general too rich spectra. Mixtures of both HAC and PAHs appear instead better. It is important to stress that, so far, we have discussed data obtained in absorption comparing them with observations of features which are detected in emission in celestial sources.

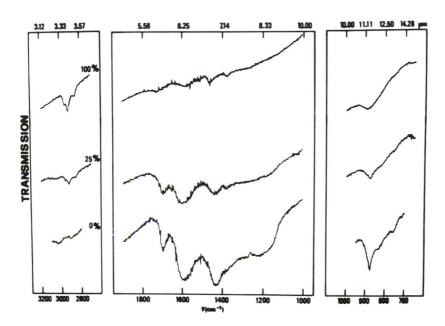

Figure 2: Same as figure 1 for HAC+PAH460 samples.

We used transmission data of our mixtures to evaluate the emission spectra at an equilibrium temperature of 300 K. In figure 3, the results obtained for the HAC+PAH460 sample are compared with the observed spectrum of NGC 7027. We limit here our comparison only to this source since:

a) it is characterized in the NIR by a 300 K thermal emission spectrum

b) at present IR absorption and emission properties of HAC and PAHs not available at T$\neq$300 K.

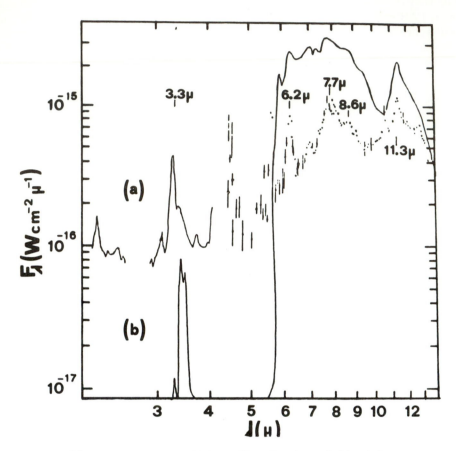

Figure 3: (a) Observed spectrum of NGC 7027 (Russell et al. 1977); (b) emission spectrum of HAC+PAH460 mixture (see text) calculated at 300 K.

As it can be seen, a good similarity between the two spectra is evident and, if we limit our attention only to the feature position, we see that the peak wavelengths of our mixture fit quite well those of the observed UIR bands. At present we do not think it is very significant to extend the comparison beyond the matching of the UIR peak position. For example, according to Léger and d'Hendecourt (1987), we tried to determine the temperature of NGC 7027 by using three couples of features attributed respectively to CH (3.3 μm, 11.3 μm), CH$_3$ (3.4 μm, 6.9 μm) and C=C (6.2 μm, 7.7 μm) bonds and our experimental results (see table 2).

Table 2: Emission temperatures for NGC 7027 deduced for different materials and using different couples of bands (λ_i, λ_j). The temperatures concerning Typical Compact PAHs have been obtained using experimental data reported by Léger and d'Hendecourt (1987).

| Material | (λ_i, λ_j) | | |
	3.3/11.3	3.4/6.9	6.2/7.7
HAC	—	570	—
PAH330	—	955	81
PAH460	905	—	—
HAC+PAH330	—	1240	72
HAC+PAH460	745	1345	—
Typical compact PAHs	730	—	304

There exist discrepancies which can be mainly attributed to:

a) none of the material listed in table 2 is the responsible for NGC 7027 bands

b) the identification of the functional groups which produce the different couples of bands may be not correct or misleading

c) the theoretical approach used for the computations may be not appropriate

d) the emission in the UIR bands comes from regions of the source at different temperatures or by components at different temperatures within the same zone.

In order to clarify some of the above mentioned problems it is also necessary to extend the comparisons between laboratory data and observed spectra of high temperature sources. This means, by an experimental point of view, that substancial work is needed in laboratory to measure the spectral characteristics at different temperatures and conditions of both neutral and ionized PAHs. Following this line of research we started a sistematic laboratory work to determine the temperature dependance of PAHs absorption properties. Very preliminary results concerning coronene are shown in figure 4. As can be seen the behaviour of the absorption bands is clearly dependent on temperature and far to be uniform. At the moment we think that there may be two main reasons for this:

a) the functional groups responsible for the various bands can undergo different chemical reactions with the matrix used for the pellet (KBr in our case);

b) the functional groups really behave differently as temperature increases.

Further work is obviously needed to clarify such problems.

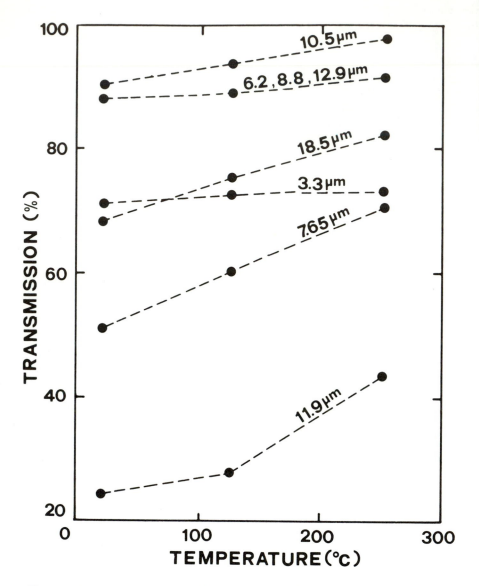

Figure 4: Transmission at peak position of the main infrared features of coronene as a function of temperature.

Ackowledgements

The authors thank the continuous technical support provided by Mr. M. Scalzo in the laboratory. This work has been partially supported by Ministero della Pubblica Istruzione (MPI) and by Consiglio Nazionale delle Ricerche (CNR).

References

Aitken, D.K.: 1981, in Infrared Astronomy, eds. Wynn Williams C.G. and Cruikshank, D.P., p. 207, Reidel, Dordrecht.

Allamandola, L.J.: 1984, in Galactic and Extragalactic Infrared Spectroscopy, eds Kessler, M.F. and Phillips, J.P., p. 5, Reidel, Dordrecht.

Allamandola, L.J., Tielens, A.G.G.M., Barker J.R.: 1987, in Polycyclic Aromatic Hydrocarbons and Astrophysics, eds. Léger, A., d'Hendecourt, L. and Boccara, N., p.255, Reidel, Dordrecht.

Borghesi, A., Bussoletti, E., Colangeli, L., Minafra, A., Rubini, F.: 1983, Infrared Phys., **23**, 85.

Borghesi, A., Bussoletti, E., Colangeli, L.: 1985, Astr. Astrophys. **142**, 225.

Borghesi, A., Bussoletti, E., Colangeli, L.: 1987, Astrophys J. **314**, 422.

Bregman, J.D., Dinerstein, H.L., Goebel, J.H., Lester, D.E., Witteborn, F.G., Rank, D.M.: 1983, Astrophys. J. **274**, 666.

Cohen, M., Allamandola, L.J., Tielens, A.G.G.M., Bregman, J., Simpson, J.P., Witteborn, F.C., Wooden, D., Rank, D.: 1986, Astrophys. J., **302**, 737.

de Muizon, M., Geballe, T.R., d'Hendecourt, L.B., Baas, F.: 1986, Astrophys. J. Lett., **306**, L105.

Goebel, J.H.: 1987, in Polycyclic Aromatic Hydrocarbons and Astrophysics, eds. Léger, A., d'Hendecourt, L., Boccara, N., p. 329, Reidel, Dordrecht.

Léger, A., d'Hendecourt, L.: 1987, in Polycyclic Aromatic Hydrocarbons and Astrophysics, eds. Léger, A., d'Hendecourt, L., Boccara N., p.223, Reidel, Dordrecht.

Morterra, C., Low, M.J.D.: 1983, Carbon 21, 283.

Russell, R.W., Soifer, B.T., Willner, S.P.: 1977, Astrophys. J. Lett., 217, L149.

Simpson, J.P., Bregman, J.D., Cohen, M. Witteborn, F., Wooden, D.M.: 1984, Bull. AAS, **16**, 523.

Willner, S.P.: 1984, in Galactic and Extragalactic Infrared Spectroscopy, eds. Kessler, M.F., Phillips, J.P., p. 37, Reidel, Dordrecht.

IR SPECTROSCOPY OF ACID INSOLUBLE RESIDUES OF CARBONACEOUS CHONDRITES

T.J. Wdowiak[1], G.C. Flickinger[1], J.R. Cronin[2]

[1]Physics Department University of Alabama at Birmingham, Birmingham, U.S.A.

[2]Department of Chemistry and Center for Meteorite Studies Arizona, State University Tempe, U.S.A.

1 Introduction

Léger and Puget (1984) have suggested polycyclic aromatic hydrocarbons (PAHs) as being responsible for the unidentified infrared emission bands (UIRs). The idea that PAHs are a component of the interstellar medium originated with Donn (1968). Because of this interest, insoluble carbon of carbonaceous chondrites obtained for the purpose of ^{13}C NMR spectroscopy (Cronin et al., 1987) appears to be due to its aromatic content an attractive material to investigate with infrared spectroscopy. This insoluble carbon is a residue of HF and HCl dissolution of the meteorite. Sample prepared from the Orgueil (C1), Murchison (CM2), and Allende (CV3) carbonaceous chondrites were examined with FTIR spectroscopy in bulk form in KBr pellets, as films evaporated under vacuum onto KBr disks, and as the residues of those evaporations in KBr pellets. The ^{13}C cross-polarization, magic angle spinning sample nuclear magnetic resonance (CP-MASS NMR) data indicates two major peaks centered at 29 and 133 ppm relative to tetramethylsilane (Cronin et al., 1987). The strength of the 133 ppm peak due to ^{13}C in aromatic structures, relative to the 29 ppm ^{13}C peak from aliphatic constituents, increases in the sequence of Allende, Murchison, and Orgueil (Figure 1). The ^{13}C aromatic peak of the Orgueil residue is of particular interest in that its lack of structure normally suggestive of carboxyl or methoxy groups, and narrowness indicative of a homogeneity of the aromatic species consistent with large PAH structures, is what would be expected of interstellar PAHs (Wdowiak 1986, Tielens et al., 1987). The 11.3 μm UIR band mandates a family of PAHs that have single hydrogens per peripheral ring as a major interstellar PAH constituent. Molecules such as coronene (Léger and Puget 1984) and hexabenzocoronene (Hendel et al., 1986) are precluded from being major interstellar PAH species because they have two and three hydrogens per peripheral ring, and hence no 11.3 μm feature exists.

145

E. Bussoletti et al. (eds.), Experiments on Cosmic Dust Analogues, 145–152.
© *1988 by Kluwer Academic Publishers.*

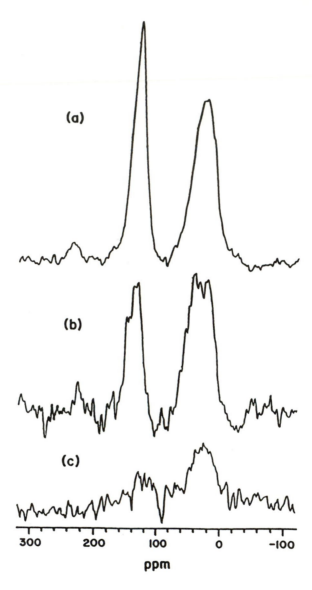

Figure 1: ^{13}C CP-MAS NMR spectra of the insoluble carbon of the Orgueil, Murchison and Allende carbonaceous chondrites. (a) Orgueil (C: 68.46%) 32,200 scans; (b) Murchison (C: 6.65%) 104,040 scans; (c) Allende (C; 15.26%); 45,000 scans (From Cronin et al., 1987).

2 Sample preparation for spectroscopy

Initial spectra were obtained of bulk residue in the amounts of 0.5 to 2 mg mixed into 100 mg of KBr, ground in an agate vibrating ball mill, and pressed into a pellet having a cross section of 0.5 cm^2. Although run against a pure 100 mg KBr pellet background, the sample pellets exhibited a higher absorbance continuum at shorter wavelengths due to scattering. To obtain spectra of the volatile components of the residues, material was placed in a 2mm diameter tube rolled from molybdenum foil capped with graphite and alumina plugs, and resistance heated at $< 10^{-5}$ torr allowing vapor to escape through a pinhole and strike a KBr disk 2 cm away. Temperature within the tube was monitored with a chromel-alumel thermocouple. The non-volatile fraction was removed from the tube after volatiles have been exhausted, and was pressed into KBr pellets. All spectra were obtained with a Mattson Polaris FTIR instrument.

3 Spectra

Because of its ^{13}C NMR spectrum the acid insoluble black residue of the Orgueil C1 carbonaceous chondrite was considered to be an attractive candidate for FTIR spectroscopy. The spectrum of the bulk residue in a KBr pellet exhibits a complex of features including aliphatic C-H stretch at 2940 cm^{-1} (3.4 μm), a feature at 1620 cm^{-1} (6.2 μm) due to C=C stretch, and features at 1710 cm^{-1} (5.85 μm), 1458 cm^{-1} (6.86μm), 1380 cm^{-1} (7.24 μm), 1235 cm^{-1} (8.1 μm), 1042 cm^{-1} (9.6 μm), 885 cm^{-1} (11.3 μm), 830 cm^{-1} (12.0 μm), 760 cm^{-1} (13.2 μm), 620 cm^{-1} (16.1 μm) and 510 cm^{-1} (19.6 μm). Heating of bulk residue in vacuum at various increased temperatures resulted in diminution of material and change in the spectrum. Most noticeable was reduction of the 2940 cm^{-1} (3.4 μm), 1710 cm^{-1} (5.85 μm), and 1042 cm^{-1} (9.6 μm bands) (Figure 2). Material evaporated during heating in vacuum at temperatures in excess of 200°C and condensed on KBr disks exhibits 2940 cm^{-1} (3.4 μm) absorption. In the case of a sample heated to dull red heat (500°C) the condensate spectrum resembled that of the bulk material. The spectra of Orgueil material heated in vacuum have a strong resemblence to the Raman spectrum of Interplanetary Dust Particle (IDP) Essex and the infrared emission spectrum from the Orion Bar (Figure 3)(Allamandola et al., 1987; Wopenka, 1987; Bregman et al., 1984). All samples of unheated and heated material have an 885 cm^{-1} (11.3 μm) band that is very evident in those heated to higher temperatures (Figure 4). Weaker features are at 830 cm^{-1} (12.0 μm) and 760 cm^{-1} (13.2 μm). These features are indicative of C-H out-of-plane bending of single, double, and triple hydrogens attached to an aromatic ring. The 11.3 μm UIR band is consistent with the idea that interstellar PAHs are dominated by structures having single hydrogens on peripheral rings and "corner" rings having two or three hydrogens (Cohen et al., 1985; Wdowiak, 1986; Tielens et al., 1987). Because the vacuum heated Orgueil C1 acid insoluble residue has the 11.3 μm, 12.0μm, and 13.2 μm bands in decreasing strength and a strong resemblance to the Orion Bar emission spectrum, serius consideration can be given to its having the same nature as interstellar PAHs.

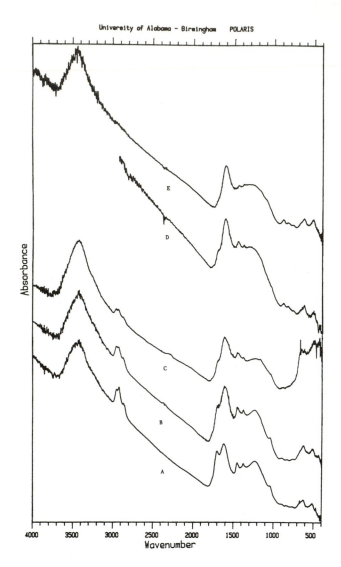

Figure 2: FTIR spectra of the acid insoluble residue of the Orgueil C1 carbonaceous chondrite (A), absorbance x 1.13, and the residue after heating in vacuum to temperatures of : (B) 195°C, x 1.63; (C) 300°C, x 5.63; (D) 415°C, x 1; and (E) 500°C, x 1.75. The broad feature at 3450 cm^{-1} is due to H_2O in the KBr pellet.

The acid insoluble residue of the Murchison CM2 carbonaceous chondrite has a high mineral content (76.5%) that is reflected in its infrared spectrum and grey color as opposed to the black Orgueil and Allende residues. The contribution of organic matter to the infrared spectrum was determined by ashing a sample and substracting its mineral spectrum from that of the bulk residue using KBr pellets.

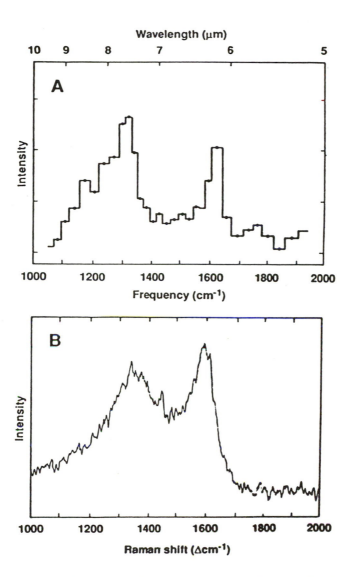

Figure 3: Comparison between (A) infrared emission spectrum from the Orion Bar and (B) Raman spectrum of the IDP Essex. (From Allamandola et al., 1987).

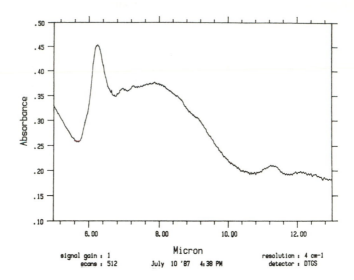

Figure 4: 5 μm to 13 μm region of the FTIR KBr pellet spectrum of the acid insoluble residue of the Orgueil C1 carbonaceous chondrite after heating in vacuum to 500°C.

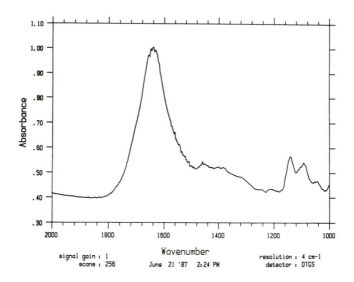

Figure 5: 2000 cm⁻¹ to 1000 cm⁻¹ region of the FTIR KBr pellet spectrum of the acid insoluble residue of the Murchison CM2 carbonaceous chondrite versus the FTIR KBr pellet spectrum of the mineral ash of the residue.

A 1620 cm^{-1} (6.2 μm) peak with a hint of the "hump" seen in the Orgueil spectrum are the only evident features (Figure 5). The Allende CV3 carbonaceous chondrite has an acid insoluble residue with a sharp 1580 cm^{-1} (6.3 μm) band and broad feature peaking at 1160 cm^{-1} (8.6 μm), both similar to the UIR emission bands at those wavelengths (Figure 6).

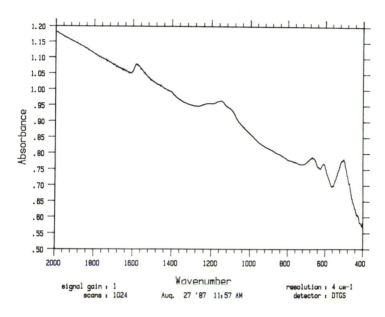

signal gain : 1
scans : 1024

Wavenumber

Aug. 27 '87 11:57 AM

resolution : 4 cm-1
detector : DTGS

Figure 6: 2000 cm^{-1} to 400 cm^{-1} region of the FTIR KBr pellet spectrum of the acid insoluble residue of the Allende CV3 carbonaceous chondrite.

4 Conclusions

Orgueil C1 acid insoluble residues heated in vacuum to temperatures in excess of 200°C (473 K) and as high as 500°C (773 K) have infrared spectra similar to the emission spectra of interstellar matter considered to be polycyclic aromatic hydrocarbons (PAHs). It is tempting to consider the hypothesis that heating has evaporated an aliphatic component leaving a non-volatile aromatic (PAH) that is intrinsic to the acid insoluble residue and interstellar in nature. The difficulty with this idea is that the temperatures necessary for removing aliphatic material as shown by condensate spectra, is in a temperature range ($>$ 200 C) at which pyrolysis can occur resulting in a PAH composition not intrinsic to the residue as extracted from the meteorite. This difficulty does not preclude the non-volatile component from being the aromatic fraction seen in the ^{13}C NMR spectrum. Also the black color of the residue before and after vacuum heating and transparent nature of condensates are consistent with the non-volatile material being intrinsic and unchanged. Large PAH molecules such as hexabenzocoronene can be heated to red heat without decomposition. It is prudent at this time to temper speculation by concluding that it is not yet established that the aromatic fraction of the Orgueil C1 carbonaceous chondrite acid insoluble residue is interstellar and predates the formation of the solar system. The spectra obtained in this study provide support for interstellar PAHs being responsible for the UIR emission bands observed for many celestial objects by demonstrating that unaltered and thermally altered acid insoluble residues of carbonaceous chondrites have infrared features at UIR wavelengths. Further study is necessary to determine if the discussed materials are of interstellar or solar nebula origin.

This work was supported by NASA grants NAGW-749 at UAB and NSG-7255 at ASU.

References

Allamandola, L.J., Sandford, S.A., Wopenka, B.: 1987, Science, **237**, 56.

Bregman, J., Allamandola, L.J., Simpson, J., Tielens, A., Witteborn, F.: 1984, in NASA/ASP Symposium, "Airborne Astronomy", NASA/Ames Research Center (NASA CP-2353).

Cohen, M., Tielens, A.G.G.M., Allamandola, L.J.: 1985, Ap.J., **299**, L93.

Cronin, J.R., Pizzarello, S., Frye, J.S.: 1987, Geochim. Cosmochim. Acta, **51**, 299.

Donn, B.: 1968, Ap., **152**, L129.

Hendel, W., Zhan, Z.H., Schmidt, W.: 1986, Tetrahedron, **42**, 1127.

Léger, A. Puget, J.L.: 1984, Astron. Astrophys., **137**, L5.

Tielens, A.G.G.M., Allamandola, L.J., Barker, J.R., Cohen, M.: 1987, in "Polycyclic Aromatic Hydrocarbons and Astrophysics", A. Léger, L. d'Hendecourt, and N. Boccara ed. (D. Reidel, Dordrecht).

Wdowiak, T.J.: 1986, Bull. Am. Astron. Soc., **18**, 1030.

Wopenka, B.: 1987, in "Lunar and Planetary Science XVIII", (Lunar and Planetary Institute, Houston), 1102.

LIGHT STANDING-WAVES IN AGGREGATES OF AMORPHOUS CARBON GRAINS

V. Capozzi[1], C. Flesia[2], A. Minafra[1]

[1]*Dipartimento di Fisica, Università di Bari, Bari, Italy*
[2]*Département de Phisyque Théorique, Université de Genève, Genève, Switzerland*

1 Introduction

The phenomenon of localization of light by disorder is a result of the interference of electromagnetic waves scattered by the random medium across which they propagate [1]. Up to now, only precursive effects of optical localization in the form of an enhanced backscattering of visible light by aqueus suspension of polystyrene spheres have been observed [2]. One of the crucial problem to measure the strong localization of light is that the localization length for classical waves is quite large, compared to that of quantum waves of a particle. A recent result [3] shows that the most important mechanism needed to reduce the localization length is to have a medium whose index of refraction very often changes randomly from a small value to a large one. This is possible if the distribution probability function $\rho(n)$, describing the random refraction index n, is uniform between $(1,1+\delta_1)$ U$(1+W-\delta_2,1+W)$, where W measures the degree of disorder of n and δ_1, δ_2 are the widths of the two distributions. The relation between the three parameters δ_1, δ_2, W defining $\rho(n)$ is crucial for the description of the propagation of light in the random medium. The limit W=$\delta_1 + \delta_2$ corresponds to a disorder that simply allows fluctuations of the refraction index around some value. In this case, the localization length is very large and the strong localization regime will be difficult to reach.

If W> $\delta_1 + \delta_2$, the behaviour of the localization length as a function of W is not monotonic at all, but shows a succession of peaks and dips and becomes very small for some values of the disorder W [3]. This implies that the only knowledge of the width W of the distribution does not characterize completely the trasmission properties of the system. In this work we present extinction spectra of amorphous carbon grains in the range 100-800 nm. With respect to the extinction spectrum of the amorphous carbon film (see figure 1), the spectra measured on sample having a granular structure show a few extinction dips at energies higher than that of the plasmon resonance mode [4,6]. On the basis of the model discussed in ref.[3], we can explain these results in terms of strong localization effects which in the granular samples are due to the random fluctuations of

153

E. Bussoletti et al. (eds.), Experiments on Cosmic Dust Analogues, 133–160.
© *1988 by Kluwer Academic Publishers.*

the refraction index of the amorphous carbon grains.

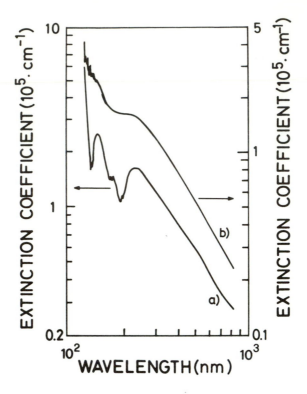

Figure 1: Extinction spectra measured at room temperature of (a) an amorphous carbon grain sample; (b) an amorphous carbon film.

2   Experimental

We prepared a few samples of amorphous carbon grains having a particular spongy structure which were obtained by an arc discharge between two amorphous carbon electrodes in a controlled Ar atmosphere at 1 Torr of pressure. Lithium Fluoride (LiF) substrates were used to collect the solide particles. The grains were simultaneously collected onto microscope grids to perform morphological analysis by means of trasmission electron microscopy. The samples appear as in figure 2, composed by grains having spheroidal shape. Diffraction methods have shown the absence of any ordered lattice structure, confirming that the material is mainly amorphous. The mean radius $< r >$ of the grains is about 50 Å. Further details on the morphological properties of our amorphous carbon grains are reported in [5].

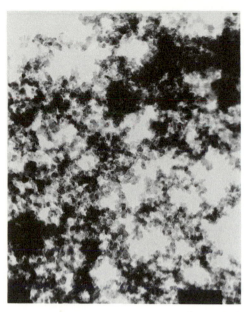

Figure 2: Electron microscope picture of a sample of amorphous submicronic carbon grains. The particles appear to have spherical shape. Note the presence of lower and higher density of grains, together with empty regions.

The ultraviolet part of the extinction spectra (100-300 nm) was performed at room temperature, at the PULS-synchrotron radiation facility of the Frascati National Laboratories, Rome. The long wavelength region (200-800 nm) of figure 1 was measured by using a double spectrophotometer. The two spectral ranges overlap to allow matching of experimental data. The synchrotron light emitted by "Adone" storage ring was analyzed by a 1 m spectrometer operating in the range 100-300 nm. The samples were placed in a vacuum chamber (10^{-9} Torr). The trasmitted light was detected by a photomultiplier tube having a CsTe photocathode and then, the signal was amplified by using standard loock-in techniques. The trasmission data were sent to an on-line computer for storage and processing.

3 Results and discussion

In figure 1, the extinction spectrum (a) shows the following features: a wide band (ω_p) centred at about 240 nm, two dips at 138 nm and 199 nm, and a steep increase of extinction at shorter wavelength. On the contrary, the spectrum (b), at wavelength shorter than ω_p shows only a continuous increase of the extinction coefficient. In according to the refs.[4,6] the extinction band at 240 nm is attributed to a surface plasmon mode and the steep increase of the absorption obseved at the shortest wavelengths in figure 1 is due to the electronic transitions to excited states of the amorphous carbon.

In order to explain the above absorption dips in terms of strong localization of light, we consider a 1-dimensional system of length L of a disordered chain consisting of N boxes of length a_i (i=1,N), in the vacuum. Each box is characterized by a constant refraction index, which can vary randomly at the junction between two boxes. The refraction index in each box is considered as independent random variable with a common distribution probability. This model can be experimentally justified, because we only detect the light which is trasmitted along the direction of the incident beam. The light scattered by the grains in this direction can be completely neglected. In fact, we also performed angle resolved diffusion measurements by rotating the detector around the sample and we found that the scattered light intensity is a few order of magnitude weaker than the trasmitted light intensity.

Consider the stationary equation of an electromagnetic wave of frequency ω along the x direction:

$$\Delta\phi(x) = -\frac{\omega^2}{c^2}n^2(x)\phi(x) \tag{1}$$

where c is the light velocity in vacuum. By using the method of ref.[7], one can show that, with probability one, the trasmission coefficient $T_L(\omega)$ of this wave through the disordered region, will decay exponentially as $T_L(\omega) \approx exp(-\gamma L)$ where γ is the reciprocal of the localization length which depends on the frequency ω and on the disorder parameters δ_1, δ_2 and W. γ is the smallest Liapunov exponent [7] of the product $\tau = t_N t_{N-1}...t_1$ of the random transfert matrices t_i of the disordered system and can be identified with the extinction coefficient if the intrinsic absorption is neglected. In fact, we are only interested to the extinction of light due to the interference effects in the disordered structure of our samples. In figure 1, this extinction is practically given by the difference between the two curves (a) and (b).

The above probability distribution function of n is determined by experimental data as in the following. In our samples, the carbon density changes randomly between a minimun and a maximun value, corresponding to the lowest and the highest density of the grains, respectively. Consequently, the corresponding refraction index can vary between 1 (vacuum value) and a maximun value n_c for amorphous carbon film for a given wavelength λ. The existence and the spectral position of the trasmission maxima, corresponding to the resonant frequencies in the disordered chain mentioned above, depends on the intrinsic disorder of the material. The harmonic condition must be satisfied simultaneously by the two experimental wavelength $\lambda_1 = 138$ nm and $\lambda_2 = 199$ nm at which we observe the two trasmission peaks (Figure 1a). The presence of these two trasmission peaks for the same sample determines in an univocal way the values of the disorder parameters defining the distribution probability. Moreover, to calculate the values of these parameters, we consider an ideal sample in which the regions of carbon film (with a constant density), takes turn with vacuum cavities. The stationary condition for the two wavelength λ_1 and λ_2 is

$$\frac{m_1\lambda_1}{2n_{\lambda_1}} = \frac{m_2\lambda_2}{2n_{\lambda_2}} \qquad m_1, m_2 \in N \tag{2}$$

where n_{λ_1} and n_{λ_2} are the experimental refraction index for the amorphous carbon film measured in ref.[8], correspondingly at the wavelength λ_1 and λ_2. The relation (2) is satisfied for $m_2/m_1 = 1.5$. From the dispersion relation $K = \omega n/c$, where K is the wave number, and from the cavity resonance condition $ka = m\pi$ we straightforward obtain

$$a = m\lambda/2n \tag{3}$$

which is the size of the box where the refraction index will be assumed to be constant in our model.

If we assume that $m_1 = 3$ and $m_2 = 2$ in order to have $m_2/m_1 = 1.5$, we obtain

$$a_1 = \frac{m_1\lambda_1}{2n_{\lambda_1}} = 106.7 \ nm \quad ; a_2 = \frac{m_2\lambda_2}{2n_{\lambda_2}} = 103.6 \ nm \tag{4}$$

and

$$a_{v_1} = \frac{m_1\lambda_1}{2} = 207 \ nm \quad ; a_{v_2} = \frac{m_2\lambda_2}{2} = 199 \ nm \tag{5}$$

which are the lengths of the vacuum resonant cavities. Now, in the real case, for an amorphous carbon grain sample, the width a of the boxes and the material density are not strictly constant. These structural random fluctuations may be expressed in terms of random fluctuations of the refraction index (see eq. (3)). Moreover, in the real case (see figure 2), we assume that the distribution probability of the refraction index takes only two ranges of values, corresponding to the very weak carbon density region ($n \approx 1$) and to the highest carbon density region ($n \approx 2$) [8].

Then, the distribution probability function will be uniform between

$$(1 + \delta_1)U(1 + W - \delta_2, 1 + W) \tag{6}$$

where

- δ_1 takes into account the variation of n due to very low density regions of the sample and the random fluctuations of the width a_v of the vacuum resonant cavities (see eq. (3)).
- δ_2 takes into account the variation of n due to the highest carbon density regions and the random fluctuations of the width a of the real box (see eq. (3)).
- W is the width of the distribution.

To estabilish the values of the disorder parameters δ_1, δ_2 and W, we consider the size a of the real box to be constant and of the same order of magnitude of the ideal box. Then, we calculate the corresponding values of the refraction index in order to satisfy the harmonic conditions for the two wavelengths λ_1 and λ_2. For $a = 100$ nm and from the eq.(3), we obtain for $n_1 = n$ (138 nm) = 2.07 and $n_2 = n$ (199 nm) = 1.99, if we take $m_1 = 3$ and $m_2 = 2$. These two values are in agreement with the refraction index of amorphous carbon measured by Duley [8]. The choice of the second and the third harmonic is therefore a resonable assumption because heigher harmonics are much more sensible to the destructive interferences. According to the above definition of the probability distribution function for the refraction index, we take $\delta_2 = 0.080$. The same calculation for a width $a_v = 199$ nm of the vacuum resonant cavities, gives $n_{v_1} = n_v$ (138 nm) = 1.04 and $n_{v_2} = n_v$ (199 nm)=1. We have then $\delta_1 = 0.04$ and $W = 1.07$. To confirm these estimated values of δ_1, δ_2, W, we have computed numerically γ as a function of δ_1, for the two wavelength $\lambda + 1$ and λ_2, by using the previous values of W and δ_2. From these calculations we have obtained the common value of δ_1 which optimizes the trasmission for the two wavelength λ_1 and λ_2: it is $\delta_1 = 0.045$, (see figures 3a and 3b).

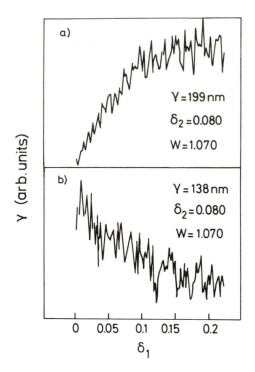

Figure 3: Calculated extinction coefficient γ as the function of the disorder parameter δ_1 for the wavelength (a) 199 nm (b) 138 nm. For the meaning of δ_1, δ_2 and W see the text.

With a good agreement, the value $W = 1.07$ optimizes the calculated trasmission with respect to the two above resonant frequencies. In fact, figure 4b shows that the minimum of the curve γ as a function of W, computed for $\lambda_1 = 138$ nm, $\delta_1 = 0.045$ and $\delta_2 = 0.080$, occurs just at $W = 1.07$. This value is also in agreement with the results of figure 4a, where the curve γ as a function of W was calculated for $\lambda = 199$ nm and for the same values of the parameters δ_1 and δ_2 of figure 4. In fact, we observe that for $W = 1.07$, the corresponding value od γ (see figure 4a) is very close to the minimum of the curve.

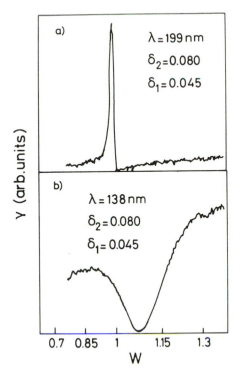

Figure 4: Calculated extinction coefficient γ as a function of W for the two wavelengths (a) 199 nm and (b) 138 nm. For the meaning of δ_1, δ_2 and W see the text.

We remark that the spectral position of the trasmission peaks and the shape of the calculated extinction spectrum reported in figure 5 is very sensible to the choice of the fitting parameters δ_1, δ_2, W as shown in figure 4. This sensibility fixes the choice of the disorder parameters: in fact, a very small variation of each single parameter changes completely the form of the spectrum (i.e. new trasmission peaks appears) and a global readjustment of the parameters will give non-physical values for the estimated refraction index of the amorphous carbon grain samples.

With this choice of the parameters for the probability distribution of $n(x)$, we have made a simulation of the experimental extinction spectrum reported in figure 1a. Figure 5 shows the calculated extinction spectrum, where we remark two minima at 135nm and 202 nm respectively which are in good agreement with the experimental dips of figure 1a. Thus, the 1-dimensional model proposed in [3] can well explain the two absorption dips of figure 1a in terms of strong localization effects of the light.

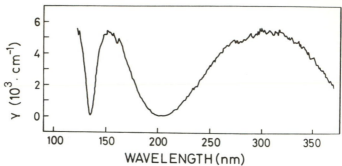

Figure 5: Calculated extinction spectrum γ as a function of wavelength by using the fitting parameters $W = 1.07$, $\delta_1 = 0.045$, $\delta_2 = 0.080$.

In conclusion, the particular spongy structure of the carbon grain sample determines random fluctuations of the refraction index. When a light beam crosses sample having a granular structure, the fluctuations of n produce optical resonance modes which are localized (standing waves) along the disordered medium. Therefore, the trasmission is enhanced for particular resonant frequencies corresponding to the positive interferences of the light within an agglomeration of carbon grains having, in our experiment, a width of about 100 nm.

Aknowledgements

We are grateful to A. Borghesi, G. Bugmann, E. Bussoletti and L. Colangeli for helful and stimulating discussions.

References

Akkermans, E., Maynard, R.: 1985, Phys. Rev. **B32**, 7850, ref.[1].

Sheng, P., Zhang, Z.Q.: 1986, Phys. Rev. Lett. **57**, 1879, ref.[1].

Anderson, P.W.: 1985, Philos. Mag. **B52**, 505, ref.[1].

Azbel, M.Y.: 1983, Phys. Rev. **B28**, 4106, ref.[1].

John, S.: 1987, Phys. Rev. Lett. **58**, 2486, ref.[1].

Van Albada, R.P., Lagendijk, Ad.: 1985, Phys. Rev. Lett. **55**, 2692, ref.[2].

Wolf, P.E., Maret, B.: 1985, Phys. Rev. Lett. **55**, 2696, ref.[2].

Etemad, E., Thompson, R., Andrejco, M.J.: 1986, Phys. Rev. Lett. **57**, 575, ref.[2].

Flesia, C., Johnston, R., Kunz, H.: 1987, Europhys. Lett. **3**, 497, ref.[3].

Gilra, D.P.: 1971, Nature **299**, 237, ref.[4].

Borghesi, A., Bussoletti, E., Colangeli, L., Minafra, A., Rubini, F.: 1983, Infrared Physics **23**, 85, ref.[5].

Johnston, R., Kunz, H.: 1983, J. Phys. **C16**, 3895, ref.[7].

Duley, W.W.: 1984, Astrophysical J. **287**, 694, ref.[8].

HYDROGENATED AMORPHOUS CARBON-COATED SILICATE PARTICLES: DIAMOND AND GRAPHITE-LIKE CARBONS AND INTERSTELLAR EXTINCTION

A.P. Jones[1], W.W. Duley[2], D.A. Williams[1]
[1]*Mathematics Department, UMIST, Manchester, United Kingdom*
[2]*Physics Department, York University, Ontario, Canada*

1 Introduction

A new model for interstellar grains has recently been proposed (Duley, Jones and Williams, 1987) in which the bulk of interstellar extinction is due to thin mantles of hydrogenated amorphous carbon (HAC) on small silicate cores (Figure 1).

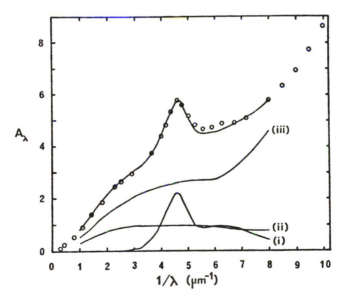

Figure 1: The fit to the average interstellar extinction made up of contributions from (i) small spherical silicate grains (radius $\leq$ 100 Å), (ii) large aspherical silicate grains (500 Å $\leq$ radius $\leq$ 2500 Å), (iii) HAC mantles on the small silicate grains ($\delta_c = 0.5$). For the extinction produced by the small core mantle grains a small particle limit theory is used.

E. Bussoletti et al. (eds.), Experiments on Cosmic Dust Analogues, 161–165.
© 1988 by Kluwer Academic Publishers.

The small particle extinction is calculated using a small particle limit theory to calculate the absorption efficiency factor, i.e.

$$Q_{abs} = 4/3\,\alpha\,a \qquad (1)$$

(van de Hulst, 1957; Platt, 1956), where α is the absorption coefficient of the grain material and a is the grain radius. The small silicate cores give rise to the UV extinction bump at 2175 Å (Steele and Duley, 1987). The large silicate particles contribute a general background extinction and are also responsible for the visual polarization. Light scattered by grains in the small particle limit makes a major contribution to the albedo determination, i.e.

$$Q_{ext} = Q_{abs} \qquad (2)$$

$$Q_{sca} = \eta Q_{abs} \qquad (3)$$

where η is a quantum efficiency factor for re-emission of radiation at the exciting wavelength. Hence the small grain albedo is $W = \eta$. Data from reflection nebulae imply that the total grain albedo in the UV is as high as 0.6 and this requires $\eta > 0.5$.

2   Consequences of the model

(a) Carbon depletions

The fractional carbon depletion, with respect to solar carbon abundance, toward a particular star can be written,

$$\delta_c = 0.07 \left[\frac{E(1250 - V)}{E(B - V)} \right] + R - 0.11 \sum_i A_{1250}(\text{silicate})_i \qquad (4)$$

For the silicate grain sizes adopted this expression reduces to

$$\delta_c = 0.07 \left[\frac{E(1250 - V)}{E(B - V)} + 1.08 \right] \qquad (5)$$

where R= 3.1 and a truncated Mathis, Rumpl and Nordsieck (1977) size distribution have been assumed. Using the above expression and the observed values of $E(1250 - V)/E(B - V)$ it is therefore a simple matter to calculate the carbon depletion along given lines of sight.

(b) 2175 Å bump strength.

From the model an expression for $E(BUMP)$ can be derived, i.e.

$$E(BUMP) = 0.6 \left[\frac{E(2175 - V)}{E(B - V)} \right] - 5.24\delta_c + 0.66 \qquad (6)$$

Figure 2 shows a plot of $E(BUMP)$ versus $E(1250 - V)$ and indicates a weak anticorrelation. This is compatible with the grain model in that carbon accreted onto small silicate grains partially quenches the silicate core absorption at 2175 Å and shorter wavelengths.

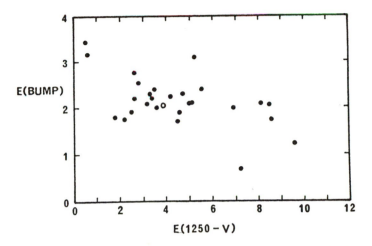

Figure 2: $E(BUMP)$ versus $E(1250 - V)$; $E(1250 - V)$ is proportional to the mass of carbon depleted into HAC mantles on small silicate grains.

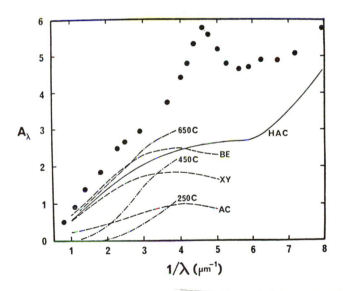

Figure 3: Extinction data for carbonaceous material ($\delta_c = 0.5$), dots – average interstellar extinction (Savage and Mathis, 1979), solid line – HAC (Duley, 1984), dashes – amorphous carbon produced by burning benzene (BE) and xylene (XY) in air, and by arcing across carbon electrodes (AC) (Borghesi, Bussoletti and Colangeli, 1985), and dash-dot – amorphous carbon deposited at 250 C and annealed at 450 C and 650 C (Smith, 1984).

3 HAC composition

Amorphous carbon extinction data presented in figure 3 shows that the optical properties of amorphous carbons are sensitive to the method of preparation and processing conditions.

It has been shown (Smith, 1984) that initial carbon condensates are rich in diamond-like carbon (i.e. rich in tetrahedrally bonded carbon atoms) which absorbs strongly in the far UV but hardly at all in the visual; whereas, thermal processing of this carbon increases the graphitic content (i.e. trignally bonded carbon component) and hence the extinction at visual wavelengths. Thus, HAC of varying composition on silicate cores can explain the observed extinction correlations (Jones, Duley and Williams, 1987).

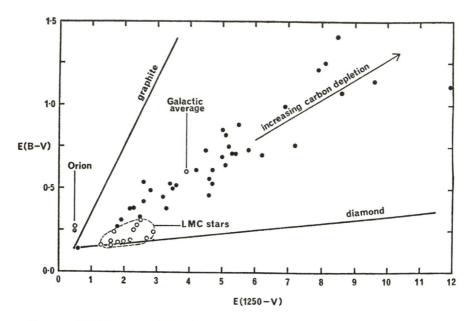

Figure 4: E(B-V) versus E(1250-V) for stars showing peculiar extinction; open circles - average for the Trapezium stars in Orion (Bohlin and Savage, 1981), LMC stars (Clayton and Martin, 1985) and the Galactic average. The labelled lines show the limiting values for HAC composed predominantly of graphitic and diamond-like forms of carbon; the intersection point of these lines corresponds to the background values for the unmantled large and small silicate particles.

Figure 4 shows the extinction data for a sample of stars with peculiar extinction; such a plot will ultimately (with additional amorphous carbon and HAC UV data) enable accurate determination of both the composition and depletion of carbon along a particular line of sight. Thus, this type of plot can provide a powerful and direct link between experimental and observational data.

References

Bohlin, R.C., Savage, B.D.: 1981, Astrophys. J. **249**, 109.

Borghesi, A., Bussoletti, E., Colangeli, L.: 1985, Astr. Astrophys., **142**, 225.

Clayton, G.C., Martin, P.G.: 1985, Astrophys. J. **228**, 558.

Duley, W.W.: 1984, Astrophys. J. **287**, 694.

Duley, W.W., Jones, A.P., Williams, D.A.: 1987, Astrophys. J. submitted.

Jones, A.P., Duley, W.W., Williams, D.A.: 1987, Mon. Not.R. astr. Soc., in press.

Mathis, J.S., Rumpl, W., Nordsieck, K.H.: 1977, Astrophys. J. **217**, 425.

Platt, J.R.: 1956, Astrophys. J. **123**, 486.

Savage, B.D., Mathis, J.S.: 1979, Ann. Rev. Astr. Ap. **17**, 73.

Smith, F.W.: 1984, J. Appl. Phys. **55**, 764.

Steele, T.M., Duley, W.W.: 1987, Astrophys. J. **315**, 337.

van de Hulst, H.C.: 1957, Light Scattering by Small Particles (New York, Dover).

AMORPHOUS CARBON AND SILICON CARBIDE GRAINS AROUND CARBON STARS

A. Blanco[1], A. Borghesi[1], E. Bussoletti[2], L. Colangeli[3], S. Fonti[1], V. Orofino[1]

[1] Physics Department, University of Lecce, Lecce, Italy
[2] Istituto Universitario Navale, Napoli, Italy
[3] ESA Space Science Department, ESTEC, Noordwwijk, The Netherlands

1 Introduction

It is well known that a large amount of dust ($\sim$ 40%) is produced in the envelopes of late-type stars (Jura, 1985). Moreover, according to Knapp and Morris (1985) a very important contribution to this amount (about 50%) comes from C-rich giant stars. Therefore, it is very interesting to study this kind of circumstellar material in order to get information on its physical, chemical and morphological properties. It is widely accepted that, in a carbon-rich stellar atmosphere, solid carbon is the dominant condensate. It is still open, however, the discussion about the structure of this material: amorphous (amorphous, glassy carbon) or crystalline (graphite). Graphite grains have been considered for quite some time (Hoyle and Wickramasinghe, 1962; Mathis et al., 1977; Mitchell and Robinson, 1980; Greenberg and Chlewicki, 1983). However the presence of this material in space has been strongly questioned because of its mechanical, thermal, optical and chemical properties, while amorphous carbon grains are more likely to exist (Czyzak and Santiago, 1973; Donn et al., 1981; Czyzak et al., 1982).

Another important component of the dust in circumstellar envelopes and planetary nebulae (Treffers and Cohen, 1974; Aitken and Roche, 1982; Cohen, 1984) seem also to be silicon carbide grains. While amorphous or graphite grains could explain the continuum emission, SiC particles should be responsible for the 11.5 μm band observed in many C-rich objects. The study of these kind of problems can be done by comparing the calculated emission spectra from the dust envelopes, obtained by means of theoretical models, with observational data. In this paper we show that the far infrared spectral trends of 78 optically thin sources as well as the full spectra of two of the most luminous optically thick sources (CIT6 and IRC+10216) are better explained by emission of amorphous carbon grains rather than graphite particles. Preliminary results on the best fit of observed spectra showing the 11.5 μm emission band seem to indicate that a mixture of carbon and silicon carbide particles can explain the infrared emission of these sources. In these cases again the state of the carbon should be amorphous, while the abundance of SiC should be about 20% of the total mass of the mixture. Experimentally determined extinction data both from amorphous carbon grains and mixtures have been used throughout this work.

E. Bussoletti et al. (eds.), Experiments on Cosmic Dust Analogues, 167–173.
© 1988 by Kluwer Academic Publishers.

2   Carbon grains continuum emission

The flux emerging from a spherical circumstellar dust cloud composed by amorphous carbon or graphite grains has been evaluated by means of a simplified model of radiative transfer widely described elsewhere (Orofino et al., 1987). The adopted values of the extinction properties of several kinds of amorphous carbon grains (experimentally obtained by Borghesi et al., 1983, 1985 and Bussoletti et al., 1987) and of graphite grains (semitheoretically deduced by Draine and Lee, 1983) are shown in figure 1.

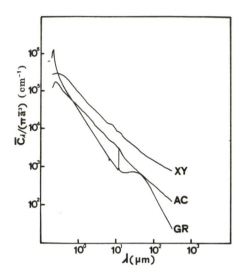

Figure 1: Extinction trends for the materials used in the model of radiative transfer. XY: amorphous carbon from xylene burning; AC: amorphous carbon from arc discharge; GR: graphite.

For sake of simplicity we have not reported the extinction trend for amorphous carbon grains produced by benzene burning (BE) since it is very similar to the curve concerning carbon particles produced by xylene burning (XY).

Besides the chemical and morphological properties of the grains in the cloud, other free parameters of the present model are the star temperature, T_*, the ratio between the cloud external radius and the stellar radius, R_E/R_*, the absorption parameter, A_c, and the density index, α. A_c takes into account the visual absorption, in magnitudes, of the stellar light by the cloud, and is defined as:

$$A_c = 1.086\, \tau_o(V)$$

where $\tau_o(V)$ is the radial optical thickness at $\lambda= 5500$ Å. The density index α determines the radial number density trend of the grains, according to the equation:

$$n_g(r) = n_o \left(\frac{R_E}{r} \right)^{\alpha}$$

The model has been used to fit the spectral trends between 60 and 100 μm of a sample of 78 optically thin sources listed by Jura (1986) and to reproduce the full spectra of two optically thick sources: CIT 6 and IRC +10216. By varying all the parameters of the model (within the limits generally reported in the literature), we obtain that the flux emerging from the model circumstellar clouds follows a power law $F_\lambda \sim \lambda^{-s}$, where the spectral index, s, ranges within the following limits

Ac carbon grains (mean radius $a= 40$ Å):	$3.7 \leq s \leq 4.3$
BE/XY carbon grains ($a= 150$ Å):	$3.5 \leq s \leq 4.3$
graphite grains ($a= 100$ Å):	$4.2 \leq s \leq 5.0$

A comparison of the above values with the observational one, $s= 3.5$ (Jura, 1986), seems to indicate that the state of the carbon around the sources considered in this work may be amorphous rather than crystalline.

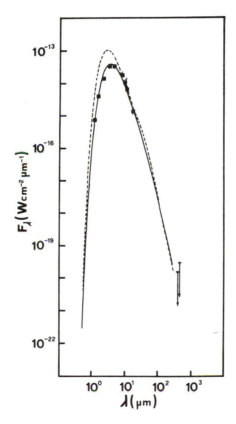

Figure 2: Comparison between observational data for CIT6 and theoretical spectra (solid line: AC amorphous carbon; dashed line: graphite). Observational data are from: Strecker and Ney, 1974 (filled squares); Sopka et al., 1985 (upper limits).

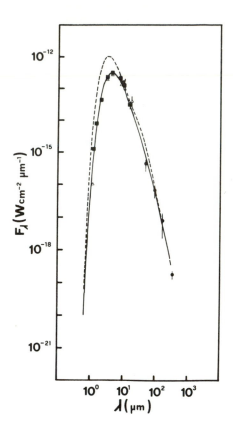

Figure 3: Same as figure 2, but for IRC +10216. Observational data are from: Becklin et al., 1969 (open triangles); Strecker and Ney, 1974 (filled squares); Campbell et al., 1976 (filled circles).

A similar indication comes from the fits of the two optically thick sources CIT 6 and IRC +10216 shown in figures 2 and 3. For both sources the best fits have been obtained for the values of the free parameters listed in table 1. They are very close to those reported in the literature (for the comparison see Orofino et al., 1987).

3   The SiC band

A broad emission band centered at about 11.5 μm is observed in the spectra of many carbon rich objects like circumstellar envelopes and planetary nebulae (Treffers and Cohen, 1974; Aitken and Roche, 1982; Cohen, 1984). This band is commonly attributed to SiC grains. In fact SiC is, after the carbon, the second material, in order of importance, which condenses in carbon-rich astronomical environments as soon as gas temperature decreases below some 1500 K (Hackwell, 1971). Here we show how a mixture of amorphous carbon and SiC particles can explain the whole spectra of some carbon stars.

Table 1: Best fit values of the model parameters for the source considered in this work.

Free parameters	Sources		
	CIT 6	IRC +10216	GL 799
T_*	1300 K	1100 K	1600 K
A_c	10.	12	0.5 (*)
α	2.	2.	—
R_E	300 R_*	400 R_*	15 R_*
T_s	—	—	800 K

(*): In this case A_c is for $\lambda = 3$ μm.

In figure 4 we report the experimental mean extinction cross sections of two mixtures, obtained in the laboratory, of βSiC grains ($a = 0.01$ μm) ans AC carbon grains.

Such cross sections have been obtained from trasmission measurements under the hypothesis of negligible matrix effects of the pellet (KBr in our case).

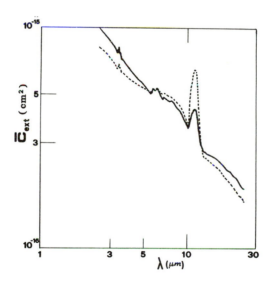

Figure 4: Experimental cross sections of AC and SiC mixtures. The two curves refer to mixtures respectively with 10% (solid line) and 25% (dashed line) of βSiC by mass. The characteristics of both materials are evident in the curves: the 11.5 μm band due to SiC grains, the 3.4 μm and 6–10 μm structures, due to AC grains.

The measured optical properties of different mixtures are being used to fit the spectra of several carbon stars showing the SiC band at 11.5 μm. We are doing this work by means of the same theoretical model (Orofino et al., 1987) used in the cases described in the previous section, except for the fact that here we have to consider a single shell model. With this modification, introduced in order to cope with the lack of UV-visible extinction data, the shell temperature T_s becomes now a new free parameter of the model.

A preliminary result, concerning the source GL 799, is shown in figure 5. The best fit has been obtained for a mixture with about 25% of βSiC by mass and for the values of the free parameters listed in table 1, which are very close to those reported by Cohen (1979). The good agreement between the observational and theoretical spectra indicates that, at least in this case, mixtures of amorphous carbon and SiC particles can explain both the 11.5 μm band and the continuum trend.

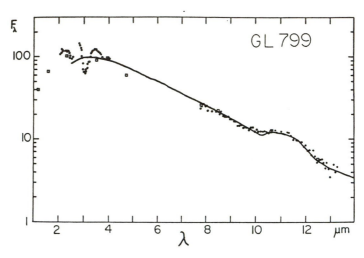

Figure 5: Comparison between the theoretical spectrum of the source GL 799 (solid line) and observational data (Cohen, 1984; dots refer to spectroscopic measurements, open squares to broadband observations). Ordinate are fluxes, F_λ, in unit of 10^{-17} W cm^{-2} μm^{-1}.

It is worthwhile to note here that on the contrary it is very difficult to fit the continuum by means of graphite grains, as already pointed out by Cohen (1984). Work is now in progress to extend this study to other C-rich sources and to verify possible differences between the experimentally obtained and semitheoretically deduced cross section of a given mixture (the latter are calculated starting from the experimental cross sections of the single components of the mixture).

Acknowledgements

The authors are grateful to Mr. M. Scalzo for his valuable assistence during the preparation of the manuscript. This work has been partly supported by Ministero della Pubblica

Istruzione (MPI) and by Consiglio Nazionale delle Ricerche (CNR).

References

Aitken, D.K., Roche, P.F.: 1982, MNRAS **200**, 217.

Becklin, E.E., Frogel, J.A., Hyland, A.R., Kristian, J., Neuegebauer, G.: 1969, Astrophys. J. Lett. **158**, L133.

Borghesi, A., Bussoletti, E., Colangeli, L., Minafra, A., Rubini, F.: 1983, Infrared Phys. **23**, 85.

Borghesi, A., Bussoletti, E., Colangeli: 1985, Astron. Astrophys. **142**, 225.

Bussoletti, E., Borghesi, A., Colangeli, L., Orofino, V.: 1987, Astron. Astrophys., in press.

Campbell, M.F., Elias, J.H., Gezari, D.Y., Harvey, P.M., Hoffmann, W.F., Hudson, H.S., Neuegebauer, G., Soifer, B.T., Werner, M.W., Westbrook, W.E.: 1976, Astrophys. J. **208**, 396.

Cohen, M.: 1979, MNRAS **186**, 837.

Cohen, M.: 1984, MNRAS **206**, 137.

Czyzak, S.J., Santiago, J.J.: 1973, Astrophys. Space Sci. **23**, 443.

Czyzak, S.J., Hirth, J.P., Tabak, R.G.: 1982, Vistas in Astr. **25**, 337.

Donn, B., Hecht, J., Khanna, R., Nuth, J., Stranz, D., Anderson, A.B.: 1981, Surface Sci. **106**, 576.

Draine, B.T., Lee, H.M.: 1984, Astrophys. J. **285**, 89.

Greenberg, J.M., Chlewicki, G.: 1983, Astrophys, J. **272**, 563.

Hackwell, J.A.: 1971, Ph. D. Thesis, Univesity College, England.

Hoyle, F., Wickramasinghe, N.C.: 1962, MNRAS **124**, 417.

Jura, M.: 1985, in "Interrelationship among circumstellar, interstellar and interplanetary dust", NASA CP-2403, eds. Nuth, J.A. and Stancel, R.E., p. 3-17.

Jura, M.: 1986, Astrophys. J. **303**, 327.

Knapp, G.R., Morris, M.: 1985, Astrophys. J. **292**, 640.

Mathis, J.S., Rumpl, W., Nordsiek, K.H.: 1977, Astrophys. J. **217**, 425.

Orofino, V., Colangeli, L., Bussoletti, E., Strafella, F.: 1987, Astrophys. Space Sci., in press.

Sopka, R.J., Hildebrand, R., Jaffe, D.T., Gatley, I., Roelling, T., Werner, M., Jura, M., Zuckerman, B.: 1985, Astrophys. J. **294**, 242.

Strecker, D.W., Ney, E.P.: 1974, Astron. J. **79**, 1410.

Treffers, R.R., Cohen, M.: Astrophys. J. **188**, 545.

GRAIN FORMATION EXPERIMENTS BY A PLASMA JET APPARATUS

T. Tanabé[1], T. Onaka[2], F. Kamijo[2], A. Sakata[3], S. Wada[4]

[1]*Tokyo Astronomical Observatory, Univ. of Tokyo, Tokyo, Japan*
[2]*Dept. of Astronomy, Fac. of Science, Univ. of Tokyo, Tokyo, Japan*
[3]*Lab. of Applied Science, Univ. of Electro-Communications, Tokyo, Japan*
[4]*Lab. of Chemistry, Univ. of Electro-Communications, Tokyo, Japan*

1 Introduction

The observational fact that the interstellar and circumstellar dust grains are not crystalline but amorphous has been accumulated and it leads to the idea that the grain formation in space is a non-equilibrium phenomenon. In order to investigate the non-equilibrium condensation, we have developed a plasma jet apparatus. The high power of the apparatus enables to dissociate molecular sample gas completely and to obtain condensates from atomized gas. The purpose of our experiments is to examine the condensates from rapidly cooled atomized gas with various abundance ratios. In the experiments, we first focused on the condensation from mixtures of H, C, O, and Si atoms. Three kinds of gas mixtures, a mixture af H and C atoms, that of H, C, and Si atoms, and that of H, C, O, and Si atoms, were chosen as reactant condensable gases. Our experiments were executed with entirely new features: (1) condensation from atomized gas and (2) condensation from mixtures of silicon, oxygen, carbon and hydrogen atoms.

2 Experimental

The plasma jet apparatus and experimental procedures are described in detail by Tanabé et al. (1986). We described them briefly here.

Figure 1 shows a schematic diagram of our experimental apparatus. The plasma jet utilizes thermal and electromagnetic pinch effects to form an arc plasma into a flame-like cone of high energy. Figure 2 shows a sectional plan of our plasma torch head. An arc produced between the cathode and the annode is carried through a nozzle (a circular opening in the annode) by a stream of fluid, resulting in a jet of hot partially ionized gas. Due to the temperature higher than 7000 K, the reactant gas is atomized completely in the plasma flame. The reactant gas is mixed with the working Ar gas on

175

E. Bussoletti et al. (eds.), Experiments on Cosmic Dust Analogues, 175–180.

the way to the torch. The used reactant sample gases are tabulated in table 1.

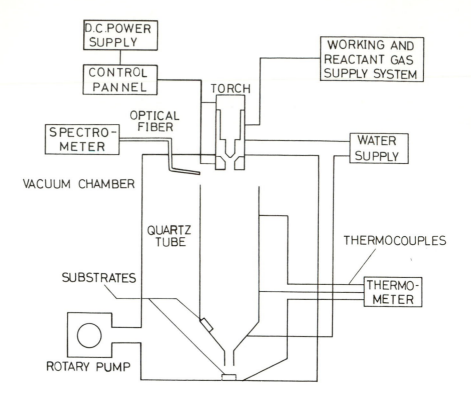

Figure 1: Schematic diagram of the plasma jet apparatus.

Table 1: Reactant gases

Name	Chemical formula
Methane	CH_4
Methane + Silane	$CH_4 + SiH_4$
Tetramethylsilane	$Si(CH_3)_4$
Tetramethoxysilane	$Si(OCH_3)_4$
Methyltrimethoxysilane	$Si(OCH_3)_3CH_4$
Tetraethoxysilane	$Si(OC_2H_5)_4$

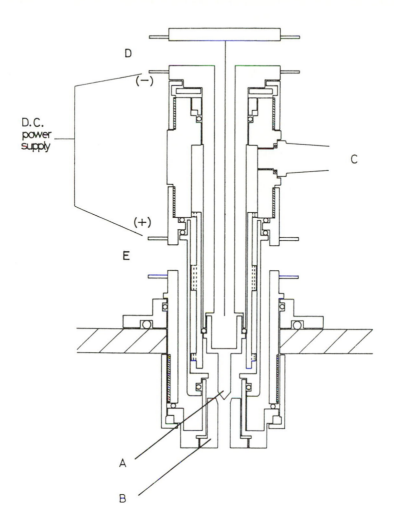

Figure 2: Sectional plan of the plasma torch head. A: cathode, B: annode, C: inlet of the working and reactant gases, D: inlet of cooling water for the cathode, E: inlet of cooling water for the annode.

The atomized gas is injected into a vacuum reactor chamber filled with Ar gas and quenched into solid particles. Experiments were performed at the ambient pressure of 200 Torr.

A photograph of the plasma flame is shown in Figure 3. The light of the plasma flame was lead to a UV-visual spectrometer through an optical fibre and emission line spectra of each species in the plasma flame were obtained. Using Ar I emission lines, we derived the excitation temperature of the plasma flame assuming a local thermodynamic equilibrium. The derived temperatures of all experiments were almost within the same range: 7000–8000 K. From the emission lines of atoms, we also obtained the relative abundance ratio of the reactant atoms of each experiment.

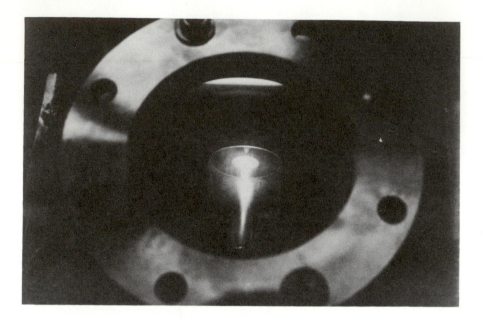

Figure 3: Photograph of the plasma flame.

The synthesized grains were trapped on the inner surface of a collection quartz tube and on the substrates placed at the end of the tube and at the bottom of the chamber. These substrates were used to measure the absorption spectra of condensed grains. Copper meshes were placed in the chamber in order to analyze the particles in the trasmission electron microscope. The collected particles were analyzed by an X-ray diffractometer.

3   Results

A morphological analysis of the synthesized grains on the electron micrograph little differences between particles. The shapes and the sizes of grains resembled one another, no matter what sample was chosen as the reactant gas. The shapes of grains were rather irregular and the sizes were smaller than 20 nm.

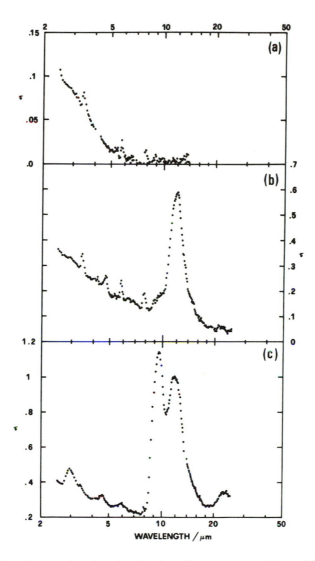

Figure 4: Absorption spectra of condensates from three reactant mixtures: (a) H and C atoms; (b) H, C and Si atoms; (c) H, C, O and Si atoms. Here, the ordinate, τ, is -1n (trasmittance).

X-ray diffraction patterns of the formed grains all showed broad features and no sharp peaks. The features of the condensates from three kinds of reactant gases are tabulated in table 2. We concluded that all of SiC, carbon, and/or SiO_2 are not in crystalline forms.

Table 2: Features appeared in the X-ray diffraction patterns and peak wavelengths of the infrared absorption bands

Mixture	X-ray diffraction feature	IR band (μm)
H, C	carbon	3.4, 5.9, (7.9)
H, C, Si	SiC, weak Si	3.4, 4.6, 5.9, 7.9, 12
H, C, O, Si	carbon or SiO-2, weak SiC	3.0, 3.4, 4.6, 5.9, 7.9, 9.4, 12, 22

The infrared absorption spectra of the synthesized grains showed several features depending upon the reactant gases. These features were all broad and band. features appeared in each condensable gas are also tabulated in table 2. The infrared absorption spectra of condensates from three mixtures are shown in Figure 4. The 9.5 and 22 μm features are assigned to Si-O bond while the 12 μm band to Si-C bond. The 3.0 and 3.4 μm bands are due to O-H and C-H bonds, respectively. One of the interesting results is that the SiO and SiC features appear together in every case of a mixture of H, C, O, and Si atoms. This result indicates that several kinds of chemical bonds were simultaneously condensed irrespective of their initial abundances. This is different from what the equilibrium calculation expects.

4   Discussion

In our experiments, the condensation took place very rapidly. We suggest from the results described above that the formation of grains in a rapidly cooling condition is dominated by the kinetic process in the condensable gas and that the number of collisions among atoms and/or molecules in the course of cooling directly controls the chemical composition and structure of grains.

The infrared spectra observed in the circumstellar regions can be compared with those obtained in our experiments. The 10 and 18 μm band features are observed in late M type stars and the 11.5 μm one in carbon stars. Two features have never been observed simultaneously in one object. However, the amorphous properties and the "dirty" characteristics observed in the astrophysical dust indicate some sort of deviation from the equilibrium. Probably the present experiments were carried out in a too rapid cooling condition and the deviation from the equilibrium was too large as a simulation of the dust formation around cool stars. Further experiments are needed to understand it.

References

Tanabé, T., Onaka, T., Kamijo, F., Sakata, A., Wada, S.: 1986, Jpn. J. Appl. Phys., **25**, 1914.

DETERMINATION OF THE U.V.-VISIBLE OPTICAL PROPERTIES OF SUBMICRONIC CARBONACEOUS PARTICLES AT HIGH TEMPERATURE FROM SCATTERING AND EXTINCTION MEASUREMENTS

B.M. Vaglieco[1], F. Beretta[2], A. D'Alessio[1]

[1]*Dipartimento di Ingegneria Chimica, Università di Napoli, Napoli, Italy*
[2]*Istituto di Ricerche sulla Combustione, C.N.R., Napoli, Italy*

1 Introduction

It is well known that both the U.V. extinction and the I.R. emission spectra measured in interstellar media could be partially attributed to carbon compounds (Huffman, 1977; Allamandola et al., 1987). Furthermore it is normally assumed that the carbon is present in the form either of submicronic solid particles or of large aromatic molecules, mainly polycyclic aromatic hydrocarbons (PAH). Therefore the problem of modelling the observed spectra is partly a problem of light scattering by small particles composed by a medium showing a dispersion of the optical properties and partly a problem of emission, absorption and fluorescence spectroscopy of large molecules. A surprising similar problem is present in the optical diagnostics of combustion of hydrocarbons in defect of oxygen; in fact also in this case carbonaceous particles (soot) and PAH are produced into the flame and often survives into the exhausts of engines or furnaces. This group of research has developed in the last few years different optical methods for sizing soot particles in flames (D'Alessio et al., 1975; D'Alessio et al., 1977) and detecting PAH molecules by laser excited fluorescence (D'Alessio, 1981; Beretta et al., 1985). In this field a specific line of research has been devoted to the measurement of the scattering and extinction coefficients inside rich flames in the wavelength range 250 nm to 550 nm (Menna and D'Alessio, 1983; Vaglieco et al., 1986) and the present communication intends to report on the evaluation of the complex refractive index of soot particles from those measurements. This approach presents the advantage of being an "in situ" technique which gives directly the optical properties of the submicronic soot particles whereas the conventional methods of determination of the optical properties from bulk samples suffer from the uncertainties of the procedure of sampling and compacting the original aerosols.

2 Theoretical background

The measured scattering coefficient $Q_\lambda(\theta)$, at the scattering angle θ and wavelength λ, is the sum of the true scattering due to soot particles Q_λ^s and fluorescence due to polycyclic

E. Bussoletti et al. (eds.), Experiments on Cosmic Dust Analogues, 181–190.
© *1988 by Kluwer Academic Publishers.*

aromatic hydrocarbons and/or flame emission Q_λ^f

$$Q_\lambda(\theta) = Q_\lambda^s(\theta) + Q_\lambda^f(\theta) \tag{1}$$

Scattering measurements in different polarization planes V and H allow to distinguish between the strongly polarized scattering and the unpolarized fluorescence.

$$Q_\lambda^V(\theta) = Q_\lambda^{Vs}(\theta) + Q_\lambda^{Vf}(\theta) \tag{2}$$

$$Q_\lambda^H(\theta) = Q_\lambda^{Hs}(\theta) + Q_\lambda^{Hf}(\theta) \tag{3}$$

at the scattering angle $\theta = 90°$ it was found experimentally that soot particles scatter much more in the vertical planes and the polarization ratio $\gamma_\lambda^s = Q_\lambda^{Hs}/Q_\lambda^{Vs}$ is around $4 \cdot 10^{-2}$ (D'Alessio et al., 1977; D'Alessio, 1981), whereas $\gamma_\lambda^f = Q^{Hf}/Q^{Vf}$ is equal to one. The measured extinction coefficient can equally be considered as the sum of the extinction due to soot particles and a molecular contribution due to absorption of molecular species (Di Lorenzo et al., 1981).

$$k_\lambda = k_\lambda^s + k_\lambda^m \tag{4}$$

Only when the molecular contribution is negligible both for scattering and absorption the measured coefficients can be interpreted in the framework of the electromagnetic theory of scattering of small particles:

$$Q_\lambda(\theta) = Q_\lambda^s(\theta) = N \cdot C_{scatt}^s(D, n_\lambda, k_\lambda) \tag{5}$$

$$k_\lambda = k_\lambda^s = N \cdot C_{ext}^s(D, n_\lambda, k_\lambda) \tag{6}$$

Equations 5 and 6 assume that the particles are monodisperse spheres in a single scattering regime so that the soot scattering and extinction coefficients are proportional to their number concentration N and to the relevant cross sections, which depend upon the particle size D and the material refractive index $m = n - ik$. When the particle size is much smaller than the wavelength the cross sections are given by the Rayleigh dipole limit expressions and the unpolarized scattering and extinction coefficients are given by:

$$Q_\lambda^s(\theta) = N\alpha^6 \left| \frac{m^2 - 1}{m^2 + 2} \right|^2 \frac{1 + \cos^2\theta}{2} \frac{\lambda^2}{4\pi^2} \tag{7}$$

$$k_\lambda^s = N\alpha^3 Im \left(\frac{m^2 - 1}{m^2 + 2} \right) \frac{\lambda^2}{4\pi} \tag{8}$$

where $\alpha = \pi D/\lambda$. If measurements of the scattering and extinction coefficients are available at i-wavelengths, and the medium has a dispersion of the optical properties, the system of the equations 7 and 8 depends upon $2i+2$ unknown quantities n_{λ_i}, k_{λ_i}, N and D. Therefore N and D have to be known by indipendent measurements of alternatively the values n_{λ_0} and k_{λ_0} at a reference wavelength λ_0.

3 Experimental apparatus and results

The scheme of the optical apparatus is reported is reported in figure 1. The measurements were carried out on primixed flames generated on flat flame porous plate burner of 50 mm diameter. Xenon continuos lamp of 450 W was used as light source since it has a good light emission also in ultraviolet.

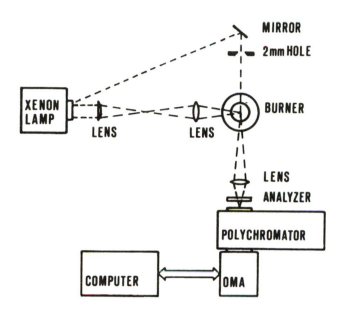

Figure 1: Experimental set-up.

The scattered light was focused with a $f=75$ mm fused silica lens on the entrance slit of a liminous, low resolution polychromator with $f/3.8$ equipped with a grating with 150 groves/mm and blazed at 450 nm. Extinction measurements were carried out using the same lamp as a source, deriving part of its light with a set of mirrors or with an optical fiber. The spectral image formed in the polychromator exit plane was detected with a photodiode array and the signals processed with an optical multichannel analyzer, and it was possible to explore a spectral range of 400 nm in a single measurement. All the signals were then stored on a microcomputer for further mathematical analysis. Typical curves of the scattering and extinction coefficients inside the soot formation region of rich premixed flames are shown in figures 2 and 3, where the results obtained for a C_2H_4/O_2 flame with C/O ratio equal to 0.73 are reported. Both the unpolarized scattering and the extinction coefficients increase with the height above the burner because the measurements were obtained in flame regions where soot formation is still going on.

The main characteristics of the extinction and scattering spectra are the pronounced peak around 300 nm, and the decrease of the scattering and extinction coefficients at longer wavelengths. It is worthwhile to remark that the scattering coefficient declines more rapidly with λ than the extinction one.

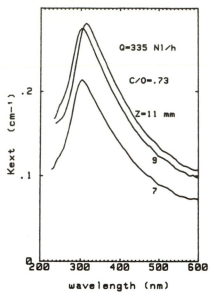

Figure 2: Scattering coefficients versus wavelength at different heights for a C_2H_4–O_2 flame at scattering angle $\theta = 90°$.

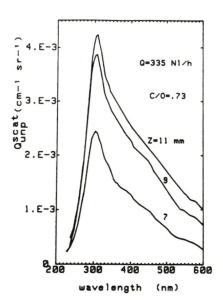

Figure 3: Extinction coefficients versus wavelength at different heights for a C_2H_4–O_2 flame.

It appears also from figure 2 that the slope of the curves of scattering coefficients versus λ is smaller in the region 500–600 nm than in the region 400–500 nm and this effect is more evident in the early zone of the flame where the soot concentration is low. This observation suggests that other emission mechanisms, in addition to the scattering due to soot particles, have a relevance in those conditions. In order to verify this hypothesis the scattering measurements were repeated employing a polarizer in the scattering beam measuring also the polarization ratio $\gamma_\lambda = Q_\lambda^H / Q_\lambda^V$ as a function of the wavelength at the scattering angle $\theta = 90°$.

The results, reported in figure 4, show that between 300 and 400 nm the polarization ratio is always very low, around $5 \cdot 10^{-2}$, whereas higher values are measurred both in the ultraviolet and in the red particularly near in the early soot formation zone. The minimum of the polarization ratio is almost coincident with the value expected from soot particles, as it has been discussed in section 2, and consequently the measurerd signals are always determined by true scattering in the 300–400 nm range. On the contrary the scattering in the region between 500–600 nm at z=7 mm is weakly polarized thus demonstrating the relative importance of unpolarized emission. In the final part of the flame (z=11 mm) the polarization ratio is almost constant, with values around $5 \cdot 10^{-2}$, thus indicating that soot scattering prevails over emission and fluorescence in the whole specttral region. A more quantitative analysis of the scattering curves in the visible showed that the scattering coefficient does not follow the λ^{-4} dependence as predicted by equation 7. This discrepancy occurs when the particles size is comparable with the Rayleigh approximation breaks down, or when there is a dispersion of the optical properties. The size of the soot particles have been determined from the ratio of the measured scattering and extinction coefficients. In fact these ratio is proportional to the volume of the scatterers, as it appears from equations 6 and 7.

In the present experiments the α-parameter in the visible is always found to be less than 0.3, which is the upper limit for the validity of the Rayleigh theory. Therefore the discrepancy between the experimental spectral scattering curves and the prediction of the theory has to be attributed to a dispersion of the optical properties. In order to study more quantitatively the dispersion effects the normalized experimental scattering curves were compared with those predicted by the Rayleigh theory using for the refractive index the U.V.-visible data determined for graphite (taft and Philip, 1965), glassy carbon (Williams and Arakawa, 1972) and for an amorphous film (Duley, 1984). This procedure was necessary because the lack of experimental determination of the optical properties of soot or carbon black in the ultraviolet. The results reported in figure 5 clearly show that the scattering due to submicronic graphite particles has a peak much more shifted toward the U.V. and its maximum intensity is almost thirty times higher than that found experimentally. On the order extreme the scattering computed using the amorphous carbon film data does not exhibit any peak in the near U.V. region. The scattering due to glassy carbon has a spectral maximum more shifted toward the red and a lower intensity respect to that of graphite, however they predict a more pronounced maximum more shifted toward the U.V. than the experimental one.

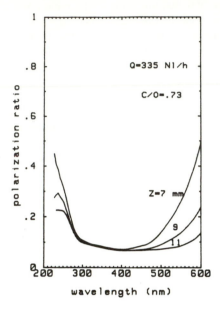

Figure 4: Polarization ratio at $\theta = 90°$ versus wavelength at different heights for a C_2H_4–O_2 flame.

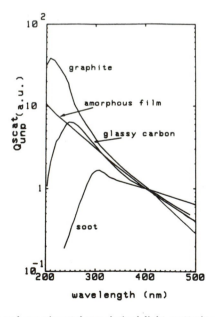

Figure 5: Computed and experimental unpolarized light scattering coefficients. For comparison they are normalized at $\lambda = 400$ nm.

4 Evaluation and discussion of the optical properties of soot

The results previously illustrated put in evidence that the optical properties of soot are different from those found for other carbonaceous compounds. The spectral behaviour of $m_\lambda = n_\lambda - ik_\lambda$ can be directly evaluated by solving numerically equation 7 and 8, assuming different values for the refractive index $m_{\lambda_0} = n_{\lambda_0} - ik_{\lambda_0}$ at the reference wavelengths $\lambda_0 = 488$ nm, according to the procedure outlined in section 2. The numerical analysis showed that it was possible to determine m in the whole spectral range only in the upper region of the $n_{\lambda_0} - k_{\lambda_0}$ plane as it is shown in figure 6.

Different experimental determination of the optical properties of soot are also reported in figure 6 (Dalzell and Sarofim, 1969; Charalmpopoulos and Felske, 1987; Erickson et al., 1964; Bockhorn et al., 1981) and it appears that only the data obtained by Lee and Tien (Lee and Tien, 1981) and by Chippet and Gray (Chippet and Gray, 1978) are located in the upper part of the $n_{\lambda_0} - k_{\lambda_0}$ plane and allow to deduce the optical properties of soot down to 250 nm.

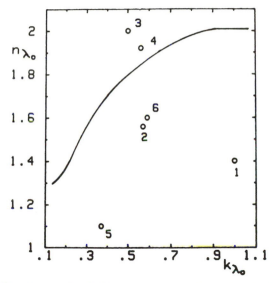

Figure 6: Possible values $n_{\lambda_0} k_{\lambda_0}$ for the determination of n_λ and k_λ in the ultraviolet. 1) Erickson and al., 1964; 2) Dalzell and Sarofim, 1969; 3) Chippet and Gray, 1978; 4) Lee and Tien, 1981; 5) Bockhorn, 1981; 6) Charalmpopoulos and Felske, 1987. a $C_2H_4-O_2$ flame.

The spectral behaviour of n_λ and k_λ computed assuming as starting values $n_{\lambda_0} = 1.92$ and $k_{\lambda_0} = 0.56$ are reported in figure 7 together with the optical data for graphite, glassy and amorphous carbons which have been used for the scattering computations of figure 5. The absorption index k_λ evaluated for soot remains almost constant in the visible, in agreement with the behaviour presented by the other materials, and have a value lower than that found for graphite and glassy carbon. The soot absorption index falls off in the ultraviolet after 300 nm while, in the same spectral region, graphite has a sharp maximum, centered at 260 nm, glassy carbon exhibits a hump, followed by a fall off beyond 250 nm, and the value of amorphous carbon film remains constant. The real part of the refractive index of soot particles n_λ shows a strong dispersion declining continuously from the visible to the U.V.. Graphite, on the contrary, has a constant value in the visible followed by a sharp fall off in the U.V.; glassy carbon has a slight dispersion in the visible and a more pronounced one in the U.V., while the carbon film data remain almost constant down to 220 nm.

Graphite is a well ordered crystal and its sharp resonance in the optical properties in the near U.V. has been attributed to a single electron transition between the π valence and conduction bands (Taft and Philip, 1965). Glassy carbon is an amorphous material produced by pyrolysis of resins or hard chars with a structure composed by small packets of hexagonal planes with very little graphitic orientation between the planes (Williams and Arakawa, 1972) defined as turbostratic structure.

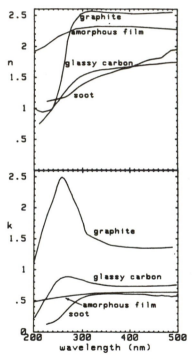

Figure 7: Soot refractive index compared with glassy and amorphous carbon data.

Its optical properties in the near U.V. can still be attributed to transitions of electrons in π orbitals but the resonance is broadened and shifted toward the red, due to a more disordered spatial structure combined with an enlargement of the typical scale size of polycyclic ring systems (Marchand, 1986). The film of amorphous carbon, whose optical properties are reported here (Duley, 1984), has been formed by direct condensation from a carbon vapor consisting primarily of atomic and simple linear molecular species. Its

U.V. optical properties do not exhibit any resonance, consequently there is no evidence of the presence of π electrons and this material is mainly composed by fourfold coordinated carbon bounded within a tetrahedral structure typical of diamond.

The fall off of k_λ shown by our data on soot indicates that this material is more similar to graphite and glassy carbon than to carbon film and therefore its U.V. optical properties are related more to the contribution of π electrons than to σ ones. In addition, the strong dispersion of n_λ is a typical feature which appears when carbon films, originally formed by tetrahedrical structures, are heated up at temperature higher than 700°C (McKenzie et al., 1983; Smith, 1984) and this effect has been attributed to the formation of graphite-like bounding. On the other side, it is well known that a dispersion of n_λ is related to an absorption band. Therefore it appears that there is a continuous transition from graphite, which exhibits a sharp resonance band around 260 nm; to glassy carbon, with a hump in the absorption spectrum and a moderate dispersion in the refractive index, down to soot where the resonance is much more broadened toward the visible. The broadening and red shifting of the U.V. absorption bands have been found in the spectra of aromatic molecules with an increase number of aromatic rings and explained with the increased delocalization of π electrons (Ademson, 1979). Therefore the surface structure of soot particles seems to consist of a large polycyclic network, with hydrogen atoms inserted in it, which does not exhibit a long range order.

5   Conclusions

The first conclusion drawn from this work is the possibility to determine the spectral behaviour of the extinction and scattering coefficients in clouds of submicronic unaggregated carbon particles from measurements inside the flames, avoiding the uncertainties due to sampling. A second conclusion is that the spectral shape of the scattering and extinction coefficients in the ultraviolet and visible indicate clearly the difference among the optical properties of these particles and those observed for other containing materials as graphite, glassy and amorphous carbon.

The "in situ" determination of the refractive index of soot particles from scattering and extinction data qualifies this statement. The resonance due to π electrons is much larger and broadened in this material than in other form of carbon because it is formed through a condensation and dehydrogenation processes at high pressure in a temperature range between 1500 and 1900 K, as a final stage of the formation of large aromatic polycondensed molecules. The differences between the spectral data found here and those observed in the interstellar dust region can probably be explained by different routes of formation of carbon compounds in that widely different conditions of density and energy. The present "in situ" method for determining both the scattering and extinction spectral curves has to be explored in a larger range of operating conditions in order to establish more detailed correlations between the optical properties of carbonaceous particles, their physical and chemical structure and their formation processes.

References

Ademson, A.W.: 1979, "A Text Book of Physical Chemistry", Academic Press, N.Y..

Allamandola, L.J., A.G.G.M., Tielens, J.R., Barker, A.G.G.M.: 1987, in "Polycyclic Aromatic Hydrocarbons and Astrophysics", Léger et al. (eds), 255.

Beretta, F., Cincotti, V., D'Alessio, A., Menna, P.: 1985, Comb. Flame **61**, 211.

Bockhorn, H., Fetting, F., Meyer, U., Reck, R.,Wannemacher, G.: 1981, XVIII (Int.) on Combustion, The Combustion Institute, 1137.

Charalmpopoulos, T.T., Felske, J.D.: 1987, Comb. Flame, **68**, 283.

Chippet, S., Gray, W.A.: 1978, Comb. Flame, **31** 149.

D'Alessio, A., Di Lorenzo, A., Sarofim, A.F., Beretta, F., Masi, S., Venitozzi, C.: 1975, XV (Int.) on Combustion, The Combustion Institute, 941.

D'Alessio, A., Di Lorenzo, A., Borghese, A., Beretta, F., Masi, S.: 1977, XVI (Int.) on Combustion, The Combustion Institute, 695.

D'Alessio, A.: 1981, in "Particulate Carbon: Formation during Combustion", D.C. Siegla and G.W. Smith (eds.), Plenum Press, New York, 207.

D'Alessio, A., Beretta, F., Cavaliere, A., Menna, P.: 1983, in "Soot in Combustion Systems and its Toxic Properties", J. Lahaye, G. Prado (eds.), Plenum Press, New York, 355.

Dalzell, W.H., Sarofim, A.F.: 1969, J. Heat Transfer, **91**, 100.

Di Lorenzo, A., D'Alessio, A., Cincotti, V., Masi, S., Menna, P., Venitozzi, C.: 1981, XVIII (Int.) on Combustion, The Combustion Institute, 485.

Duley, W.W.: 1984, Astrophys. J., **287**, 694.

Erickson, W.A., Williams, G.C., Hottel, H.C.: 1964, Comb. Flame, **8**, 127.

Huffman, D.R.: 1977, Adv. Phys., **26**, 129.

Jenkins, G.M., Kawamura, K.: 1976, "Polymeric Carbons - Carbon Fibre, Glass and Char", Cambridge University Press, New York.

Lee, S.C., Tien, C.L.: 1981, XVIII (Int.) on Combustion, The Combustion Institute, 1159.

Marchand, A.: 1987, in "Polycyclic Aromatic Hydrocarbons and Astrophysics", Léger et al. (eds.), Reidel Publishing Company, 31.

McKenzie, D.R., McPhedran, R.C., Savvides, N.C., Botten, C.: 1983, Phyl. Mag., B. **48**, 341.

Menna, P., D'Alessio, A.: 1983, XIX (Int.) on Combustion, The Combustion Institute, 1421.

Smith, F.W.: 1984, J. Appl. Phys., **55**, 764.

Taft, A., Philip, H.R.: 1965, Physical Rev. A **197**, 138.

Vaglieco, B.M., Beretta, F., D'Alessio, A.: 1986, in AEROSOLS: Formation and Reactivity, 2nd Int. Aerosol Conf., Berlin, 976, Pergamon Press, Oxford.

Williams, M.W., Arakawa, E.T.: 1972, J. Appl. Phys. **43**, 3460.

EXPERIMENTAL STUDIES OF CIRCUMSTELLAR, INTERSTELLAR AND INTERPLANETARY REFRACTORY GRAINS ANALOGS

J.A. Nuth[1], R.N. Nelson[1,2], M. Moore[1,3], B. Donn[1]
[1]Code 691, NASA/GSFC, Greenbelt., U.S.A.
[2]Chemistry Dept., Georgia Southern College, Statesboro, U.S.A.
[3]Chemistry Dept., U. of Maryland, College Park, U.S.A.

1 Introduction

The history of refractory matter in the universe can be extremely complex. After synthesis in stellar interiors and migration to the stellar surface, refractory atoms and molecules can be expelled in a variety of ways as part of a circumstellar outflow. If conditions are right, nucleation occurs and circumstellar grains begin to grow via a series of potentially independent chemical pathways. Depending upon conditions within the circumstellar shell, numerous changes might occur in the structure of the grain prior to its ejection into the interstellar medium. Once in the interstellar medium, grains can be destroyed via shock processes or could survive and accrete the refractory vapor liberated in the destruction of other grains. This materials could then be metamorphosed via UV, X-ray or proton bombardment. During the collapse of a giant molecular cloud to form a protostar, a new set of metamorphic influences might be encountered including thermal annealing, partial melting, distillation, hydrous alteration, oxidation (or reduction), growth and coagulation.

We have carried out experiments which examine several aspects of the life cycle of cosmic dust including nucleation, thermal metamorphism, hydrous alteration and the irradiation of refractory ice mantles. We will discuss several new studies which we have recently initiated. These initiatives include the construction of a differentially pumped cluster beam-mass spectrometer apparatus with which we expect to determine the relative stability of small clusters in the systems Fe-Mg-Si-O and Ca-Ti-Al-O as a function of temperature and pressure; the construction of a "Smoke Generator" with which to produce gram quantities of amorphous grains for use in thermal annealing and hydrous alteration studies of smokes containing Fe-Mg-Al-Ti-Ca-Si-O and Cl; the use of a low temperature cryostat to observe the processing of ices containing $Fe(CO)_5$, SiH_4 and H_2O by 1 MeV protons and ultraviolet radiation in order to simulate refractory grain growth in the interstellar medium; trapping of rare gases during the condensation of Si_2O_3 smokes; and oxygen isotopic fractionation during the condensation of grains at temperatures in excess of 500 K from a gaseous mixture of SiH_4, $Fe(CO)_5$, $Al(CH_3)_3$ and

E. Bussoletti et al. (eds.), Experiments on Cosmic Dust Analogues, 191–207.

O_2 in a H_2 carrier gas.

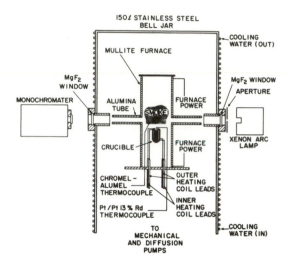

Figure 1: Schematic diagram of the older nucleation system in which a crucible is used to produce a known concentration of refractory vapor inside a well defined temperature gradient. Nucleation produces very fine grained smoke particles ($\sim$ 50 nm radius) when the critical supersaturation is exceeded. The onset of particle formation is detected optically.

2 Experimental apparatus

2.1 Refractory grains

Two experimental systems have been used in our laboratory to produce refractory grains. The older bell-jar based apparatus is shown in figure 1 and consist of an outer furnace which controls the ambient temperature and one or more crucibles from which refractory vapors diffuse through an ambient gas (usually H_2). The system has been described by Nuth and Donn (1982, 1983a); the major problem with this system is that complete mixing for even two component systems is rarely achieved. This system was used in the experimental study of the absorption efficiency of rare gases into Si_2O_3 smokes which will be described in section 3.3.

Refractory smokes can also be produced by passing volatile metal containing precursors through a hot ($T \geq 1500$ K) furnace tube and allowing the reactants to condense in a lower temperature environment (Figure 2). Metallic reactants are premixed before entering the furnace to ensure a homogeneous gas phase. Particles produced in this system are collected on copper collectors downstream from the furnace and are typically macroscopic aggregates of ten nanometer sized particles. A variety of experiments have been run to date. The goals of these experiments are: to derive the crystal structure and morphology of newly condensed refractory materials of cosmic composition, to determine the manner in which such smokes would change under various metamorphic influences, to assess the ease with which such changes occur and to estabilish diagnostic observational signatures by which the structure and composition of circumstellar and interstellar grains could be discerned.

We have produced mixed oxide smokes containing Fe, Al, and Si and have made some preliminary measurements of the rate and nature of the infrared spectral changes induced by vacuum thermal annealing and hydrous alteration.

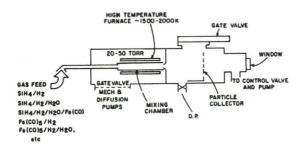

Figure 2: Schematic diagram of the flow condensation apparatus in which amorphous smokes are produced at relatively high temperature from a mixture of refractory gas precursors (such as SiH_4, $Fe(CO)_5$, $TiCl_4$, $Al(CH_3)_3$) and other reactive gases (e.g. O_2, H_2S or C_2H_2) in a flow of H_2. Particles are collected at approximately room temperature.

These experiments will require a considerable amount of analysis before we can apply our results directly to astronomical observations as we previously did with the less complicated MgSiO system (Nuth, Donn and Nelson, 1986). In addition, we will soon produce refractory smokes containing Ti oxides as well as Fe, Al, and Si and we will again study the spectral properties of the grains during metamorphism via vacuum annealing or hydrous alteration. Over the next 3 years similar studies of grains containing Na, K, Ca, Mg, S and C in various combinations with O, Ti, Fe, Al, and Si will also be attempted. Preliminary studies of the infrared spectra of grains condensed in this system are discussed in section 3.1. Results of studies of isotopic fractionation will be described in section 3.4.

2.2 Refractory ices

A helium cooled cryostat cell is available which can reach temperatures in the vicinity of 10 K and which is installed in the beam line of a 1 MeV Van de Graaff proton accelerator. The system has previously been used to study the proton irradiation of volatile ices such as CH_4, NH_3 and H_2O (Moore et al., 1983) or the irradiation of sulfur containing compounds applicable to the surface of Io (Moore, 1984). By turning the cryostat $180°$ away from the proton beam line, it is possible to obtain an infrared spectrum of the irradiated ice deposit. It is typical that several cycles of irradiation followed by spectroscopy are possible without loss of a sample. A mass spectrometer is available to monitor gas emitted during warm up of the sample. The temperature of the sample can be controlled to within 2 K over the range 15 K$\leq$ T $\leq$ 300 K. This permits very controlled reaction of radiation produced radicals trapped within the ice matrix.

Recent calculations of the efficiency with which interstellar grains may be destroyed in shocks (Seab and Shull, 1985) have indicated that typical grains may be destroyed, on the average, approximately 10 times between their formation in a circumstellar environment and their incorporation into a protostellar nebula. Such high destruction efficiencies, if correct, would indicate that a considerable mass of the interstellar dust grew in situ from the vapor, probably after deposition on the surfaces of surviving grains. In such a scenario, refractory species such as silicon would more likely react with hydrogen atoms on grain surfaces than with oxygen or other metals [Tielens, personnel communication]. The result would be hydrogenated species such as SiH_4. AlH, FeH, etc. In addition, other molecules such as H_2O, CO, CH_4 or NH_3 would co-exist as ices with the metal hydrides on the surface of the grain. These refractory and ice coatings would be subject to both UV and cosmic ray bombardment as well as to a certain degree of thermal processing during temperature fluctuations.

We have carried out preliminary studies of the processing of refractory ice mixtures in systems containing SiH_4-$Fe(CO)_5$ and H_2O and have monitored changes during processing by both mass spectrometry and infrared spectroscopy. We have observed IR features which may be observable in the spectra of interstellar materials and which might be diagnostic of such ices. We gradually warm up the ices after irradiation and record the infrared spectrum of the refractory residue. Preliminary results from these studies will be briefly discussed in section 3.2.

2.3   Refractory clusters

A schematic diagram of the Cluster Beam System is shown in figure 3.

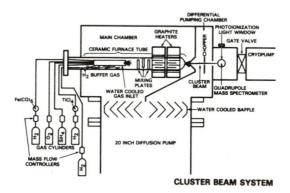

Figure 3: Schematic diagram of the cluster beam apparatus in which refractory clusters will be produced in low concentration in a H_2 buffer gas from a variety of refractory gas precursors. The clusters will be detected by a quadrupole mass spectrometer as a function of the temperature of the reaction tube and the partial pressure of the reactants.

The system consists of a vacuum chamber, mounted on top of a 20" diffusion pump, into which a ceramic furnace tube is inserted. A small hole is bored into the end of the tube, which contains small quantities of reactive metal precursors (such as SiH_4, $Fe(CO)_5$, $Al(CH_3)_3$, or $TiCl_4$) plus oxygen diluted in a large bath of H_2. The tube can be heated to $T \sim 2000$ K and the equilibrium cluster distribution which is estabilished within the heated portion of the furnace will be measured by a quadrupole mass spectrometer after the gas has expanded cooled and been differentially pumped to $\sim 10^{-5}$ Torr. If the cluster concentration is sufficiently dilute, no cluster growth will occur during the adiabatic cooling of the gas, and the true equilibrium cluster ratios will be measured.

a. *Silicate condensation*: Previous studies have indicated that condensation of Mg-SiO smokes at high temperatures is independent of the concentration of Mg in the gas phase (Nuth and Donn, 1983b), while nucleation at $T < 925$ K can be effected by the presence of magnesium. It was suggested that mixed Mg-SiO clusters are stable at lower temperatures, whereas only pure $(SiO)_x$ clusters are stable at high temperatures. This seemed consistent with the work of Stephens and Bauer (1981) which showed that the pressure of Fe had very little or no effect on the nucleation of Fe-SiO systems at 1250 $K \leq T \leq 4400$ K in a shock tube. We will measure the relative abundances of the precondensation clusters in hydrogen gas to which a very small quantity of SiH_4 and O_2 have been added as a function of temperature, reactant partial pressure and total pressure. Similar measurements will be made for a dilute mixture of SiH_4, $Fe(CO)_5$ and O_2. We will then modify the system by the addition of a crucible in the side of the furnace which will be used to introduce Mg into the gas system. In this way, cluster distributions in the systems Mg-Si-O and Fe-Mg-Si-O can be measured, and the reaction path through the more stable silicate clusters can be plotted from simple molecules through complex solids.

b. *CAI nucleation*: No previous work on the nucleation of these systems has been carried out despite the fact that calcium, aluminum and titanium oxides are extremely refractory and likely to be among the first thermodinamically stable condensates from a cooling vapor of cosmic composition. We will measure the relative abundances of the precondensation clusters in hydrogen gas to which a very small quantity of $Al(CH_3)_3$ and O_2 has been added as a function of ambient temperature, reactant partial pressure and total pressure. Similar measurements will be made for a dilute mixture of $Al(CH_3)_3$, $TiCl_4$ and O_2. We will then use the modified system described above to introduce Ca vapor from a crucible into the gas stream and study the systems Ca-Al-O, Ca-Ti-O and Ca-Al-Ti-O.

3   Results and discussion

3.1   Refractory grains

The structure and composition of pre-solar grains can greatly effect the thermal and dynamic history of the solar nebula, yet the only way to study the modern equivalents of such materials in circumstellar outflows and dense interstellar clouds is via remote sensing techniques. Interpretation of the data obtained by these techniques usually requires calibration against 'known' standards which have similar properties. Nuth and Donn (1984) have previously argued that simple silicate smokes are good analogs to more complex natural grains and that studies of such simple particles can be extremely helpful in understanding the processing of grains in various astrophysical environments. The morphology and crystal structure of amorphous Mg-SiO smokes annealed at 1000 K in vacuo for varying times has been studied by Rietmeijer et al. (1986). Figure 4a is a scanning electron microscope (SEM) image of a Mg-SiO smoke annealed in vacuo at 1000 K for 4 hours (the scale bar is 1 micron). Figure 4b is an SEM image of an interplanetary dust particle (IDP) collected in the stratosphere (Bradley et al., 1983); note that the scale bar in this image is also 1 micron.

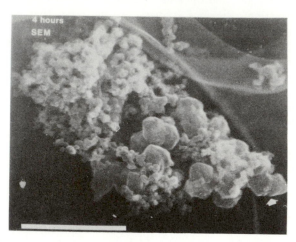

Figure 4a: SEM image of an amorphous MgSiO smoke particle produced in our laboartory after 4 hours of vacuum annealing at 1000 K (from Rietmeijer et al., 1986). Note 1 micron scale bar.

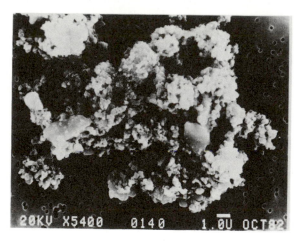

Figure 4b: SEM image of a chondritic IDP collected in the earth's stratosphere (from Bradley et al., 1983). Note 1 micron scale bar and the fact that the analog material in 4a could represent an enlargement of any number of regions of this IDP.

The morphology of the artificial smoke is quite similar to that of the natural material even though the size of the IDP is significantly larger than that of the MG-SiO analog.

    One of the most diagnostic spectral regions for the analysis of minerals is the infrared. We have prepared a series of smokes using the Flow Condensation Apparatus and recorded the infrared spectra of the initial condensates. These condensates range in composition from relatively pure SiO_x, FeO_x, and AlO_x (the spectra of which are shown in figure 5a) to more complicated mixtures of iron, aluminum and silicon oxides (these spectra are shown in figures 5b-f).

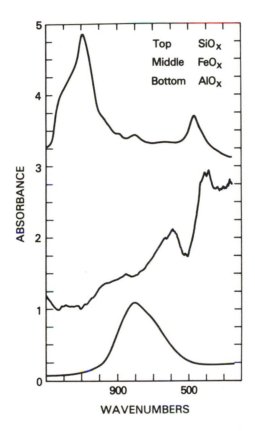

Figure 5a: Infrared absorbance spectra of the major features in SiO_x, FeO_x and AlO_x smokes over the wavenumber range from 1300–200 cm^{-1}.

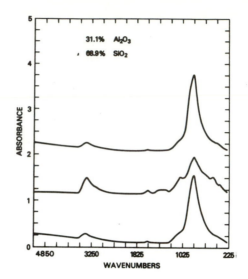

Figure 5b-f: Infrared spectra of amorphous smokes produced in the flow condensation apparatus. The spectrum of the initial condensate is given as the bottom curve; that of a sample heated for 5 days at 378 K in liquid water is the middle curve; and that of a sample heated in vacuo at 1200 K for 16 hours is the top curve. All absorbance have been scaled to one milligram of sample and the middle and top curves have been displaced upwards by 1 and 2 absorbance units, respectively. Note that a scale change occurs at 2000 cm^{-1} which results in a compression of the region from 5250–2000 cm^{-1} by a factor of 2 compared to the region from 2000 cm^{-1} to 225 cm^{-1}.

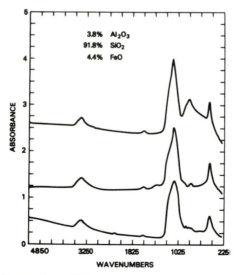

Figure 5c: For caption see figure 5b-f.

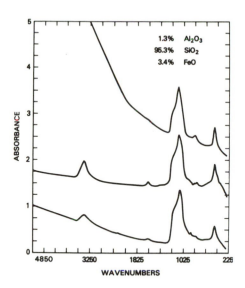

Figure 5d: For caption see figure 5b-f.

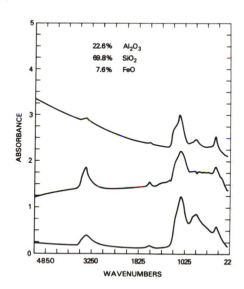

Figure 5e: For caption see figure 5b-f.

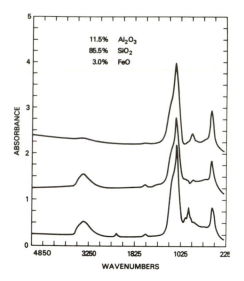

Figure 5f: For caption see figure 5b-f.

Because it is unlikely that all grains which formed in circumstellar outflows would survive passage through the interstellar medium and into the solar nebula without first being subjected to conditions under which some degree of metamorphism might occur, we have investigated the effects of both thermal annealing in vacuo and hydrous alteration in liquid water on the infrared spectra of our mixed oxide condensates. In the spectra shown below, the lower curve is the absorbance of the initial condensate, the middle curve is the absorbance of the sample after heating for 5 days at 378 K in liquid water and the upper curve is the absorbance of the sample after heating in vacuo for 16 hours at 1200 K. In all cases the composition of the initial condensate is given at the top of the figure. Note that in these spectra a scale change occurs at 2000 cm^{-1}; at lower energy there is one tic mark every 200 cm^{-1}, at higher energy, there is one tic mark every 400 cm^{-1}. All absorbances in figures 5b-f have been scaled to one milligram of sample. Each hydrated spectrum is displaced upward by the addition of one absorbance unit, each annealed spectrum is displaced by the addition of two absorbance units. From figure 5a it is obvious that SiO_x is characterized by absorbance peaks at $\sim$ 1100, 800 and 475 cm^{-1} with a minor shoulder at $\sim$ 950 cm^{-1}. FeO_x shows minor peaks at $\sim$ 600 and 400 cm^{-1} on top of a relatively flat background. AlO_x has only one peak centered at $\sim$ 800 cm^{-1} which is quite broad and strong. In figures 5c-f, no peaks attributable to FeO_x are seen, even though some iron is present in each sample. However, in figures 5c and d, annealing at 1200 K appears to have increased the absorbance (or scattering) at higher energies. This is not observed in samples containing little or no iron and may be due to a broad absorption band of FeO_x in the near infrared. Further work is needed to confirm this hypothesis. In figures 5b and 5e, one can observe that hydrous alteration of samples containing AlO_x significantly reduces the intensity of the 800 cm^{-1} peak. However, since samples were treated in a closed system and the water was drawn off via freeze drying, no AlO_x was removed from the sample. Obviously, the alumina was converted to an alternate structure which has little or no infrared signature in the spectral region we have studied. Characterization of these samples must await detailed analysis via electron microscopy. Comparison of figures 5b, c and e reveals no consistent pattern for the changes observed in the relative intensities of the 1100 cm^{-1} SiO_x and 800 cm^{-1} AlO_x peaks for samples annealed at 1200 K for 16 hours. In sample 5e it appears that SiO_x increases with annealing whereas in sample 5c it is obvious that the intensity of the AlO_x feature increases relative to that of SiO_x. Little change can be noted in either feature from the curves in figure 5b. Figure 5e are the spectra of samples produced at room temperatures via a series of explosive reactions in the Flow Condensation Apparatus. Despite the fact that the sample contains more than 11% AlO_x, no trace of the 800 cm^{-1} feature is seen in any of the tracings. This should be compared to figure 5b (only 1.3% AlO_x) in which the 800 cm^{-1} featute became stronger with annealing. Again detailed study via electron microscope will be needed to unravel the mineralogy of these particles. However, two points *should* be made about interpretations of the spectra of complex grains. First, it is difficult to determine the composition and structure of such materials solely from their infrared spectra. Second, the infrared spectrum of a particular material can vary drastically depending upon the details of the processing which the grain may have undergone and the manner in which it was formed.

3.2 Refractory ices

The infrared spectrum of the mixture of SiH_4, $Fe(CO)_5$ and H_2O ices at 15 K prior to irradiation is shown as the bottom curve in figure 6. Major, broad features near 3, 6 and 12.5 μm are due to H_2O; sharper features at 4.6 and 11.1 μm are due to the SiH stretching and bending modes in SiH_4 and the feature at 5 μm is due to the CO stretch in $Fe(CO)_5$. Changes in the infrared spectrum of this ice mixture due to proton bombardment are evident in figure 6; most noteworthy is the general blending of the initially sharper ice features of SiH_4 and $Fe(CO)_5$ near 5 μm and of SiH_4 and H_2O near 12 μm. One can also observe the growth of a shoulder near 9–10 μm, presumably due to the formation of SiO, the growth of a features near 17 μm, possibly due to SiO polymerization and the appearance of a sharp feature at 4.25 μm due to CO_2. As the ices were warmed from 15 K to $\sim$ room temperature, the water gradually evaporated, as did the SiH_4 and $Fe(CO)_5$, while the 9–10 μm feature became more pronounced and a broad feature began to grow at 20 μm. It is interesting to note that features due to SiH_4 at 4.6 μm and $Fe(CO)_5$ at 5 μm are still visible after the sample has been warmed in a vacuum to 250 K.

Figure 7 shows the infrared spectrum of the residue left after the sample was warmed to 300 K and left in a vacuum for 72 hours, then removed from the cryostat and inserted into the reflection attachment of a Perkin Elmer grating infrared spectrometer. Strong bands are still evident at 4.5 μm and 5.0 μm due to SiH_4 and $Fe(CO)_5$, respectively. Additional strong features are evident at 3 and 6.2 μm (H_2O) and at 10 and 22 μm (silicate).

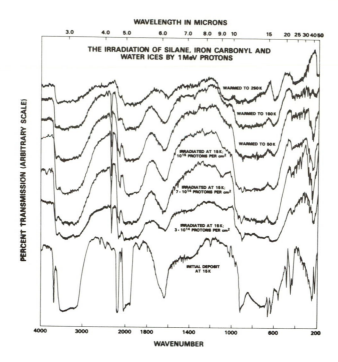

Figure 6: Infrared spectra of mixed ices of SiH₄, Fe(CO)₅ and H₂O over the wavelength range from 4000 cm⁻¹ to 200 cm⁻¹. The bottom curve is the initial spectrum of the ice mixture taken after deposition at 15 K. The next 3 curves up from the bottom were taken at 15 K after irradiation for total doses of $3 \cdot 10^{14}$, $7 \cdot 10^{14}$ and $1 \cdot 10^{15}$ protons per cm², respectively. The final 3 curves are the spectra taken after warmup to 50 K, 150 K, and 250 K, respectively.

A weak shoulder is evident near 7 μm, as is a weak band at ∼ 11 μm; this latter feature might be the remnant of the SiH bending mode. Upon warming the sample at 400 K for 30 hours in vacuo (top curve in figure 7), the peak at 5 μm (Fe(CO)₅) has disappeared, the feature at 7 μm has become more pronounced; the 6.2 μm feature (formerly part of a doublet) has disappeared while the 6.4 μm portion of the doublet remains strong. The major silicate features at 10 and 22 μm appear unchanged. The most interesting results of this processing is that the absorption feature at ∼ 4.5 μm are still evident even after heating to 400 K.

Lacy et al. (1984) have observed features at 2135 cm⁻¹, 2140 cm⁻¹ and 2165 cm⁻¹ at high resolution that had previously been detected by Soifer et al. (1979). Lacy et al. (1984) proposed that the features at 2135 cm⁻¹ and 2140 cm⁻¹ were due to solid CO and to CO complexed to other molecules. They identified the broad feature at 2165 cm⁻¹ (4.6 μm) as due to the presence of various cyano groups of the general form

XCN within ice mantles on the grains. They also noted that the feature at 2165 cm^{-1} appeared to be decoupled from the features at 2135 cm^{-1} and 2140 cm^{-1} and so were not likely to be related directly to the presence of CO.

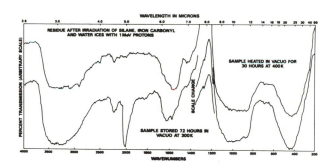

Figure 7: Infrared spectra of the residue left after evaporation of the volatile ices which remained after proton bombardment and warmup of a mixture of SiH$_4$, Fe(CO)$_5$ and H$_2$O to 300 K for 72 hours in vacuum (bottom) and after vacuum annealing of this same residue at 400 K for 30 hours (top curve).

We propose an alternate explanation for the 2165 cm^{-1} feature; namely, that this feature is due to the presence of SiH bonds either within an ice mantle or on free silicate surfaces. This idea will be explored in more detail in the near future once results from experiments on the irradiation of SiH$_4$–H$_2$O ice mixture have been analyzed.

3.3 Trapping rare gases

We have measured the quantity of Ne, Ar, Kr and Xe trapped in Si$_2$O$_3$ smokes during vapor phase condensation as a function of the pressure of the individual rare gases in the ambient atmosphere (Nuth et al. 1987a). Our data is in substantial agreement with previous work by Honda et al. (1979) which showed that the quantity of rare gas trapped during condensation was approximately proportional to its partial pressure and that the quantity of gas trapped is dependent on the chemical composition of the condensate. We find that the bulk of the Ne and Ar trapped during grain growth is lost upon heating to $\sim$ 1000 K for prolonged times although a substantial portion of the trapped Kr and Xe would probably survive heating to such temperatures. Unfortunately, extrapolation of our measured trapping efficiencies to realistic condensation pressures predicts that only a trivial quantity ($\sim$ 10^{-12} ccSTP/g) of even Kr and Xe would be trapped in the first place by this process.

Because the trapping efficiency for argon was significantly greater in Si$_2$O$_3$ than would be expected from the results of Honda et al. (1979) we conclude that argon is efficiently trapped by Si$_2$O$_3$ smokes and that more efficient traps for the other noble gases may also exist. It is not apparent at this time if the argon is trapped during formation, where it would occur preferential to the other noble gases (Table 1), or if it occurred upon later exposure to atmospheric gases, in which case the other atmospheric species would not be readily observable.

Table 1:

	Ambient Gas (μm)	Trapped Gas ccSTP/g·10^8	Ambient Gas vs. Ar	Trapped Gas vs. Ar	Trapping Efficiency
Ne	282	6.0	0.43	$2.3 \cdot 10^{-4}$	$5.3 \cdot 10^{-4}$
Ar	650	26000.0	1.0	1.0	1.0
Kr	55	10.0	0.084	$4.0 \cdot 10^{-4}$	$4.8 \cdot 10^{-3}$
Xe	13	8.2	0.02	$3.2 \cdot 10^{-4}$	$1.6 \cdot 10^{-2}$

In either case, Si_2O_3 smokes represent substrates for noble gas incorporation. For this reason, we intend to investigate both the trapping and absorption efficiency of simple and mixed oxides of Ti, Al, Si and Fe. Fanale and Cannon (1972) have already shown that adsorption of rare gas occurs efficiently on silicate minerals only at low temperature, while others (Lancet and Anders, 1973; Zaikowski and Schaeffer, 1976) have shown that incorporation of rare gases during hydrothermal alteration yields abundance patterns which are inconsistent with those observed in meteorites. If we fail to observe a significant affinity for the rare gases by oxide condensates of Ti, Al, Si and Fe, then we would be forced to conclude that significant rare gas enhancements will only be found in carbonaceous grains. This is consistent with the mechanism proposed by Huss and Alexander (1987) in which the bulk of the rare gases found in the solar system materials is bound in the carbonaceous mantles of pre-solar grains which formed in the molecular cloud from which the nebula collapsed. Unfortunately, before we can draw definitive conclusions fron our work, several additional experiments must be performed in order to characterize the noble gas absorption efficiency of mixed oxide smokes.

3.4 Oxygen isotopic fractionation

We have investigated the possibility that oxygen becomes isotopically fractionated during the nucleation of refractory metal oxides (Nuth et al., 1987b). Studies have been made using two techniques for grain production: thermal vaporization/recondensation of SiO in the Bell Jar system and chemical reaction of a gas mixture in the Flow Condensation Apparatus. Nucleation in the Bell Jar system produces no measurable isotopic fractionation even though some degree of disproportionation occurs in the vapor during the process which produces an Si_2O_3 (and Si metal) condensate from a vapor of SiO. Only a small degree of Rayleigh Distillation was observed between the smokes and the SiO residue which remained in the crucible after the experiments. Smokes produced via chemical reaction from gas mixtures containing SiH_4, $Al(CH_3)_3$, $Fe(CO)_5$, O_2 and H_2 at temperatures often in excess of ~ 700 K show a significant degree of isotopic fractionation when compared to the isotopic composition of the initial O_2 gas. The composition of these smokes is shown in table 2. One should note that many of these samples contained a significant amount of excess O_2: oxygen in excess of the amount expected based on the sample weight and the known stoichiometry of the sample as determined via an electron microprobe. This excess O_2 is likely to be adsorbed either from the flow stream or from the atmosphere and the effect of such an adsorbed component would be to dilute the measured fractionation efficiencies of the refractory smokes by a significant amount in many of these samples. The degree of oxygen isotopic fractionation during the formation of our smokes is likely to be significantly greater than the values reported in table 2.

Table 2: Results of the Flow Condensation Experiments

Sample Number	1	2	3	4	5	6	7	8	9
$\delta^{18}O(o/oo)$[a]	−20.2	−16.0	−23.3	−21.9	−14.7	−29.5	−23.1	−28.7	−31.3
$\delta^{17}O(o/oo)$[a]	−9.9	−8.2	−11.7	−11.2	−7.5	−15.2	−11.9	−14.7	−15.8
Condensation Temperature (K)	550	500	1050	775	875	800	300	725	725
Chemical Composition									
Al_2O_3 (% by wt.)	0	0	0	1.3	3.8	22.5	11.5	31.0	31.0
SiO_2 (% by wt.)	100	100	0	5.3	91.8	70.0	85.5	69.0	69.0
FeO (% by wt.)	0	0	100	3.4	4.4	7.5	3.0	0	0
O_2 (μmol)	102	90	87	67	104	>280	>280	>280	33
Sample wt. (mg)	6.2	4.1	2.8	3.2	6.9	8.8	6.0	7.0	1.4
Excess O_2 (μmol)	(−1)	22	60	14	(−8)	>148	>185	>180	12
Notes				b			c		d

a.   Isotopic compositions are reported relative to that of the flow gas isotopic composition, which has $\delta^{18}O = 24.97 \pm .01$, $\delta^{17}O = 12.63 \pm .03$ with respect to SMOW.

b.   Combination of 6 separate experimental runs.

c.   Product of explosive chemical reaction at approximately room temperature (~ 300 K).

d.   Sample aliquot of run 8. This sample was vacuum annealed at 1200 K.

One sample was formed nominally at room temperature (7) in a series of explosive events. In this "experiment" the heater burned out prior to the start of the run but we estabilished the flow of reactive gases normally. We saw a slight haze, indicative of the formation of particles, soon after the flow was estabilished. At 3 to 5 minute intervals we heard loud detonations within the system accompanied by light flashes in a line along the axis of the flow tube. After each detonation a significant increase in the particle density was observed. Thermocouple readings within the flow varied from about 305 K after an explosive event to about 330 K just *prior* to a detonation. As can be seen from table 2, the isotopic fractionation during the explosive event at relatively low temperature was less than the fractionation produced at higher temperature in the steady state flow. Normally one would expect that the degree of isotopic fractionation would be enhanced at lower temperature. Fractionation probably occurs during the chemical reactions which produce the original metal oxide monomers prior to the nucleation and growth of the smokes. These reactions (e.g., between SiH and O_2) are similar to those which may occur in the expanding shells of oxygen rich stars. Therefore, it is possible that the oxygen in dust condensed in circumstellar outflows is isotopically light compared to the gas from which it formed. We plan to carry out additional experiments, using N_2O as our source of oxygen rather than O_2, in order to elucidate the possible mechanism for the fractionation and to assess its relevance to circumstellar condensates.

4   Summary

We have presented an overview of the experimental techniques by which we are attempting to understand the many processes which occur from the time a refractory grain nucleates in a circumstellar environment to the time at which it is incorporated into a new star or solar system. No result are yet available from the cluster beam system as it is still under construction. The properties (infrared spectra and oxygen isotopic fractionation) of the grains produced in the Flow Condensation Apparatus are still not completely understood and work on the material produced in this system will continue to be a major part of our research effort. We still do not have a satisfactory explanation for the remarkable sorption efficiency of Si_2O_3 smokes for Ar gas and a considerable effort is required in order to investigate the possibility that other noble gases may be preferentially adsorbed by grains of specific composition.

A new line of research which we have just begun to explore is the processing of ices containing refractory precursors such as SiH_4 or $Fe(CO)_5$ plus water. The infrared spectrum of the residue left after 1 MeV proton bombardment and warm-up to room temperature is not significantly different from that expected of an amorphous Fe–SiO smoke. The residue has major broad absorptions at 10 and 22 μm, lesser features at 3, 6.2 and 7 μm and a small feature at 4.6 μm. We have identified this latter feature as the SiH stretching vibration and have suggested that the 4.6 μm feature observed by Lacy et al. (1984) in W33A might be due to SiH rather than the cyano group compounds which they had originally proposed. More work, both observational and laboratory, is needed to evaluate the merit of this proposal.

References

Bradley, J.P., Brownlee, D.E., Veblen, D.R.: 1983, Nature **301**, 473.

Fanale, F.P., Cannon, W.A.: 1972, Geochim. et Cosmochim. Acta **36**, 319.

Honda, M., Ozima, M., Nakada, Y., Onaka, T.: 1979, EPSL **43**, 197.

Huss, G.R., Alexander, E.C.: 1987, Proc. 17th Lun. Plan. Sci. Conf. in J. Geophys. Res. (Red) **92**, E710.

Lacy, J.H., Baas, F., Allamandola, L.J., Persson, S.E., McGregor, P.J., Lonsdale, C.J., Geballe, T.R., van de Bult, C.E.P.: 1984, Ap. J. **276**, 533.

Lancet, M.S., Anders, E.: 1973, Geochim. et Cosmochim. Acta **37**, 1371.

Moore, M.H.: 1984, Icarus **59**, 114.

Moore, M.H., Donn, B., Khanna, R., A'Hearn, M.F.: 1983, Icarus **54**, 388.

Nuth, J.A., Donn, B.: 1982, J. Chem. Phys. **77**, 2639.

Nuth, J.A., Donn, B.: 1983a, Proc. 13th Lun. Plan. Sci. Conf. in J. Geophys. Res. (Red) **88**, A847.

Nuth, J.A., Donn, B.: 1983b, J. Chem. Phys. **78**, 1618.

Nuth, J.A., Donn, B.: 1984, Proc. 14th Lun. Plan. Sci. Conf. in J. Geophys. Res. (Red) **89**, B657.

Nuth, J.A., Donn, B., Nelson, R.N.: 1986, Ap. J. Lett. **310**, L83.

Nuth, J.A., Olinger, C., Garrison, D., Hohenberg, C., Donn, B.: 1987a, in Proc. 18th Lun. Plan. Sci. Conf. (Cambridge Univ. Press, Boston) accepted.

Nuth, J.A., Thiemens, M., Nelson, R.N., Donn, B.: 1987b, in Proc. 18th Lun. Plan. Sci. Conf. (Cambridge Univ. Press, Boston) accepted.

Rietmeijer, F.J.M., Nuth, J.A., Mackinnon, F.D.R.: 1986, Icarus **66**, 211.

Seab, C.G., Shull, J.M.: 1985, In Interrelationships Among Circumstellar Interstellar and Interplanetary Dust (J. Nuth and R. Stencel, eds.) NASA CP-2403 page 37.

Soifer, B.T., Puetter, R.C., Russell, R.W., Willner, S.P., Harvey, P.M., Gillett, F.C.: 1979, Ap. J. Lett. **232**, L53.

Stephens, J.R., Bauer, S.H.: 1981, personal communication.

Zaikowski, A., Schaeffer, O.A.: 1976, Meteoritics **11**, 394.

OPTICAL PROPERTIES OF GLASSY BRONZITE AND THE INTERSTELLAR SILICATE·BANDS

J. Dorschner, C. Friedemann, J. Gürtler, Th. Henning
Universitäts-Sternwarte, Jena, G.D.R.

1   Introduction

Observations of broad and structureless bands at 10 and 20 μm wavelengths in the spectra of many infrared sources gave rise to extensive laboratory work on amorphous silicates supposed to be responsible for these bands (for a recent review see, e.g., Dorschner and Henning, 1986). Up to now, no experimentally manufactured silicates have been presented that could satisfactorily reproduce the large width of the bands observed in the cosmic sources or explain the observed infrared spectra over a broad wavelength interval. Consequently, a number of attempts has been made to evaluate realistic optical properties for cosmic silicates either by modifying experimental results in such a way that the observations could be more satisfactorily described (Jones and Merrill, 1976; Papoular and Pégourié, 1982; Pégourié and Papoular, 1985; Rowan-Robinson, 1986) or by combining astronomical observations, experimental data, and theoretical relations (Draine and Lee, 1984; Henning et al., 1983).

Cosmic abundances, conditions of grain formation in circumstellar envelopes, and interstellar elemental depletions all point to magnesium-rich silicates as the dominating interstellar silicate component. Mineralogical investigations of primitive meteorites and interplanetary dust particles revealed pyroxenes and olivines as the main constituents (Dodd, 1981; Sandford and Walker, 1985). Combining this mineralogical information with the interstellar depletion pattern Jones and Williams (1987) concluded that pyroxenes should contribute significantly to the mass of interstellar grains.

Bronzite is an abundant pyroxene in the chondritic meteorites. Thus, it is of prime interest as an interstellar dust material. From a terrestrial bronzite mineral Dorschner et al. (1986) manufactured a glass with optical properties that seem very promising for the representation of the observed dust spectra in the middle infrared. It is the aim of this paper to explore the applicability of this material for modeling the spectra of cosmic infrared sources in more detail. First, the dielectric functions for glassy bronzite are derived in order to ease the application in model computations, and absorption cross-sections in the 7–40 μm range are provided as a function of grain size. Confronting the properties of bronzite grains both with other experimental results and astronomical observations corroborates the conclusions reached by Dorschner et al.

E. Bussoletti et al. (eds.), Experiments on Cosmic Dust Analogues, 209–225.
© *1988 by Kluwer Academic Publishers.*

(1986), namely that glassy bronzite has optical properties in the mid-IR that make it a promising candidate for an astronomically important silicate material.

2   Optical constants of glassy bronzite

2.1   Derivation of n and k from measured mass absorption coefficients of submicrometre-sized particles

The straight-forward way to the refractive indices of a mineral is measuring trasmissivity and reflectivity of a thin slab of bulk material. On the other hand, the mass absorption coefficient (MAC) is often determined from transmission spectra of powders embedded in some matrix (e.g., KBr for IR spectra). This is especially true for astronomically interesting dust silicates, which are often artificially produced in small amounts only and not as large enough pieces that allow thin slabs to be cut from them.

The present study is also based on trasmission spectra from which MAC were computed. As shortly described by Dorschner et al. (1986), in the present investigation trasmission spectra were taken by means of a Perkin-Elmer 457 spectrometer from KBr pellets containing fixed quantities of powdered glassy bronzite. The charges of the pellets varied from 0.3 to 1.0 mg corresponding to column densities from $9 \cdot 10^{-5}$ to $3 \cdot 10^{-4}$ g cm^{-2}. The transmissivity of the pellets was set to unity at $z = 1500$ cm^{-1}. The glassy bronzite was produced from a mineral specimen coming from Paterlestein near Kupferberg (Franconia, FRG). Chemical analysis showed a FeSiO$_3$ content of about 10 mol per cent. The raw material was melted in an arc and the droplets were effectively quenched in a basin of mercury. Repetitions of the procedure confirmed that the glasses produced in that way had the same optical properties. The glass chunks were triturated by means of an Ardenne vibrator for 2 hours and the powder was suspended in acetone for 1.5 hours. From our experience with former experiments we know that all of the particles larger than about 1 μm have been precipitated after that time. The floating fraction was regained and used for the spectroscopic work. The trasmission spectra of the KBr pellets are used to derive the wavelength dependence of tha MAC. The complex index of refraction may be determined from the MAC along the following line. Assume the particles are spherical with radius a and their material has the density δ. The MAC x is then given by

$$x = \frac{3}{4} \frac{Q_{abs}}{a\delta} \tag{1}$$

where Q_{abs} is the efficiency factor for absorption. The Mie theory yields for particles with $x = 2\pi a/\lambda \ll 1$ the expression

$$Q_{abs} = 4 \cdot Im\left(\frac{\epsilon - 1}{\epsilon + 2}\right) \tag{2}$$

Here, $\epsilon = \epsilon' + i\epsilon''$ is the complex dielectric function, which is connected with the refractive index $m = n + ik$ by the relation

$$n = \left\{\frac{1}{2}\left[\left(\epsilon'^2 + \epsilon''^2\right)^{\frac{1}{2}} + \epsilon'\right]\right\}^{\frac{1}{2}} \tag{3a}$$

$$k = \left\{ \frac{1}{2} \left[\left(\epsilon'^2 + \epsilon'^2 \right)^{\frac{1}{2}} - \epsilon' \right] \right\}^{\frac{1}{2}} \tag{3b}$$

Deriving two functions, $n(\lambda)$ and $k(\lambda)$, from only one measured function $x(\lambda)$ is possible only by means of additional information. A relation between the real and imaginary parts of ϵ in the regions of band absorption may be established by means of the dispersion theory. The Clausius-Mosotti-Lorentz-Lorenz relation states, when N oscillators contribute to the absorption, that

$$\frac{\epsilon - 1}{\epsilon + 2} = \frac{\epsilon_\infty - 1}{\epsilon_\infty + 2} + \sum_{j=1}^{N} \frac{F_j}{\omega_j^2 - \omega^2 - i\gamma_j\omega} \tag{4}$$

Here, ϵ_∞ is the dielectric function in the limit of very high frequencies, ω the angular frequency, ω_j the eigenfrequency of the j-th oscillator, and γ_j its dampimg costant. The coefficients F_j are connected with the plasma frequencies $\omega_{p,j}$ and the oscillator strenghts f_j via

$$F_j = \frac{1}{3} \omega_{p,j}^2 f_j \tag{5}$$

Thus, if F_j, ω_j, γ_j and ϵ_∞ are known, the dieletric function ϵ or n and k can be calculated. Combining eq. (1), (2), and (4) and considering that ϵ_∞ is real yields

$$x = \frac{3\omega}{\delta c} \sum_{j=1}^{N} F_j \frac{\omega\gamma_j}{(\omega_j^2 - \omega^2)^2 + \omega^2\gamma_j^2} \tag{6}$$

Eq. (6) shows that the parameters defining the dielectric function ϵ may be derived by fitting the measured $x(\lambda)$ by a superposition of N dispersion profiles and determining their parameters ω_j, γ_j, and F_j. The necessary steps for deriving n and k from MAC measurements are displayed in figure 1.

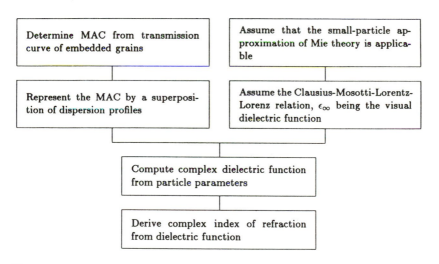

Figure 1: Steps and assumptions necessary for the determination of refractive indices from IR trasmission spectra.

2.2   Representation of MAC curves by lorentzian profiles

Figure 2 shows transmission curves for both crystalline and glassy bronzite. The trasmission spectrum for crystalline bronzite exhibits at least 12 absorption peaks within the wavelength range from 7–40 μm. The complicated structure disappears in the trasmission spectrum of glassy bronzite and only two broad absorption bands remain.

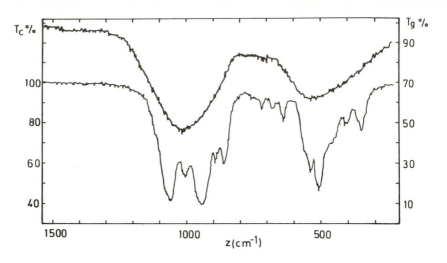

Figure 2: Trasmission curves for crystalline (T_c) and glassy (T_g) bronzite. The column density of submicrometre-sized bronzite grains in the KBr pelletes is $1.59 \cdot 10^4$ g cm^2.

The MAC was determined from transmission spectra of pellets with different charges of submicrometre-sized bronzite particles. For the following an average of these individual MACs was used. The influence of the KBr matrix on the transmission spectrum of the embedded grains was corrected according to the procedure described by Dorschner et al. (1986). The appearance of the spectrum for crystalline bronzite suggests to assume at least 12 contribution oscillators with lorentzian profiles. The detailed analysis revealed that 16 lorentzian profiles give a satisfactory representation (Figure 3a). The parameters of the profiles are summarized in table 1. Further increasing the number of profiles does not improve the accuracy of the representation. Although the spectrum of glassy bronzite display only two broad structureless absorption peaks, the asymmetry of both bands suggests that two dispersion profiles will not give a fully satisfactory representation of the MAC curve (Figure 3b). For want of anything better and being anxious to avoid arbitrariness we took the 16-profile approximation of the crystalline bronzite as a basis for the representation of the MAC of glassy bronzite. All profile parameters $(\omega_j, \gamma_i, \gamma_j, F_j)$ are allowed to be varied for achieving an optimal fit. The results are listed in table 1.

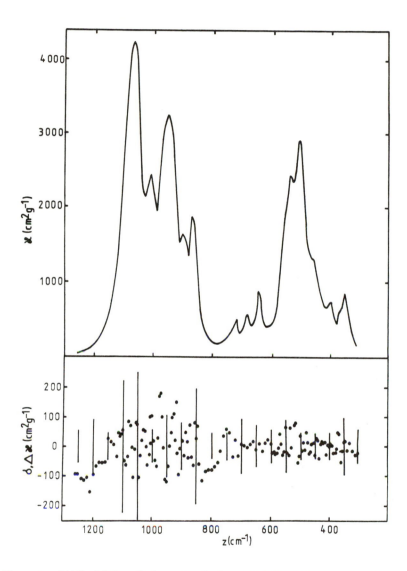

Figure 3a: MAC $x(z)$ for submicrometre-sized grains of crystalline bronzite. The curve shown is the representation by a superposition of 16 lorentzian profiles of the measurements corrected for KBr influence. (For the profile parameters see table 1). The lower panel shows the difference Δx between the measured MAC and the 16-profile approximation (dots) and the standard deviation σ of the measured MAC (bars).

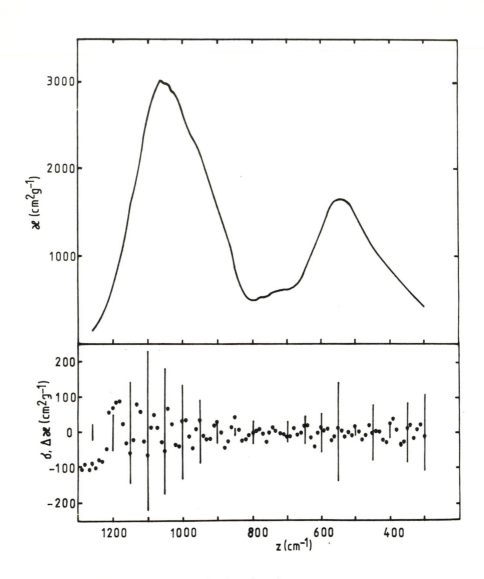

Figure 3b: The same as figure 3a for glassy bronzite.

Table 1: Parameters of the dispersion profiles for the representation of for crystalline and glassy bronzite.

j	Crystalline bronzite				Glassy bronzite			
	z_j (cm^{-1})	ω_j $(10^{14}s^{-1})$	γ_j $(10^{12}s^{-1})$	F_j $(10^{26}s^{-2})$	z_j (cm^{-1})	ω_j $(10^{14}s^{-1})$	γ_j $(10^{12}s^{-1})$	F_j $(10^{26}s^{-2})$
1	1127.0	2.1240	3.357	0.5532	1150.0	2.1680	11.445	3.1250
2	1090.0	2.0550	7.941	5.8810	1099.0	2.0720	11.647	5.8000
3	1063.0	2.0040	7.621	7.5600	1057.0	1.9920	11.300	6.0000
4	1013.0	1.9090	5.504	2.1320	1015.0	1.9130	11.898	5.9050
5	949.5	1.7900	13.080	13.4500	971.8	1.8320	11.830	4.1220
6	893.2	1.6840	2.492	0.4152	932.0	1.7570	12.507	3.6950
7	863.2	1.6270	5.325	2.3480	882.9	1.6640	15.166	3.7190
8	723.8	1.3640	4.450	0.4738	763.6	1.4390	11.945	0.5818
9	684.6	1.2900	3.766	0.4404	710.0	1.3380	17.737	1.4300
10	644.9	1.2160	4.599	0.9533	614.2	1.1580	16.870	3.1920
11	560.4	1.0560	5.167	1.8710	562.7	1.0610	134.941	3.8940
12	539.9	1.0180	3.926	1.6030	522.3	0.9845	12.486	2.9030
13	507.7	0.9570	7.515	6.1780	481.0	0.9067	14.005	2.6410
14	458.7	0.8646	11.947	3.3620	428.3	0.8073	16.998	2.8500
15	402.4	0.7585	5.022	0.6272	363.7	0.6818	19.698	2.7370
16	349.6	0.6590	7.640	1.8310	276.0	0.5202	14.663	1.1920

If the profile parameters for crystalline bronzite are compared with those for glassy bronzite the following conclusion can be drawn:

1) The peak wavelengths of the main absorption components agree within 5 percent with each other, whereas some of the weaker components show a rather large shift. As a common trend the shift of a component increases with its increasing distance from the peak centre. The components within the range of the 20 μm band of glassy bronzite underwent shifts toward longer wavelength if compared with their peak wavelength in the crystalline case (the only exception is component $j = 11$) while the component within the 10 μm band underwent shifts toward shorter wavelengths (the only exception is component $j = 3$).

2) With the exception of the component $j = 5$, all lorentzian profiles became broader, but the factor of broadening varies from 1.4 to 5.

3) The strengths F_j of most components are langer for glassy bronzite than for crystalline bronzite. The range of the band strengths is narrower for the glassy substance than for the crystalline one. Whereas for crystalline bronzite the band strengths F_j are proportional to $\gamma_j^{2.5}$, no relation was found for glassy bronzite.

It is not clear whether the behaviour described is due to the transition from the crystalline to the glassy state of the lattice or originates from the fitting procedure alone.

2.3   Numerical results for n and k

The parameters of the dispersion profiles determine the dielectric function ϵ with the exception of the constant ϵ_∞ (see eq. (4)). We suppose that ϵ measured in the visual part of the spectrum is a good approximation for ϵ_∞. Rösler (1981) gives $n_{vis} = 1.66$, hence $\epsilon_\infty = 2.756$. The resulting values of n and k for glassy bronzite are presented in table 2.

Table 2: Real part n and imaginary part k of the index of refraction and efficiency factors for absorption, Q_{abs}, and extinction Q_{ext} for particles of glassy bronzite of different radii a.

$\lambda(\mu m)$	n	k	a = 0.1 μm		a = 1.0 μm		a = 3.0 μm	
			Q_{abs}	Q_{ext}	Q_{abs}	Q_{ext}	Q_{abs}	Q_{ext}
6.0	1.50	0.093	1.95–3	1.98–3	2.86–2	2.76–1	1.32–1	3.42
6.4	1.46	0.013	2.56–3	2.58–3	3.54–2	2.07–1	1.66–1	3.02
6.8	1.42	0.018	3.45–3	3.46–3	4.53–2	1.60–1	2.05–1	3.37
7.2	1.37	0.025	4.84–3	4.85–3	6.02–2	1.32–1	2.43–1	1.72
7.6	1.30	0.039	7.32–3	7.33–3	8.58–2	1.26–1	3.11–1	1.18
8.0	1.20	0.067	1.27–2	1.27–2	1.39–1	1.55–1	4.32–1	7.62–1
8.04	1.05	0.157	3.10–2	3.10–2	3.03–1	3.11–1	6.96–1	8.25–1
8.8	1.00	0.39	7.70–2	7.70–2	6.97–1	7.34–1	1.12	1.53
9.0	0.97	0.51	1.02–1	1.02–1	8.85–1	9.45–1	1.23	1.78
9.2	1.05	0.64	1.21–1	1.21–1	1.08	1.16	1.37	2.07
9.4	1.10	0.73	1.32–1	1.32–1	1.20	1.30	1.44	2.53
9.6	1.21	0.80	1.29–1	1.29–1	1.24	1.37	1.51	2.40
9.8	1.27	0.84	1.27–1	1.28–1	1.27	1.40	1.55	2.49
10.0	1.39	0.88	1.15–1	1.15–1	1.22	1.36	1.58	2.60
10.2	1.44	0.86	1.07–1	1.07–1	1.16	1.29	1.59	2.61
10.4	1.52	0.90	1.00–1	1.00–1	1.13	1.26	1.61	2.69
10.6	1.59	0.89	9.15–2	9.15–2	1.06	1.19	1.62	2.73
10.8	1.68	0.91	8.40–2	8.40–2	1.01	1.13	1.63	2.79
11.0	1.79	0.85	7.17–2	7.17–2	8.88–1	1.00	1.63	2.84
11.2	1.84	0.81	6.45–2	6.45–2	8.07–1	9.12–1	1.62	2.85
11.4	1.92	0.78	5.71–2	5.71–2	7.31–1	8.31–1	1.62	2.91
11.6	2.02	0.68	4.59–2	4.59–2	6.02–1	6.93–1	1.60	2.99
11.8	2.04	0.53	3.56–2	3.56–2	4.67–1	5.44–1	1.54	3.07
12.0	2.00	0.42	2.86–2	2.86–2	3.68–1	4.32–1	1.45	3.08
12.2	1.93	0.34	2.46–2	2.46–2	3.08–1	3.60–1	1.36	2.98
12.4	1.86	0.30	2.25–2	2.26–2	2.76–1	3.18–1	1.28	2.78
12.6	1.79	0.28	2.22–2	2.22–2	2.66–1	3.00–1	1.21	2.52
12.8	1.73	0.28	3.31–2	2.31–2	2.71–1	3.01–1	1.16	2.28
13.0	1.69	0.30	2.45–2	2.45–2	2.84–1	3.11–1	1.15	2.13
13.2	1.67	0.31	2.53–2	2.53–2	2.90–1	3.14–1	1.13	2.04
13.4	1.65	0.30	2.53–2	2.53–2	2.88–1	3.10–1	1.10	1.93
13.6	1.62	0.31	2.56–2	2.56–2	2.89–1	3.08–1	1.07	1.82
14.0	1.57	0.32	2.74–2	2.74–2	3.04–1	3.20–1	1.06	1.68
14.4	1.54	0.33	2.80–2	2.80–2	3.07–1	3.20–1	1.03	1.56
15.0	1.44	0.34	3.00–2	3.00–2	3.21–1	3.40–1	9.85–1	1.37
16.0	1.28	0.51	4.69–2	4.69–2	4.84–1	4.93–1	1.19	1.51
17.0	1.25	0.69	6.08–2	6.08–2	6.24–1	6.36–1	1.39	1.80
18.0	1.28	0.91	7.32–2	7.32–2	7.60–1	7.76–1	1.61	2.17
19.0	1.36	1.06	7.28–2	7.28–2	7.69–1	7.86–1	1.73	2.39
20.0	1.49	1.16	6.44–2	6.44–2	6.93–1	7.08–1	1.78	2.51
21.0	1.59	1.24	5.80–2	5.81–2	6.29–1	6.43–1	1.80	2.54
22.0	1.71	1.27	4.99–2	4.99–2	5.43–1	5.55–1	1.75	2.46
23.0	1.75	1.29	4.63–2	4.63–2	5.03–1	5.12–1	1.72	2.37
24.0	1.83	1.36	4.23–2	4.23–2	4.60–1	4.69–1	1.68	2.31
25.0	1.94	1.39	3.73–2	3.73–2	4.06–1	4.13–1	1.59	2.17
26.0	1.99	1.38	3.41–2	3.41–2	3.70–1	3.77–1	1.50	2.02
27.0	2.02	1.41	3.24–2	3.24–2	3.50–1	3.56–1	1.44	1.91
28.0	2.08	1.48	3.02–2	3.02–2	3.27–1	3.32–1	1.38	1.82
29.0	2.18	1.54	2.72–2	2.72–2	2.94–1	2.99–1	1.30	1.70
30.0	2.31	1.56	2.39–2	2.39–2	2.59–1	2.63–1	1.19	1.54
31.0	2.41	1.52	2.12–2	2.12–2	2.30–1	2.33–1	1.08	1.39
32.0	2.47	1.46	1.94–2	1.94–2	2.10–1	2.13–1	9.88–1	1.26
33.0	2.48	1.43	1.86–2	1.86–2	1.99–1	2.02–1	9.23–1	1.15
34.0	2.46	1.43	1.82–2	1.82–2	1.95–1	1.97–1	8.84–1	1.09
35.0	2.46	1.49	1.80–2	1.80–2	1.92–1	1.94–1	8.60–1	1.05
36.0	2.48	1.58	1.75–2	1.75–2	1.87–1	1.89–1	8.34–1	1.01
37.0	2.56	1.68	1.64–2	1.64–2	1.76–1	1.77–1	7.94–1	9.57–1
38.0	2.68	1.77	1.48–2	1.48–2	1.59–1	1.61–1	7.29–1	8.91–1
39.0	2.85	1.83	1.31–2	1.31–2	1.41–1	1.42–1	6.77–1	8.16–1
40.0	3.00	1.84	1.17–2	1.17–2	1.26–1	1.27–1	6.19–1	7.44–1

We have used these optical constants to compute the efficiency factors for absorption and extinction. The results for three different grain sizes are listed in table 2, too. The estimated errors of the refractive indices are of about 1–5 percent. These values are the standard deviations of the mean for n and k derived from the individual MAC curves. The deviations of the analytical approximation from the mean MAC are smaller than the standard deviations due to errors in the measurements and in the nominal charges of the pellets.

3 Discussion

3.1 Properties of glassy bronzite grains

In figure 4 the efficiency factor for absorption is plotted for various particle sizes in the wavelength region 7–40 μm.

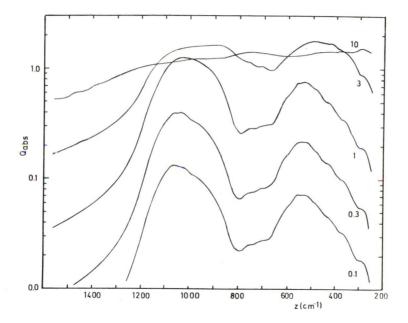

Figure 4: Efficiency factor for absorption, $Q_{abs}(z)$, for grains of glassy bronzite with radii 0.1, 0.3, 1, 3, and 10 μm.

The influence of the particle size on the appearence of the spectrum can be clearly seen. Characteristic parameters, namely the centres and halfwidths of the 10 and 20 μm bands, the ratio of the peak absorption of both bands, and the absorption in the "trough" between the bands relative to the 10 μm band peak absorption are summarized in table 3. It is common to all of these parameters that they are size-independent for particles with radii ≤ 0.3 μm and increase strongly with the grain size for radii beyond that limit.

Next, we will compare our results for glassy bronzite with the infrared properties of other experimentally produced amorphous silicates. Figure 5 summarizes the normalized MACs for glassy bronzite, amorphous Mg_2SiO_4 synthesized by Day (1979), irradiated olivine and olivine smoke (Krätschmer and Huffman, 1979).

Table 3: Properties of the 10 and 20 μm bands for grains of glassy bronzite.

a (μm)	10 μm band λ_c (μm)	FWHM (μm)	trough x_{rel} at 12.6 (μm)	13.3 (μm)	20 μm band λ_c (μm)	FWHM (μm)	x_{rel}
0.1	9.52	2.48	0.17	0.19	18.5	9.5	0.55
0.3	9.52	2.50	0.17	0.19	18.5	9.5	0.55
1.0	9.80	2.80	0.20	0.23	18.5	10.0	0.60
1.5	10.10	3.21	0.30	0.32	18.6	11.6	0.74
2.0	10.70	3.65	0.42*	0.43	19.4	12.7	0.94
3.0			0.60	0.67	20.6		1.10

* deepest point, at 14.8 μm

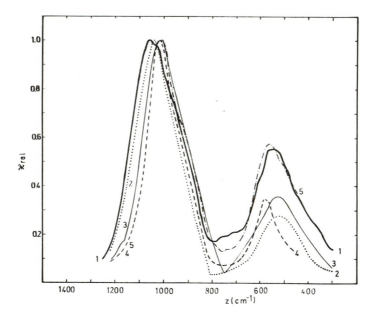

Figure 5: Experimentally derived MAC for amorphous silicates normalized at the peak of the 10 μm band: 1) glassy bronzite (this paper), 2) amorphous $MgSiO_3$ (Day, 1979), 3) amorphous Mg_2SiO_4 (Day, 1979), 4) irradiated olivine (Krätschmer and Huffmann, 1979), 5) olivine smoke (Krätschmer and Huffmann, 1979).

Table 4 gives characteristic parameters of the MAC curves. The data in figure 5 and table 4 all refer to particles of radius 0.1 μm.

Table 4: Characteristic parameters for MAC curves of experimentally produced amorphous silicates.

Silicate	References	10 μm band		13.5 μm	Trough at	20 μm band		
		λ_c (μm)	FWHM (μm)	x_{max} (cm^2 g^{-1})	x_{rel}	λ_c (μm)	FWHM (μm)	x_{rel}
Bronzite	this work	9.5	2.5	3000^a	0.19	18.5	9.9	0.55
MgSiO$_3$	Day (1979)	9.6	2.0	2920^a	0.045	19.2	8.4	0.275
Mg$_2$SiO$_4$	Day (1979)	9.9	2.2	2300^b	0.045	19.0	9.7	0.36
Irrad. olivine	Krätschmer & Huffman (1979)	9.8	1.7	3900^b	0.07	17.4	4.2	0.345
Olivine smoke	Krätschmer & Huffman (1979)	9.9	2.0	1750^c	0.14	17.8	9.7:	0.575

a $\delta = 3.3$ g cm^{-3}
b $\delta = 3.5$ g cm^{-3}
c $\delta = 3.0$ g cm^{-3}

The comparison of the MAC curves in the 10 μm band region reveals that two groups may be distinguished. The silicates of comporison X_2SiO_4 peak at about 9.9 μm, whereas those of composition $XSiO_3$ have their maximum absorption at a clearly shorter wavelength. As for the widths of the 10 μm band, glassy bronzite shows the widest bands, the particles of silicate smoke give a width of about 2 μm, and the particles of irradiated olivine have the most narrow band. The same trend can be seen for the 20 μm band. The differences of the parameters of the 20 μm band among the various experimental results show no clear dependence on the chemical composition. The irradiated olivine deviates noteworthy from the other substances by exhibiting the shortest peak wavelength as well as the smallest band width.

Concerning the relative band strength of the 20 μm band, there are two groups with $x_{rel} \approx 0.33$ and $x_{rel} \approx 0.55$, resp. The interpretation is not quite clear. An uncertainty in the appropriate choise of the continuum necessary for the determination of the MAC may decrease our x_{rel} at the peak of the 20 μm band by 10 percent at most. An earlier study of protosilicates (Dorschner et al., 1980) resulted in similar values for x_{rel} as the olivine smoke of Krätschmer and Huffman (1979) and our glassy bronzite. The crystalline raw material of our glassy bronzite shows $s_{rel} \approx 0.8$. Such high values are typical of crystalline silicates (cf., e.g., Friedemann et al. 1979). The depth of the trough between the bands shows the same trend as x_{rel} of the 20 μm band. This depth is significantly smaller for bronzite and olivine smoke than for Day's silicates and irradiated olivine. Our crystalline bronzite sample is also more trasparent than the glassy form. The uncertainty in the position of the continuum has more severe consequences here, but may deepen the trough by 30 percent at most, so that the difference cannot be removed in that way.

Most confrontations of observations with experimental work have taken Day's measurements as a standard MAC or a starting point at least. Comparing Day's results for MgSiO$_3$ with our bronzite curve, we see general agreement in the region of the 10 μm band, where bronzite has the advantage of an even larger bandwidth. Marked differences occur in the region beyond 12 μm. After the optical data for glassy bronzite had become available, the unreflected use of Day's results is not justified any longer. Both sets of optical data must be considered in the interpretation of the astronomical observations without a priori preference.

3.2 Confrontation with the observations

Confrontation of experimental MAC with the observations cannot be performed without modeling the infrared sources the spectra of which are to be compared with. As a matter of fact, two ways have been gone:

1) The wavelength dependence of the MAC was extracted from the spectra of optically thin sources. These sources are envelopes around late-type giants and supergiants the only exception being the Ney-Allen nebula in the Orion Trapezium region (Forrest et al., 1975; Henning et al., 1983; Papoular and Pégourié, 1983; Pégourié and Papoular, 1985).

2) The experimentally determined grain properties were used to model the sources and the predicted spectra are then compared with the observations (e.g., Jones and Merrill, 1976; Rowan-Robinson and Harris, 1982; Rowan-Robinson, 1982a,b; Gürtler and Henning, 1986).

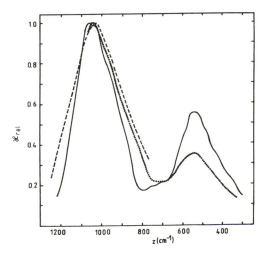
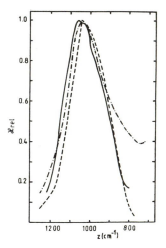

Figure 6a: Comparison of the normalized MAC of glassy bronzite grains with a = 0.1 μm (full line) with the wavelength dependence of the MAC derived from the infrared emission in the Orion Trapezium region (dashed line, Forrest et al., 1975) and the "astronomical silicate" (dotted line) by Draine and Lee (1984).

Figure 6b: Comparison of the normalized MAC of glassy bronzite grains with a = 0.1 μm (full line) with the mean T Tauri emission profile (dashed line) given by Cohen (1980) and the MAC profile derived by Chini et al. (1986) for the dust in NGC 7538 (E) (dashed-dotted line).

The wavelength position of the band centre depends rather weakly on model assumptions. There exist, however, two groups of astronomical spectra distinguished by the peak position of the 10 μm band. These differing spectra are associated with well distinguished classes of infrared sources. Among the sources showing the 10 μm peak at 9.5–9.7 μm are young objects, e.g., T Tauri stars, BN objects, and the Ney-Allen nebula (Figures 6a,b). The spectra of evolved sources, i.e., shells around oxygen-rich late-type giants and supergiants, exhibit the 10 μm band at 9.9–10.0 μm (Figure 7).

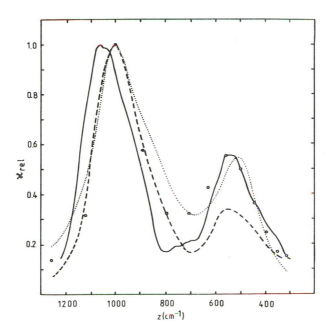

Figure 7: Comparison of the normalized MAC of glassy bronzite grains with a = 0.1 μm (full line) with tha MAC profiles derived for circumstellar dust around late-type giants and supergiants: dashed line - Pégourié and Papoular (1985), dotted line - Henning et al. (1983), circles - Rowan-Robinson (1986).

Reasoning that these differences in the peak wavelengths may reflect true chemical/mineralogical differences in the silicate component responsible for the respective spectra, Gürtler and Henning (1986) introduced the distinction of O-type (from olivine, X_2SiO_4) and P-type (from pyroxene, $XSiO_3$) silicates without claiming a mineralogical identification of the respective silicate. The differences may be due to (1) different grain size distributions, (2) different grain shapes, and (3) different chemical structure of the grains. Adopting our glassy bronzite as typical of P-type silicates, then the representation of the observed 10 μm bands in late-type stars by means of P-type silicates would require a grain size of ≈ 1 μm. However, particles are unlikely to grow up to such large sizes according to theoretical modeling of grain condensation in stellar atmospheres (e.g., Gail and Sedlmayr, 1986; Fadeyev and Henning, 1987). Moreover, particles of a ≈ 1 μm produce broader bands than these observed in spectra of red giants (cf. table 3).

Also in the case of optically thin sources the extraction of the wavelength dependence of the MAC from the infrared spectra is not possible without model assumptions on the emitting dust. The most decisive parameter is the temperature of the grains. The traditional procedure for the derivation of the Trapezium emissivity often used as a kind of standard for the optical properties of cosmic dust in the 10 μm range assumes 250 K as temperature for these grains. T Tauri stars form another class of young sources including members with the silicate feature in emission (Cohen, 1980; Cohen and Witteborn, 1985). Up to now, MAC curves have not yet been derived because suitable shell models are lacking. The shape of the mean emission profile deduced by Cohen (1980) is very similar of the Trapezium profile as far as the wavelength position of the emission peak is concerned. The width is, however, significantly smaller (cf. figure 6).

From modeling dust emission spectra of star forming regions Chini et al. (1986) derived the refractive indices n and k within the 10 μm band for the source NGC 7538(E). The absorption coefficient calculated from these optical data is shown in figure 6b. The peak absorption agrees with that of the "T Tauri silicate". The profile longward from the peak is very similar to that of glassy bronzite, whereas in the 11–14 μm range the "star-forming region silicate" deviates from both of the other profilies.

The large width (3.5 μm) of the Trapezium profile was attributed to particles larger than those in the general interstellar medium by Rowan-Robinson (1975) and Rouan and Léger (1984). If grains of our glassy bronzite are to be used for modeling the width of the Trapezium profile, grain radii of about 1 μm are needed (cf. table 3). On the other hand, the peak absorption of such large grains is shifted to 9.9–10.0 μm contrary to the observations. Thus, we conclude that interpretations other than grain size effects, e.g., additional emission by carbon grains (Tielens and de Jong, 1979) should be considered seriously.

Draine and Lee (1984) have compiled the optical properties of an "astronomical silicate" in a very broad spectral range by combining various observational and experimental data with theoretical arguments. It is widely used as a kind of standard model of cosmic silicate dust. Wavelength positions and widths of the 10 and 20 μm bands were modeled in accordance with the observation of the Trapezium region (for the 10 μm band) and optically thin circumstellar shells around latetype stars (for the 20 μm band). The peak absorption ratio was taken as 0.4 being the mean value of the range (0.3–0.5) deduced by McCarthy et al. (1980) for the Galactic Centre. Incidentally, that value is also the mean of our result for glassy bronzite and Day's for amorphous MgSiO$_3$.

Up to this point, we have discussed the optical properties of silicate dust associated with young objects. A more extensive observational data base exists for silicate dust around evolved oxygen-rich stars. The circumstellar shells around these objects have the additional advantage of consisting of silicate dust only whereas in the general interstellar medium an admixture of carbon grains may be present.

Henning et al. (1983) as well as Pégourié and Papoular (1985) derived efficiency factors for circumstellar silicate grains from mid-IR spectra of late-type stars with optically thin dust shells. As mentioned above, model assumptions on the dust must be made to obtain the temperature of the emitting grains. The assumptions made may be reduced to fixing the peak ratio of both silicate bands. Henning et al. (1983) assumed a value of 0.55 suggested by the experimental results for glassy bronzite whereas Pégourié and Papoular (1985) started from the experimental data of Day (1979) for Mg_2SiO_4, that means a peak ratio of 0.3. Independent support for the higher ratio used by us is provided by a model fit of the observations of μ Cephei by Rogers et al. (1983) based on both UV and IR spectral data.

Support for a peak ratio as suggested by our experimental results comes also from the work of Rowan-Robinson and his collaborators. Rowan-Robinson (1986) gives the absorption efficiency for an "amorphus silicate" in a broad wavelength region. Using these data, it was possible to reproduce the spectra of dust shells around late-type stars (Rowan-Robinson et al., 1986) as well as the interstellar extinction curve of the infrared background radiation observed by IRAS (Rowan-Robinson, 1986). In the mid-IR these efficiencies are essentially based on the optical constants measured by Penman (1976) for the Vigarano and Murchison meteorites but are modified in the 8–10 μm range for a better fit of the observations. The ratio of both absorption peaks is 0.55 in excellent agreement with the ratio measured by us. Obviously, the agreement does not include the wavelength position of the 10 μm peak. A peak ratio of 0.5 would be just compatible with the estimate by McCarthy et al. (1980).

From the foregoing comparison of our experimental data with the observations we conclude that glassy bronzite is a promising candidate for the silicate component associated with molecular clouds and young stellar objects. We point out that bronzite is a main mineral of the chondrites and a major fraction of collected interplanetary dust particles consists of pyroxenes (Sandford and Walker, 1985).

Although our measurements provide us with information on the MAC up to 40 μm only, the analytical representation in terms of a superposition of dispersion profiles implies a λ^{-2} dependence of the MAC in the far infrared independent of whether the particles are of glassy or crystalline bronzite. Experimental and theoretical results seemed to indicate a wavelength dependence $\lambda^{-\cdot}$ with $0.8 \leq p \leq 3$ for various silicates (Aannestad, 1975; Hasegawa and Koike, 1984). Observations had been unable to produce decisive results so that the issue remained controversial. Recent observations by Chini et al. (1986a,b) on the submillimetre emission spectra of compact HII regions gave clear evidence in favour of $p = 2$. This lends confidence in using our superposition ansatz for describing the optical properties of silicate grains in the far IR.

4   Conclusions

Pyroxenes belong to the main minerals of the silicate component of meteorites and interplanetary dust particles and are therefore expected to contribute to interstellar silicate dust, too. For this reason, we produced a glass from the most abundant pyroxene, bronzite and manufactured submicrometre-sized grains from it. Mass absorption coefficients in the spectral range from 7 to 40 μm were determined from transmission spectra of the bronzite powder embedded in KBr and used to derive the complex indices of refraction by means of dispersion theory. Comparison of the emission properties of glassy bronzite grains with the observations showed that there is a surprisingly good agreement with the dust emissivity in infrared sources associated with evolutionarily young objects (e.g., T Tauri stars and BN objects). This agreement includes the position of the absorption peaks of the 10 and 20 μm bands as well as the widths of these bands. Thus, glassy bronzite seems to be a suitable P-type silicate according to the classification by Gürtler and Henning (1986). It obviously fails to reproduce the infrared spectra of oxygen-rich late-type giants and supergiants in the range of the silicate bands.

Submicrometre-sized grains of glassy bronzite have the widest 10 μm band of all amorphous silicates experimentally studied so far. Nevertheless, their widths are not quite sufficient to reproduce the notoriously broad profile of the μm band silicate emission in the Orion Trapezium region. A number of reasons suggest, however, that special conditions may hold in this region.

Among the various experimentally manufactured amorphous silicates the substances synthesized by Day (1979) play a central role. A significant difference to our glassy bronzite is the peak ratio of the bands. It amounts to 0.33 for his $MgSiO_3$, whereas our glassy bronzite has a ratio of 0.55. A number of arguments speak in favour of the higher value of the band ratio. This and the broader profile of the 10 μm band give rise to serious doubt in Day's amorphous silicate as a kind of standard for the optical properties of interstellar dust.

The dispersion ansatz used in deriving the optical constants of glassy bronzite predicts a λ^{-2} law for the grain emissivity in the far infrared. Recent observations confirmed this law.

References

Aannestad, P.A.: 1975, Astrophys. J. **200**, 30.

Chini, R., Kreysa, E., Mezger, P.G., Gemünd, H.P.: 1986, Astron. Astrophys. **154**, L8.

Chini, R., Krügel, E., Kreysa, E.: 1986, Astron. Astrophys. **167**, 315.

Cohen, M.: 1980, Monthly Notices Roy. Astron. Soc. **199**, 499.

Cohen, M., Witteborn, F.: 1985, Astrophys. J. **294**, 345.

Day, K.L.: 1979, Astrophys. J. **234**, 158.

Dodd, R.T.: 1981, Meteorites, Cambridge Univ. Press, Cambridge.

Dorschner, J., Friedemann, C., Gürtler, J., Duley, W.W.: 1980, Astrophys. Space Sci. **68**, 159.

Dorschner, J., Friedemann, C., Gürtler, J., Henning, Th., Wagner, H.: 1986, Monthly Notices Roy. Astron. Soc. **218**, 37P.

Dorschner, J., Henning, Th.: 1986, Astrophys. Space Sci. **128**, 47.

Draine, B.T., Lee, H.M.: 1984, Astrophys. J. **285**, 89.

Fadeyev, Yu., Th.: 1987, on preparation.

Forrest, W.J., Gillett, F.C., Stein, W.A.: 1975, Astrophys. J. **195**, 423.

Forrest, W.J., McCarthy, J.F., Houck, J.R.: 1979, Astrophys. J. **233**, 611.

Friedemann, C., Gürtler, J.,, Dorschner, J.: 1979, Astrophys. Space Sci. **60**, 297.

Gail, H.P., Sedlmayr, E.: 1968, Astron. Astrophys. **166**, 225.

Gillett, F.C., Forrest, W.J., Merrill, K.M., Capps, R.W., Soifer, B.T.: 1975, Astrophys. J. **200**, 609.

Gordon, M.A., Jewell, P.R., Kaftan-Kassim, M.A., Salter, C.J.: 1986, Astrophys. J. **308**, 288.

Gürtler, J., Henning, Th.: 1986, Astrophys. Space Sci. **128**, 163.

Hasegawa, H., Koike, Ch.: 1984, Occ. Rep. Roy. Obs. Edinburgh No. 12, p. 137.

Henning, Th., Gürtler, J., Dorschner, J.: 1983, Astrophys. Space Sci. **94**, 333.

Jones, T.W., Merrill, K.M.: 1976, Astrophys. J. **209**, 509.

Jones, A.P., Williams, D.A.: 1987, Monthly Notices Roy. Astron. Soc. **224**, 473.

Krätschmer, W., Huffmann, D.R.: 1979, Astrophys. Space Sci. **61**, 195.

McCarthy, J.F., Forrest, W.J., Briotta, D.A., Houck, J.R.: 1980, Astrophys. J. **242**, 965.

Papoular, R., Pégourié, B.: 1983, Astron. Astrophys. **128**, 335.

Pégourié, B., Papoular, R.: 1985, Astron. Astrophys. **142**, 451.

Penman, J.M.: 1976, Monthly Notices Roy. Astron. Soc. **175**, 149.

Rösler, H.J.: 1981, Lehrbuch der Mineralogie. VEB Deutscher Verlag für Grundstoffindustrie, Leipzig.

Roger, C., Martin, P.G., Crabtree, D.R.: 1983, Astrophys. J. **272**, 175.

Rouan, D., Léger, a.: 1984, Astron. Astrophys. **132**, L1.

Rowan-Robinson, M.: 1975, Monthly Notices Roy. Astron. Soc. **172**, 109.

Rowan-Robinson, M.: 1982a, Monthly Notices Roy. Astron. Soc. **201**, 281.

Rowan-Robinson, M.: 1982b, Monthly Notices Roy. Astron. Soc. **201**, 289.

Rowan-Robinson, M.: 1986, Monthly Notices Roy. Astron. Soc. **219**, 737.

Rowan-Robinson, M., Harris, S.: 1982, Monthly Notices Roy. Astron. Soc. **200**, 197.

Rowan-Robinson, M., Lock, T.D., Walker, D.W., Harris, S.: 1986, Monthly Notices Roy. Astron. Soc. **222**, 273.

Sandford, S.A., Walker, R.M.: 1985, Astrophys.J. **291**, 838.

Tielens, A.G.G.M., de Jong, T.: 1979, Astron. Astrophys. **75**, 326.

PIROXENE GLASSES - CANDIDATES FOR INTERSTELLAR SILICATES

J. Dorschner, J. Gürtler, C. Friedemann, Th. Henning
Jena University Observatory, Jena, G.D.R.

It is an old working hypothesis that solar system solids mineralogy could be of importance for understanding the nature of interstellar dust (Dorschner, 1968). Recently, a new approach to this subject by Jones and Williams (1987) considered interplanetary material as a guide to mineralogical modeling of the composition of interstellar grains.

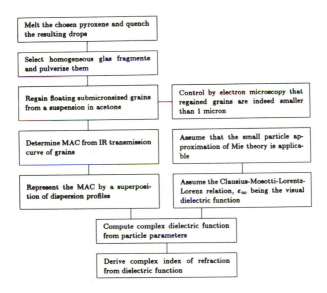

Figure 1: Experimental steps and theoretical background for deriving complex refractive indices.

Most primitive solids of the solar system, chondrites, IDP's, and cometary dust, which are assumed to be representative for condensed phases forming in molecular-cloud gas during star formation, contain Mg-silicates of olivine and orthopyroxene type.

E. Bussoletti et al. (eds.), Experiments on Cosmic Dust Analogues, 227–230.

General considerations of cosmic abundance of elements, interstellar depletion (Whittet, 1984) and modeling of dust condensation in O-rich late-type stellar winds
point to Mg-silicates with stoichiometric relations of olivines and pyroxenes, but with
lacking long-range lattice order ("amorphous silicates", cf. Dorschner & Henning, 1986).

IR spectroscopy of silicate features of very young objects (BN objects, Orion
Trapezium nebula, NGC 7538(E), T Tauri shells) shows clearly that this interstellar
silicate is rather of pyroxene stoichiometry ("P-type dust" according to classification of
Gürtler & Henning, 1986) than of olivine one.

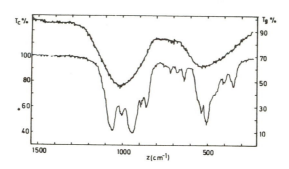

Figure 2: Transmission curves of crystalline (T_c) and glassy (T_g) bronzite. The column
density of submicrometre-sized grains in the KBr pellets was $1.59 \cdot 10^{-4}$ g cm^{-2}.

Pyroxene-type dust has been simulated by laboratory experiments at Jena
University Observatory. The programme started with bronzite, $(Mg,Fe)SiO_3$ (Dorschner
et al., 1986). The experimental procedure and the philosophy of deriving optical data
from the particulate is schematically shown in figure 1. For details cf. Dorschner et al.
(1987).

Figure 2 demonstrates the influence of the vitrification on the appearance of
the IR spectrum of crystalline bronzite, generating striking resemblence to the interstellar
silicate features.

Figure 3 represents the derived complex refractive indices from 7 to 40 μm
and figures 4 and 5 compare the relative mass absorption coefficient rel of small grains
of glassy bronzite with the observations of very young objects. The profile of the 10 μm
band of glassy bronzite is in good agreement with the observations of T Tauri stars and
IR sources in star-forming regions. The deviation of the latter profile beyond 11 μm is no
serious objection against this statement because the optical data derived by Chini et al.
(1986) are of less accuracy at these wavelengths. The Trapezium dust profile is broader
than that of small grains of glassy bronzite. If, however, particle radius increases from
0.1 to 1 μm, then the absorption peak shifts from 9.5 to 9.8 μm and FWHM becomes 2.8
instead of 2.5 μm. It is noteworthy that the peak ratio (20 μm)/(10 μm) = 0.55 of glassy
bronzite is considerably larger than that of the "astronomical silicate" by Draine & Lee
(1984), which was based on Day's (1979) experimental data. In agreement with Rowan-
Robinson (1986) who used a peak ratio of about 0.5 we conclude that Day's optical data
should no longer be used for the representation of observed silicate spectra in the range

around 20 μm.

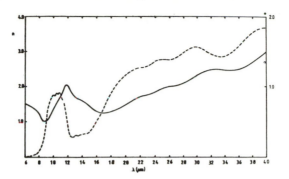

Figure 3: Real (n, solid line) and imaginary (k, dashed line) part of refractive index of glassy bronzite.

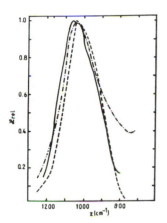

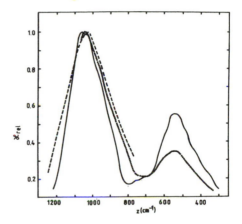

Figure 4: Comparison of the normalized MAC of glassy bronzite grains with a = 0.1 μm (solid line) with the mean T Tauri star silicate emission profile (Cohen, 1980, dashed line) and the MAC calculated with the optical data drived by Chini et al. (1986) for dust in NGC 7538(E) (dashed-dotted line).

Figure 5: Comparison with the profile of the silicate emission feature of the Orion Trapezium region (dashed line, Forrest et al., 1975) and the "astronomical silicate" (dotted line) by Draine and Lee (1984).

References

Chini, R., Krügel, E., Kreysa, E.: 1986, Astron. Astrophys. **167**, 315.

Cohen, M.: 1980, Monthly Not. Roy. Astron. Soc. **199**, 499.

Day, K.L.: 1979, Astrophys. J. **234**, 158.

Dorschner, J.: 1968, Astron. Nachr. **290**, 171.

Dorschner J., Henning Th.: 1986, Astrophys. Space Sci. **128**, 47.

Dorschner, J., Friedemann, C., Gürtler, J., Henning, Th., Wagner, H.: 1986, Monthly Not. Roy. Astron. Soc. **218**, 37P.

Dorschner, J., Friedemann, C., Gürtler, J., Henning, Th., Wagner, H.: 1987 submitted to Astron. Astrophys.

Draine, B.T., Lee, H.M.: 1984, Astrophys. J. **285**, 89.

Forrest, W.J., Gillett, F.C., Stein, W.A.: 1975, Astrophys. J. **195**, 423.

Gürtler, J., Henning, Th.: 1986, Astrophys. Space Sci. **128**, 163.

Jones, A.P., Williams, D.A.: 1987, Monthly Not. Roy. Astron. Soc. **224**, 473.

Rowan-Robinson, M.: 1986, Monthly Not. Roy. Astron. Soc. **219**, 737.

Whittet, D.C.B.: 1984, Monthly Not. Roy. Astron. Soc. **210**, 479.

SILICATE GRAINS IN SPACE: LABORATORY RESULTS

F. Leucci[1], L. Colangeli[2], E. Bussoletti[3], W. Krätchmer[4]
[1]*Physics Department, University of Lecce, Lecce, Italy*
[2]*ESA Space Science Department, ESTEC, Noordwijk, The Netherlands*
[3]*Istituto Universitario Navale, Naples, Italy*
[4]*Max-Planck-Institut für Kernphysik, Heidelberg, F.R.G.*

1 Introduction

Since their discovery in the spectra of some strong IR sources (Stein et al., 1969; Woolf and Ney, 1969; Forrest et al., 1979), the bands at around 10 μm and 20 μm have been respectively attributed to typical Si-O stretching and O-Si-O bending modes in the SiO_4 tetrahedra of the silicate structure. The presence of silicate grains around "O" stars ($[O]/[C]\gg1$) and in the interstellar medium (ISM) is also supported by theoretical computations on the grain condensation mechanism in space conditions. In fact, in the "O" stars envelopes CO is very likely to exist in the gas phase (Salpeter, 1977), as it is also confirmed by astronomical observations (Knapp and Morris, 1985; Zuckerman and Dyck, 1986). The oxygen in excess takes part to the condensation in solid grains which involve also Si, Fe, Al, ecc. (see, for instance, Hackwell, 1971).

Today it is commonly accepted that the silicate grains are one of the "main components" of cosmic dust, but their actual chemical, physical and structural properties are still not well defined. By analyzing a great number of astronomical spectra from different reddened stellar objects and interstellar nebulae, Merrill (1979), in agreement with other authors, concluded that the profile of the 9.7 μm band was essentially always the same. On the contrary, Papoular and Pegourié (1983), on the base of more recent and detailed observations, have identified three different classes of objects. All these sources show the 10 μm band, but slight differences in the profile have been evidenced. Semi-regular super-giants, like μ-Cep and RW Cyg, present a sharp and intense feature; cool M stars, like W Hya, show a smooth and weak band; M giants, like oCet and R Cas, have an "intermediate" band profile.

In the past, simulation of the bands has been often performed on theoretical computations based on Mie theory whose validity for small grains (radius a $\lesssim$ 0.1 μm) is today doubtfull. In addition, when experimental data from dust silicate samples have been used, not always the internal consistency of the laboratory data has been "a priori" checked before performing astrophysical applications.

To overcome these problems, we have started a research program aimed to compare theoretical results with experimental data obtained from materials whose presence in space is suggested by physical and astrophysical considerations. In particular, in

231

E. Bussoletti et al. (eds.), Experiments on Cosmic Dust Analogues, 231–238.
© 1988 by Kluwer Academic Publishers.

our experiments the ambient conditions are carefully controlled and the morphological, structural and optical properties of the dust samples are characterized with best attention. This allows to obtain homogeneous and reproduceable laboratory data. Here we present some preliminary laboratory results on silicate grains. We have also performed a comparison between grain properties deduced according Mie computations and those measured in the laboratory in order to check the limits of validity of this "classic" theory to silicate small grains.

2   Sample

Due to the large variety of silicate compounds present on the earth it is necessary to select some possible candidates to be "analogues" of cosmic dust according to the observational constraints. In particular, the observed bands at ~10 and 20 μm appear mainly smooth in profile and featureless so that they are expected to be produced by amorphous materials. Furthermore, the actual thermodynamic interstellar conditions do not allow for the condensation of "too complex" chemical species so that simple silicate compounds must be the most abundant. Finally, the identification of different classes of features (Papoular and Pegourié, 1983) suggests that several are the chemical compounds present in space.

Table 1: IR Extinction properties.

Material	SiO$_2$ wt (%)	λ_p (μm) I_p (cm^{-1}) W (μm)	= peak wavelength = $[Q_{ext}/a](\lambda_p)$ = full width at half maximum				
CR SiO$_2$	100	8.6	12.8		18.9	20.8	
		1.7 E5	6 E3		2 E4	9 E3	
		0.25	0.45		0.75	0.70	
AM SiO$_2$	100	8.8	12.8			20.5	
		8 E4	3 E3			2 E4	
		0.35	1.0			0.9	
OBSIDIAN 1	76.20	9.0	14.5			21.0	
		3 E4	2 E3			1 E4	
		0.9	–			1.8	
OBSIDIAN 2	73.45	9.0	14.5			20.8	
		3 E4	2 E3			9 E3	
		0.9	–			1.8	
ANDESITE	54.15	9.0		16.5	18.2		23.0
		2 E4		5 E3	4 E3		3 E3
		1.3		1.7	1.4		3.2
BASALTIC GLASS	53.45	9.4				20.4	
		2 E4				4 E3	
		1.5				4.9	
BASALT	53.25	9.0		16.5	18.0		23.0
		2 E4		5 E3	4 E3		3 E3
		0.9		1.9	2.0		2.9

In the computations we have considered seven materials with different $[SiO_2$ wt. %] content (see table 1) whose experimentally measured optical constants (n, k) have been reported in tabular form by Bussoletti and Zambetta (1976):

a) Crystalline quartz (SiO_2):

the "dispersion parameters" reported by Spitzer and Kleinman (1961) have been used to deduce the optical constants for this material. Two different sets of values exist, for light polarization parallel and perpendicular to the optical axis of the crystal. Due to the lattice symmetry, the two contributions must be weighted by factors 1/3 and 2/3, respectively.

b) Amorphous quartz (SiO_2):

(n, k) for this material have been published by Steyer et al. (1974) in the wavelength range 7–25 μm.

c) Obsidian 1 and 2:

the optical constants of these rocks have been measured by Pollack et al. (1973) between 0.2 μm and 50 μm. These silicates come respectively from Lake County (Oregon) —Obsidian 1— and from Little Glass Mt. (Calif.) —Obsidian 2—.

d) Andesite, Basalt, Basaltic Glass:

for these materials too, the optical constants have been obtained by Pollack et al. (1973).

For the laboratory measurements, we have selected three different submicron dust samples: amorphous quartz, plagioclase and amorphous olivine. The main characteristics of these materials are reported in table 2.

3 Results and discussion

3.1 Mie theory computation

For all the silicate compounds listed in table 1, the extinction factor, $[Qext/a](\lambda)$, has been computed between 5 and 25 μm, by assuming a grain radius a = 0.1 μm, in vacuum (n_o= 1.0). In table 1 the peak positions of the main IR features, λ_p, their intensities, $I_p=[Qext/a](\lambda_p)$, and the full widths at half maximum, W, are reported. Three main broad bands are commonly found at around 10, 13 and 20 μm (see figure 1). Some qualitative, more than quantitative, evidences may be identified on both the behaviour of the peak position and the band profile:

i) λ_p (10 μm) shifts towards longer λ and W (10 μm) rises as: 1) the amorphous structure prevails, for the same material (AM SiO_2 vs. CR SiO_2 and Basaltic glass vs. Basalt); 2) the $[SiO_2$ wt %] decreases, for a similar amorphous degree (AM SiO_2, Obsidian 1–2 and Basaltic glass). On the contrary, I_p (10 μm) is more sensible to $[SiO_2$ wt %] than to the crystalline degree (Basalt vs. Basaltic glass) and lowers as $[SiO_2$ wt %] decreases.

ii) I_p (13 μm) reduces and W (13 μm) rises as the $[SiO_2$ wt %] decreases. Furthermore, for $[Sio_2$ wt %]< 54% the band disappears, in agreement with its typical SiO_2 origin (Nuth and Donn, 1983).

iii) For amorphous materials a single band appears at $\sim 20\ \mu$m. In this case, λ_p is essentially independent of [SiO_2 wt %] while I_p and W follow the same behaviour observed for the $\sim 10\ \mu$m band.

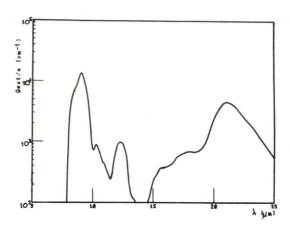

Figure 1: Extinction efficiency for a typical amorphous SiO_2 dust sample.

Table 2: Laboratory dust samples.

Material	Origin	Average radius (μm)
AMORPHOUS SiO_2*	industrial	0.02
PLAGIOCLASE	rock $Na(AlO_2)(SiO_2)_3$ $Ca(AlO_2)_2(SiO_2)_3$	$\sim$0.5
AMORPHOUS OLIVINE	20 keV Ar^+ ions bombardment	0.03

(*) 99.8% purity.

3.2   Laboratory measurements

For each of the three samples listed in table 2, we have measured the trasmittance, $T(\lambda)$, in the wavelength range 2.5–50 μm, by means of the classic KBr pellet technique. The

$[Qext/a](\lambda)$ was calculated from $T(\lambda)$ data by using the formula:

$$\frac{Qext}{a}(\lambda) = \frac{4}{3}\frac{\rho S}{M}ln\frac{1}{T(\lambda)}.$$

Here S is the cross-section of the KBr pellet, M is the dust mass in the sample and ρ is the mass density of the grain material.

The trasmission spectra measured in the wavelength range 5–25 μm are reported in figures 1 to 3. The main extinction features are summarized in table 3.

Table 3: Extinction features in KBr.

Material	λ_p (μm)	I_p (cm^{-1})/W (μm)		
AM SiO$_2$	9.05	12.35		21.19
	1.4 E4	1.0 E3		4.7 E3
	1.05	1.1		2.9
PLAGIOCLASE	10.0		17.3	18.5
	5.7 E3		2.5 E3	2.2 E3
	2.3		(*)	(*)
AM OLIVINE	10.4			19.3
	3.2 E3			1.8 E3
	2.7			~5

(*) difficult to determine

The behaviour of λ_p, I_p, W follow some of the qualitative seen in trends the Mie computations. In particular we observe that, as [Sio$_2$ wt %] decreases λ_p (10 μm) shitf towards longer λ, I_p (10 μm) reduces and W (10 μm) increases; I (20 μm) reduces and W (20 μm) increases, while the (13 μm) band tends to disappear.

3.3 Matrix effect

To use our experimental results for astrophysical applications it is necessary to correct extinction data for possible "matrix effects". These are mainly due to the interaction of the "small" grains with the embedding matrix and affect the λ_p, I_p and W behaviour. In table 4 the results of Mie computations and laboratory measurements for amorphous SiO$_2$ grains, embedded in different matrices, are reported. The comparison between the two sets of data allows for the following conclusions:

i) The values of λ_p, I_p and W obtained according to Mie theory seem not completely in agreement width obtained on dust samples for the same material in a selected matrix;

ii) The experimental trend of matrix effects on λ_p and I_p, is in good agreement with the Mie predictions: λ_p shifts towards longer λ and I_p rises as n_o increases;

iii) the behaviour for W is not clear; in fact, experimental data show always a reduction of W as n_o reduces, while Mie computations predict a decrease, an increase and stability respectively for the ~ 10 μm, ~ 20 μm and ~ 13 μm bands;

iv) Mie data show a strong linear, y=mx+b, correlation between $y = \lambda_p, I_p, W$ and $x = n_o$ (except for the 13 μm band).

Table 4: Matrix effects for AM SiO_2.

Matrix	λ_p (μm)/I_p (cm^{-1})/W (μm)					
	Laboratory data			Mie Theory		
vacuum				8.78	12.45	20.5
(n_o=1.0)				8.3 E4	3.0 E3	1.3 E4
				0.40	0.95	0.8
KBr	9.05	12.35	21.19	8.93	12.49	20.9
(n_o=1.5)	1.4 E4	1.0 E3	4.7 E3	1.3 E5	4.4 E3	3.2 E4
	1.05	1.1	2.9	0.25	1.0	1.1
CsI	9.09	12.42	21.30	9.01	12.51	21.0
(n_o=1.7)	1.7 E4	1.2 E3	5.7 E3	1.4 E5	4.8 E3	3.3 E4
	1.03	0.92	2.5	0.20	1.0	1.15

Table 5: Extinction properties extrapolated to vacuum.

Material	λ_p (μm)/I_p (cm^{-1})/W (μm)		
AM SiO_2	8.95	12.18	20.92
	6.5 E3	4.8 E2	2.2 E3
	1.10	1.6	3.9
PLAGIOCLASE	9.8	17.0	18.2
	2.6 E3	1.2 E3	1.0 E3
	2.4	–	–
AM OLIVINE	10.3		19.0
	1.5 E3		8.5 E2
	2.8		6.0

On the base of this last evidence, we have fitted our experimental data in KBr and CsI with linear laws in order to reduce them to vacuum. The results of this extrapolation are summarized in table 5.

4 Conclusions

As already mentioned in Par. 1, many galactic sources show peaks at $\lambda_p = 9.7$ μm with observed value of $W = 2.5$–3 μm and at $\lambda_p \sim 19.0$ μm.

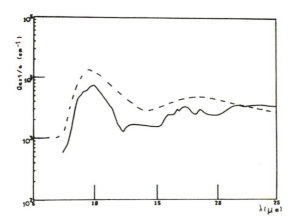

Figure 2: Extinction efficiency for Plagioclase dust sample (full line). The spectrum for the "astronomical silicate" synthesized by Draine (1985) is also reported for comparison (dashed line).

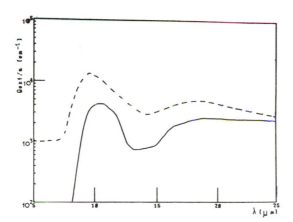

Figure 3: Extinction efficiency for Amorphous Olivine dust sample (full line). The spectrum for the "astronomical silicate" synthesized by Draine (1985) is also reported for comparison (dashed line).

The experimental results reported in table 5 show that:

- AM SiO_2 does not fit astronomical data both in λ_p and in W;
- Plagioclase fits well both λ_p and W at 10 μm but shows a 19 μm band too rich in structures, probably due to some degree of crystallinity of the material (Figure 2);
- Amorphous olivine fits well W at 10 μm and λ_p at 20 μm but some discrepancy exists in λ_p at 10 μm (Figure 3)

In conclusion, our results seem to confirm that amorphous SiO_2 cannot be a main component of interstellar silicate grains. Further measurements must be addressed to oder amorphous materials, with low ($< 50\%$) [SiO_2 wt %] content, and with chemical composition similar to olivine and plagioclase compounds. At the end, we would stress that present results may be also interesting in view of an interrelationship between "interstellar" and "cometary" dust. Olivine and plagioclase have been just proposed among the possible dust components of Halley's comet on the base of "in situ" PUMA 1 and 2 measurements (Sagdeev et al., 1986).

Acknowledgements

This work has been partially supported by Ministero Pubblica Istruzione and by Consiglio Nazionale delle Ricerche under the contracts CNR–85.00281, PSN–85.0018 and PSN–85.0019.

References

Bussoletti, E., Zambetta, A.M.: 1976, Astron. Astrophys. Suppl. Ser. **25**, 549.

Bohren, C.F., Huffmann, D.R.: 1984, "Absorption and Scattering of light by small particles" (Wiley & Sons, New York).

Draine, B.T.: 1985, Astrophys. J. Suppl. Ser. **57**, 587.

Forrest, W.J., McCarthy, J.F., Houck, J.R.: 1979, Astrophys. J. **233**, 611.

Hackwell, J.A.: 1971, Ph.D. Thesis, University College, London.

Knapp, G.R., Morris, M.: 1985, Astrophys. J. **292**, 640.

Merrill, K.M.: 1979, IAU Circ. N. 3444.

Nuth, J.A., Donn, B.: J. Geophys. Research 88, A847.

Papoular, R., Pegourié, B.: 1983, Astron. Astrophys. **128**, 335.

Pollack, J.B., Tonn, O.B., Khore, B.N.: 1973, Icarus **19**, 372.

Sagdeev, R.Z., Kissel, J., Evlanov, E.N., Mukhin, L.M., Zubkov, B.V., Prilutskii, O.F., Fomenkova, M.N.: 1986, in 20th ESLAB Symp. on the Exploration of Halley's Comet, Heidelberg, 27–31 October 1986, ESA SP-250 (December 1986).

Salpeter, E.E.: 1977, Ann. Rev. Astron. Astrophys. **15**, 267.

Spitzer, W.G., Kleinman, D.A.: 1961, Phys. Rev. **121**, 1324.

Stein, W.J., Gaustad, J.E., Gillett, F.C., Knacke, R.F.: 1969, Astrophys. J. **155**, L3.

Steyer, T.R., Day, K.L., Huffman, D.R.: 1974, Appl. Optics **13**, 1586.

Woolf, N.J., Ney, E.P.: 1969, Astrophys. J. **155**, L181.

Zuckerman, B., Dyck, H.M.: 1986, Astrophys. J. **304**, 394.

LABORATORY STUDIES OF EXTRATERRESTRIAL CHEMISTRY INITIATED BY ATOMIC SILICON IONS IN THE GAS PHASE

D.K. Bohme, S. Wlodek, A. Fox
Department of Chemistry and Centre for Research in Experimental Space Science, York University, Ontario, Canada

1 Introduction

Gas-phase chemistry initiated by atomic silicon ions occurs in a variety of extraterrestrial environments including diffuse interstellar clouds,[1] dense interstellar clouds,[2] and the atmospheres of certain stars.[3] Reliable observational evidence for extraterrestrial molecules containing Si exists for SiO, SiS and SiC_2 but is limited otherwise, in contrast to evidence which has been obtained for larger numbers of molecules containing less abundant elements. This lack of observational evidence is partly due to the lack of experimental spectral information for relevant silicon-containing molecules. Such information is difficult to obtain in the terrestrial laboratory. Also, one needs to know *which* silicon-containing molecules to search for. Suggestions have been made in the past,[4] but the basis of these suggestions is somewhat obscure. Here we explore experimentally the kinds of silicon-containing molecules which may be produced via the chemistry initiated by atomic silicon ions in the gas phase.

Experimental results are reported for gas-phase ion/molecule reactions initiated by ground-state $Si^+(^2P)$ ions. A variety of interstellar and circumstellar molecules have been chosen as neutral reactants including H_2, CO, H_2O, H_2S, CH_4, C_2H_2, NH_3, amines, cyanides, dimethyl ether, alcohols and carboxylic acids. The silicon may appear either in the neutral product formed directly by the ion/molecule reaction or in the ionic product. The ionic product may neutralize in a secondary reaction, such as charge transfer or proton transfer. When electrons are present, as is likely to be the case in all natural environments, neutralization may also proceed by electron/ion recombination.

2 Experimental

This report is confined to the measurements made recently in the Ion Chemistry Laboratory of York University with the Selected-Ion Flow Tube (SIFT) technique.[5,6] With this technique atomic silicon ions, known to be in their ground state, are selected from a suitable source, and injected into flowing helium gas to which is added the desired neutral reactant.[7] The reactant and product ions are monitored with a mass spectrometer as a function of the addition of the neutral reagent. The observations provide a measurement

239

E. Bussoletti et al. (eds.), Experiments on Cosmic Dust Analogues, 239–244.
© *1988 by Kluwer Academic Publishers.*

of the specific rate and products of the reaction of the silicon ions and information about the secondary and higher ion/molecule reactions.

3 Results

The results of the measurements are discussed below for groups of related molecules. The actual values of the rate constants, k, and product distributions of the individual reactions which were determined have not been tabulated in this report. Instead, the nature of the products is indicated, and reactions are classified as fast when k was found to be equal to or greater than $1 \cdot 10^{-10}$ cm^3 molecule^{-1} s^{-1}, or as slow when k was found to be in the range from $1 \cdot 10^{-10}$ and $1 \cdot 10^{-12}$ cm^3 molecule^{-1} s^{-1}. Unless indicated otherwise, the reaction is said not to occur when k was found to be less than $1 \cdot 10^{-12}$ cm^3 molecule^{-1} s^{-1}. The actual values which were determined for the rate constants can be found in the references. The measurements were carried out at 296 K and moderate pressures (ca. 0.35 Torr) of helium. The rapid ion/molecule reactions with bimolecular channels which are reported should be essentially independent of temperature and pressure. Sometimes the bimolecular channel was observed to compete with adduct formation. Adduct formation is expected to be termolecular under the adopted experimental conditions (measurements were not made as a function of total pressure) and so should become less important as a competing channel at lower pressures. Adduct formation prceeding exclusively by termolecular collisional stabilization in our SIFT experiment must proceed by radiative association if it is to be efficient at very low pressures.

4 Hydrogen and carbon monoxide

Ground-state Si(^{2}P) ions have been found to be unreactive towards H$_2$.[7] In contrast, the excited ^{4}P state reacts rapidly to produce SiH$^+$, and so will be filtered out preferentially in an environment rich in molecular hydrogen. SiH$^+$ produced in this fashion does not react with H$_2$. Atomic silicon ions, and any SiH$^+$ which may be formed due to the presence of excited silicon ions, will therefore be available for chemical reaction with other constituents in hydrogen environments, even when molecular hydrogen predominates. Ground-state silicon ions have been observed to be unreactive toward CO.[7]

5 Water and hydrogen sulphide

Silicon ions have been observed to react rapidly with both H$_2$O and H$_2$S to produce protonated silicon monoxide and silicon monosulphide:[7,8]

$$Si^+ + H_2O \rightarrow SiOH^+ + H$$
$$Si^+ + H_2S \rightarrow SiSH^+ + H$$

Molecular orbital calculations have shown that protonation occurs at the O and S sites rather than at the Si atom.[8] The proton may be lost through proton-transfer reactions with molecules having higher proton affinities than SiO and SiS. SiOH$^+$ has been observed to lose a proton to NH$_3$ and CH$_3$CN. H$_2$O adds to SiOH$^+$ to form SiH$_3$O$_2^+$. SiSH$^+$ has been observed to lose a proton to NH$_3$, HCN and H$_2$S. H$_2$O reacts with SiSH$^+$ to

generate $SiOH^+$:
$$SiSH^+ + H_2O \rightarrow SiOH^+ + H_2S$$

6 Methane and acetylene

Methane reacts only slowly with Si^+ (^2P) to form the adduct $SiCH_4^+$.[8] The structure of this adduct is uncertain. Both Si^+CH_4 and $H\text{-}Si^+\text{-}CH_3$ are possible. The latter would be produced by C-H bond insertion. Neutralization by proton transfer or recombination with electrons may produce $SiCH_3$ or $HSiCH_2$.

Acetylene reacts primarily by association with H atom elimination:[8]
$$Si^+ + C_2H_2 \rightarrow SiC_2H^+ + H$$
$$\rightarrow SiC_2H_2^+$$

The Si_2H^+ ion may neutralize to form the SiC_2 molecule which is known to have a symmetric cyclic rather than linear ground state. Also, the further reaction of SiC_2H^+ with acetylene is worthy of note. It has been observed to react in the following manner:
$$SiC_2H^+ + C_2H_2 \rightarrow SiC_4H^+ + H_2$$
$$\rightarrow SiC_4H_3^+$$

The SiC_4H^+ ion may neutralize to form SiC_4. This interesting silatetracarbon molecule may have a structure which is linear, cyclic, or even three-dimensional (a square pyramid)!.

7 Ammonia

With ammonia we have observed the following reaction sequence:
$$Si^+ + NH_3 \quad \rightarrow SiNH_3^+ + H$$
$$SiNH_2^+ + NH_3 \rightarrow SiNH + NH_4^+$$

This sequence estabilishes hydrogen silaisonitrile, $SiNH$, the silicon analogue of hydrogen isonitrile.[9]

8 Amines

Experiments have been conducted with methyl-, dimethyl-, and trimethyl amine.[10] The reactions of all of these amines with atomic silicon ions were observed to be fast and can be understood in terms of Si^+ insertion into N-H and N-C bonds according to the following reaction:
$$Si^+ + CH_3\text{-}NR_1R_2 \rightarrow CH_3 + Si^+NR_1R_2$$

proceeding in competition with hydride transfer to form immonium ions and the neutral SiH molecule according to the reaction:
$$Si^+ + H\text{-}CH_2\text{-}NR_1R_2 \rightarrow SiH + CH_2NR_1R_2^+$$

Rapid secondary proton-transfer reactions were observed for $SiNH_2^+$ and $SiNHCH_3^+$ to produce $SiNH$ and $SiNCH_3$ molecules. The structure of $SiNCH_3$ is intriguing since the three heavy atoms in this molecule may form a ring. With methylamine $SiNH_2^+$ also appeared to produce $H_2SiNH_2^+$ which may deprotonate to form the simplest silanimine, H_2SiNH. Other minor reaction channels were observed for the reactions of Si^+ with the methylamines which lead directly or indirectly to SiCH and SiCH$_3$:

$$Si^+ + CH_3NH_2/(CH_3)_2NH \rightarrow SiCH + NH_4^+/CH_3NH_3^+$$
$$Si^+ + (CH_3)_3N \quad\quad \rightarrow SiCH_3 + CH_2NHCH_3^+$$
$$\rightarrow SiCH_2^+ + (CH_3)_2NH$$

9   Cyanides

A range of reactivity was observed for the reaction of Si$^+$ with the cyanides HCN, C$_2$N$_2$, CH$_3$CN and HC$_3$N.[8] Hydrogen cyanide reacted only slowly:

$$Si^+ + HCN \rightarrow (CHNSi)^+$$
$$\rightarrow (CNSi)^+ + H$$

Neutralization of the product ions by proton transfer or charge transfer respectively, may lead to silicon cyanide or isocyanide. The reaction with methylcyanide was much more rapid and proceeded in the following manner:

$$Si^+ + CH_3CN \rightarrow (CH_3CNSi)^+$$
$$\rightarrow SiCH_2^+ + HCN$$

The bimolecular product channel produces SiCH$_2^+$ which will neutralize to form SiCH or HSiC. Cyanogen also showed both a bimolecular and addition channel:

$$Si^+ + C_2N_2 \rightarrow (C_2N_2Si)^+$$
$$\rightarrow (CNSi)^+ + CN$$

Neutralization by charge transfer may produce several isomers of (C$_2$N$_2$Si) and (CNSi). The reaction of Si$^+$ with cyanoacetylene mimicked the reaction with acetylene in that SiC$_2$H$^+$ was the predominant product:

$$Si^+ + HC_3N \rightarrow HSiC_2^+ + CN$$
$$\rightarrow (SiHC_3N)^+$$

and so represent another source of SiC$_2$.

10   Water, dimethyl ether, alcohols and carboxylic acids

We have observed that Si$^+$ reacts with molecules containing hydroxyl groups to abstract the hydroxyl group:[7]

$$Si^+ + R\text{-}OH \rightarrow SiOH^+ + R$$

For R=H, CH$_3$, C$_2$H$_5$, HCO and CH$_3$CO. Therefore, when followed by neutralization, all of these reactions become sources of the SiO molecule. A second, less predominant, channel with methanol yielded the SiOCH$_3^+$ ion which may neutralize to form SiOCH$_2$ which has been shown to have a cyclic ground state.[11] SiOCH$_3^+$ was the predominant product in the fast reaction observed between Si$^+$ and dimethyl ether:

$$Si^+ + CH_3OCH_3 \rightarrow SiOCH_3^+ + CH_3$$

The silene cation SiOH$^+$ was found to be unreactive toward hydrogen and carbon monoxide. Five molecules of water were observed to add to SiOH$^+$. Predominant bimolecular products were observed for the fast reactions of SiOH$^+$ with methanol, ethanol, formic acid, and acetic acid. Formation of SiOCH$_3^+$ predominated with methanol as in the reaction with Si$^+$:

$$SiOH^+ + CH_3OH \rightarrow SiOCH_3^+ + H_2O$$

$SiH_3O_2^+$ was the main product ion observed for the reactions with ethanol and formic acid:

$$SiOH^+ + C_2H_5OH/HCOOH \rightarrow SiH_3O_2^+ + C_2H_4/CO$$

Neutralization of this ion may yield silanoic acid, $HSiOOH$, or silicon dihydroxide, $Si(OH)_2$. These molecules may be formed directly in the reaction observed with acetic acid:

$$SiOH^+ + CH_3COOH \rightarrow CH_3CO^+ + (SiH_2O_2)$$

Still further, higher order, chemistry was observed with methanol, ethanol and formic acid. The chemistry appeared to be directed toward increasing the silicon/oxygen coordination. One sequence of reactions initiated by Si^+ in methanol produced the following ions:

$$Si^+ \rightarrow SiOCH_3^+ \rightarrow HSi(OCH_3)_2^+ \rightarrow HSi(OCH_3)_3H^+$$

Neutralization of the molecular ions by proton transfer can yield the neutral molecules $SiOCH_2$, $Si(OCH_3)_3$ and $HSi(OCH_3)_3$. An analogous sequence was observed with ethanol:

$$Si^+ \rightarrow SiOC_2H_5 \rightarrow HSi(OC_2H_5)^+ \rightarrow HSi(OC_2H_5)_3H^+$$

In this case neutralization can produce the molecules $SiOCHCH_3$, $Si(OC_2H_5)_2$, and $HSi(OC_2H_5)_3$.

Still other reaction sequences initiated by Si^+ were identified which appeared to saturate the silicon centre. With ethanol the experimental data was consistent with the following additional bimolecular reaction sequences:

$$Si^+ \rightarrow SiOH^+ \rightarrow HSi(OH)_2^+ \rightarrow HSi(OH)_3H^+$$
$$Si^+ \rightarrow SiOH^+ \rightarrow HSi(OH)_2^+ \rightarrow HSi(OC_2H_5)OH^+$$

With formic acid complete saturation appears to be achieved in three bimolecular steps as with ethanol as follows:

$$Si^+ \rightarrow SiOH^+ \rightarrow HSi(OH)_2^+ \rightarrow HSi(OH)_3H^+$$

or in four steps as follows:

$$Si^+ \rightarrow SiOH^+ \rightarrow HSi(OH)_2^+ \rightarrow HSi(OCHO)OH^+ \rightarrow HSi(OCHO)(OH)_2H^+$$

Neutralization of some of these product ions can yield dihydroxysilene, $Si(OH)_2$, trihydroxysilane, $HSi(OH)_3$, and other substituted silanes such as $HSi(O)(OC_2H_5)$, $HSi(O)(OCHO)$, and $HSi(OCHO)(OH)_2$.

11   Summary and conclusions

The experimental study reported here has led to the identification of a large number of new molecules potentially, of interest in extraterrestrial chemistry. The gas-phase ion/molecule reactions which were identified in this study represent possible contributors to the formation of the following silicon-containing molecules:

SiH, SiO, SiS, $SiCH$, SiC_2, $SiCH_3$, SiC_4, $SiNH$, H_2SiNH, $SiNCH_3$, $SiCN$, SiC_2N_2, $SiOCH_2$, $SiOC_2H_4$, SiO_2H_2, $HSi(OH)_3$, $HSi(O)(OCHO)$, $HSi(O)(OC_2H_5)$, $HSi(OCHO)(OH)_2$, $Si(OCH_3)_2$, $Si(OCH_3)_3$, $Si(OC_2H_5)_2$, $HSi(OC_2H_5)_3$. Several isomers are possible for many of these molecules. In some cases cyclic structures may be preferred as, for example, with SiC_2, $SiNHCH_2$, $SiOCH_2$, and $SiOC(H)CH_3$. A three-dimensional square pyramid structure may be preferred for SiC_4. Several other points are worthy of emphasis:

1. Ground state atomic silicon ions are unreactive toward hydrogen and carbon monoxide and so are available in environments where these gases predominate for other chemistry with other constituents.

2. SiO and SiS are readily generated from their protonated forms which can easily be produced from reactions between silicon ions and molecules containing OH or SH, respectively.

3. In contrast to C^+, Si^+ does not react with methane in a bimolecular fashion. Si^+ reacts with acetylene in a fashion which is analogous to the reaction of C^+. It forms SiC_2H^+ which potentially is a source for silacyclopropyne. The reaction of Si^+ with cyanoacetylene produces SiC_2H^+ and so is another possible precursor for this molecule. By inference, we can also expect reactions of atomic silicon ions with polyacetylenes and cyanopolyacetylenes to be possible precursors to the formation of SiC_2.

4. The reaction of Si^+ with ammonia, when followed by proton transfer, leads *selectively* to the hydrogen silaisonitrile isomer, SiNH.

5. The hydride transfer reactions with amines may provide a unique source for SiH. Other ion/molecule reactions of Si^+ leading directly to SiH, or to SiH_2^+ which may neutralize to form SiH, have not yet been identified.

6. The results obtained with water, alcohols and carboxylic acids are striking in that they clearly indicate that the chemistry with these molecules initiated by Si^+ is inexorably directed toward increasing the silicon/oxygen co-ordination.

7. The results with water, alcohols and carboxylic acids suggest the efficient formation by gas-phase chemistry of intermediates important in silicate formation. For example, re-ionization of trihydroxysilane, followed by a dehydrative reaction with a carboxylic acid, may provide a source for tetrahydroxysilane which has been shown to be a building block for the condensational synthesis of hydrated silicate networks.[12,13]

References

Turner, J.L., Dalgarno, A.: 1977, Astrophys. J. **213**, 386.

Millar, J.: 1980, Astrophys. Space Sci. **72**, 509.

Clegg, R.E.S. van IJzendoorn, L.J., Allamandola, L.J.: 1983, Mon. Not. Astr. Soc. **203**, 125.

Lovas, F.J.: 1974, Astrophys. J. **213**, 265.

Mackay, G.I., Vlachos, G.D., Bohme, D.K., Schiff, H.I.: 1980, Int. J. Mass Spectrom. Ion Phys. **36**, 259.

Raksit, A.B., Bohme, D.K.: 1984, Int. J. Mass Spectrom. Ion Processes **55**, 69.

Wlodek, S., Fox, A., Bohme, D.K.: 1987, J. Am. Chem. Soc., in press.

Wlodek, S., Rodriquez, C.F., Lien, M.H., Hopkinson, A.C., Bohme, D.K.: 1987, Chem. Phys. Letters, submitted for publication.

Wlodek, S., Bohme, D.K.: 1987, J. Am. Chem. Soc., submitted for publication.

Hopkinson, A.C.: 1987, private communication.

Abe, Y., Mison, T.: 1984, J. Polym. Sci. Polym. Lett., **22**, 565.

Meinhold, R.H., Rothbaum, H.P., Newman, R.H.: 1985, J. Colloid Interface Sci. **108**, 234.

LIGHT SCATTERING FROM SIMULATED INTERSTELLAR DUST

J.R. Stephens
Los Alamos National Laboratory, Los Alamos, U.S.A.

1   Introduction

Significant progress has been made in recent years in restricting the range of possible compositions, shapes, and size distributions of circumstellar and interstellar dust. Much of the progress has occurred by comparing observations with results of laboratory and theoretical studies of likely dust materials based on the physical and chemical conditions existing in dust producing astronomical environments. Significant uncertainty remains, however, on the shape and size distribution of dust. Theoretical studies have usually modeled the optical properties of dust by assuming that the dust consists of spheres, infinitely long cylinders, or ellipsoids because such assumptions make detailed calculations tractable. Dust produced in laboratory studies frequently occur as tangled chains of submicron grains, variously described as "birdsnests", "aggregated" or "coagulated". The aggregated structure of laboratory dust reflect the large contribution of grain-grain collisions to grain growth in the relatively high pressure processes used to prepare the dust samples. In astronomical dust forming regions, grain growth can occur both by grain-molecule growth, which leads to compact dust, and by grain-grain collisions, which leads to low density dust consisting of strings of grains. Numerical simulations of grain-grain aggregation result in low density dust particles that can be described by their fractal dimension (Meakin, 1986).

Recent observations have promoted the development of models describing the physical and chemical conditions existing in dust forming environments, for example, in red giant stars. Knowledge of the conditions in such environments allows mechanisms of dust formation to be studied. Lattimer (1982), for example, considered the relative contribution of grain-grain aggregation and grain-molecule collisions to grain growth in supernova and nova ejecta and also in the envelopes of cool stars. He concluded that growth by aggregation can make a significant contribution to grain growth, particularly in the envelopes of cool stars where dust is accelerated by radiation pressure. Observational evidence for the existence of aggregated interstellar dust is not definitive, perhaps due to the lack of theoretical methods to accurately calculate the extinction, scattering, and absorption of irregularly shaped dusts. Jura (1980) has speculated that the dust in some dense molecular clouds may be aggregated, leading to extinction that appears to be due

245

E. Bussoletti et al. (eds.), Experiments on Cosmic Dust Analogues, 245–252.
© 1988 by Kluwer Academic Publishers.

to grains larger than the general interstellar population. Donn (1987) has modeled the formation of dust in a protosolar nebula and concluded that grain-grain aggregation may be the dominant mechanism of dust growth. Both interplanetary and cometary dust are believed to consist of low density aggregates based on sunlight scattered from zodiacal dust (Giese, et al. 1978) and data from the recent flyby of comet Halley by the Giotto spacecraft (Giese, et al. 1986). Interplanetary dust collected in the stratosphere consists predominantly of aggregates of micron and submicron grains (Zolensky, 1987 and references there in).

In order to provide data on the optical properties of coagulated dust grains composed of astronomically interesting materials a program to measure the scattering and extinction properties of laboratory prepared dusts has been initiated. The objective of the program is to measure, over the visible spectral range, the intensity of light scattered from well characterized dust materials. Reported here are color normalized scattering intensities of silicate and carbon dust materials as a function of scattering angle and wavelength. The scattering data are compared to calculated scattering data of spherical graphite and silicate particles of various size distributions.

2   Experimental

Particle clouds were prepared by vaporizing a solid target of olivine $((MG_{0.9}, Fe_{0.1})_2SiO_4)$, a silicate mineral, or spectroscopic grade graphite using a beam from a 20 W CW CO_2 laser focused onto the target. The olivine and carbon samples were vaporized in air and argon atmospheres respectively at 590 torr pressure in a 6 inch diameter cylindrical glass chamber. A schematic of the apparatus is shown in figure 1.

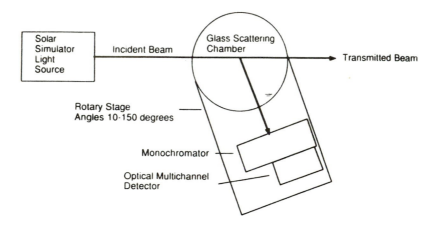

Figure 1: Schematic of dust light scattering apparatus showing the light source, dust chamber, collection optics, and monochromator/detector.

The dust particles, formed by recondensation of the vapor from the target, grew by grain-molecule condensation and grain-grain coagulation leading to 100–500 Å particles aggregated into tangled chains.

The cloud of particles, suspended by the gas in the chamber, was stable over several hours. The mass of particles settling to the bottom of the chamber was monitored using a California Measurements Quartz Crustal Microbalance. Dust column densities in the experiments were less than 1 microgram/cm^2. The cloud of particles was illuminated using a 500 W Xe light source, focused to a 10 mm diameter beam at the center of the dust cloud. The scattered light was collected by an optical system with a 6 degree acceptance angle. A polarizing beamsplitter in the collection optics allowed measuring the intensity of scattered light polarized parallel or perpendicular to the scattering plane. The scattered light was focused into the slit of a fixed monochromator with a 10 nm bandwidth and onto a Princeton Applied Research Optical Multichannel Detector/Analyzer. The detector consisted of an unamplified Reticon Detector with 1024 channels covering the spectral range of 200 to 1200 nm. Only the spectral range 400–700 nm were used for data reduction, due to the limitations in the spectral range of the polarizing beamsplitter and order overlap in the monochromator. The intensity of scattered light was measured at 10 degree increments from 10 degrees to 150 degrees scattering angle. A large dynamic range was achieved by varying the exposure time of the detector as a function of angle under computer control. The intensity of the source was measured using calibrated neutral density filters at zero degrees scattering angle.

The data was reduced by subtracting the detector dark current from the spectrum at each scattering angle using the exposure time at that angle. A correction for diffuse scattered light in the monochromator was also subtracted from each spectrum. The source spectrum was corrected to the true intensity of light incident on the cloud by correcting for the dark current and scattererd light and also multiplying by the trasmission curves for the neutral density filters used to measure the source spectrum. The corrected spectrum at each angle was then divided by the corrected source spectrum to yield the spectrum of apparent scattered intensity. To correct for possible variations in the source intensity and number of particles in the beam, each spectrum was then normalized to the center of the V band (550 nm). The resulting spectra described a color surface in color, wavelength, and angle space that is characteristic of the dust particles in the cloud.

3   Results

The scattering data for the silicate and carbon dust are shown in figures 2 and 3, respectively. The data are shown as "color surfaces", calculated as discussed above. The value of the normalized scattering intensity at 550 nm is equal to one at each angle. The data shown are for polarized scattered light perpendicular to the scattering plane with the source unpolarized.

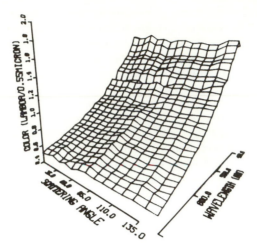

Figure 2: Measured scattering color surface of silicate dust from laser vaporized Olivine. Scattered light polarized perpendicular to the scattering plane with source unpolarized.

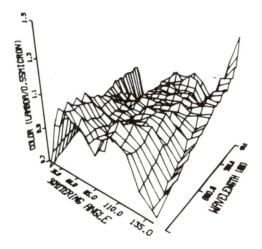

Figure 3: Measured scattering color surface of carbon dust from laser vaporized graphite. Scattered light polarized perpendicular to the scattering plane with source unpolarized.

4   Discussion

The scattering color surfaces of the silicate and carbon dust are qualitatively different. The silicate scattering data for several clouds was reproducible within 10 %. The silicate scattering surface shows a monotone increase in scattering cross section at shorter wavelengths independent of the scattering angle. The carbon color scattering surface was, however, quite variable with both the shape of the color surface and the color values varying on the order of 50% between various dust clouds. The carbon scattering surface is irregular, without the increase in scattering toward shorter wavelengths seen in the silicate data.

In order to understand the differences between the carbon and silicate scattering, scattering surfaces were calculated for various size distributions of spheres, using Mie calculations, for both materials. The optical constants of Draine (1985) for "astronomical silicate" were used to calculate the silicate scattering. The wavelength dependence of the real and imaginary parts of the index of refraction was fit to a third order polynomial for the calculations. For the carbon smoke, a wavelength independent index of refraction of 2–1i was used. To match the experimental angular resolution, the calculated scattering color was integrated over the acceptance angle of the receiving optics. Grain size distributions were included by summing the scattering for each grain size with the appropriate weighting. An exponential grain size distribution with a exponent of −3.5 was assumed in the calculations, based on the size distributions used by Mathis et al. (1977) to fit the interstellar extinction curve.

Figure 4a shows the color surface calculated for "astronomical silicate" spheres, assuming equal weighting for particles of 0.01, 0.02 and 0.03 μm radius, approximately equal to the size of the individual grains in the silicate. The resulting surface is close to that of the scattering expected for Rayleigh spheres. The shape of the calculated surface is similar to the measured scattering of the silicate dust, except the measured range of color is smaller. The smaller color contrast of the measured dust reflects the contribution of multiple scattering within aggregates.

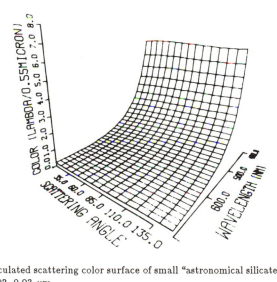

Figure 4a: Calculated scattering color surface of small "astronomical silicate" particles with radius 0.01, 0.02, 0.03 μm.

The addition of larger particles to the "astronomical silicate" size distribution should decrease the contrast of the calculated spectrum. The result of a calculation that used a Mathis size weighting with grain size weighting with grain sizes of 0.02, 0.04, 0.06, 0.08, and 0.1 μm radius is shown in figure 4b. While the contrast is decreased in the backscattering direction the shape of the surface is markedly different from the measured spectrum.

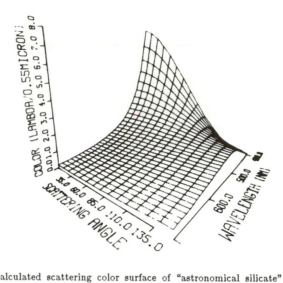

Figure 4b: Calculated scattering color surface of "astronomical silicate" particles with radius 0.02, 0.04, 0.06, 0.08, and 0.1 μm.

The addition of still larger particles to the size distribution produces an irregular surface due to scattering resonances of the grains at various wavelengths and angles that is clearly disparate from the measured color surface. The similarity of the color scattering surface calculated for small grains and the measured surface imply that the scattering from the clusters may be dominated by close particle to particle structure, rather than clusters, the diameter of which is much larger. More modeling of the scattering from nonspherical particles, including chains of grains, is needed to resolve this issue.

The calculated color surface for carbon grains of radius 0.02 and 0.03 μm is shown in figure 5. As in the silicate case, the surface is nearly the same as scattering from pure Rayleigh particles and is very different from the irregularly shaped measured color surface of carbon dust. The addition of larger grains will reduce the contrast in the calculated color surface for carbon. The variability of the measured surfaces are, however, problematic. The variability may reflect a strong sensitivity of the measured color surface to the structure of the grains in the aggregates. Since the optical depth is very low (<1 % absorption), the measured scattering must be a sum of scattering from individual clusters, and include very little cluster to cluster multiple scattering. Intracluster scattering is a much more likely explanation for the variability of the measured color surface. Due to the high absorption of carbon, multiple scattering within the clusters could give rise to a relatively flat, irregular scattering color surface. More modeling calculations are required to test this hypothesis.

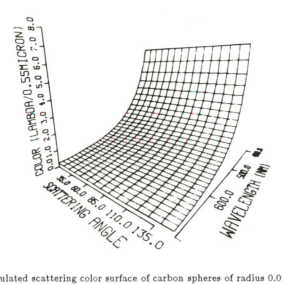

Figure 5: Calculated scattering color surface of carbon spheres of radius 0.02 and 0.03 μm.

5 Conclusions

The color scattering surfaces of silicate and carbon dust have been measured and compared with simple calculations of the scattering using various size distributions of spheres. The measured color surfaces for silicate and carbon dust are clearly different. The results may have observational implications for diagnosing the properties of astronomical dusts. The gross feature of the measured color surfaces is that a material with low optical absorption in the visible shows a strong increase in scattering cross section at shorter wavelengths, while a material with high absorption exhibits a relatively flat, irregular spectrum. Observations of scattering as a function of wavelength of optically thin dust regions may allow diagnosis of the optical properties of the grains. In regions where silicate dust is present, for example, it may be possible to infer the amount of absorbing material in the silicate, since the silicate dust is a very poor absorber in the visible.

References

Donn, B.: 1987, Proceedings of the 18th Lunar and Planetary Science Conference, p.243.

Draine, B.: 1985, Ap. J. Suppl. **57**, 587.

Giese, R.H., et al.: 1978, Astron. Astrophys. **65**, 265.

Giese, R.H., et al.: 1986, 20th ESLAB Symposium on the Exploration of Halley's Comet, V. 2, p.53.

Jura, M.: 1980, Ap. J. **235**, #1, 63.

Lattimer, J.M.: 1982, Formation of Planetary Systems, A. Brachic (Ed.), Cepadues-Editions, Toulouse, France, p.192.

Mathis et al.: 1977, Ap. J. **217**, 425.

Meakin, P., Fractals: 1986, in Physics, L. Pietronero, E. Tosatti (Eds.), North Holland, New York, p.205.

Zolensky, M.E.: 1987, Science **237**, 1466.

LABORATORY MEASUREMENTS OF LIGHT SCATTERING BY DUST PARTICLES

P. Bliek[1,2], **P. Lamy**[2]
[1] *Université de Provence, Marseille, France*
[2] *Laboratoire d'Astronomie Spatiale, Marseille, France*

1 Experimental principle and set-up

1.1 Description

The principle retained for our experimental investigation consists in generating a continuously flowing aerosol, a section of which is illuminated in order to study its scattering properties over the 0–180° interval of scattering angles. It can therefore be considered as a nephelometer whose key part is a fluidized-bed generator which receives a mixture of the dust to be studied (in the form of powder) and of "large" glass spheres having diameters of 100 to 200 μm. An air flow forces the mixture to "boil", the collisions between the glass spheres desagglomerate the powder and the liberated dust particles are preasported by the air flow. The "boiling" chamber is fed by an endless screw which continuously enriche the mixture so as to compensate for the loss of transported particles. It can be shown that the operation of the generator is governed by a very simple equation which relates the ponderal concentrations of the mixture and the flow rates to the concentration of the aerosol. After a transitory phase of approximately 10 nm, the stationary regime given by the equation is reached and was experimentally shown to be remarkably stable. A nozzle concentrates the aerosol in a column having a diameter of 5 mm; it is then allowed to flow freely over a lenght of 20 mm before being taken over by a suction device. The free part is of course used for the optical measurements. It is illuminated by a collimated beam from a quartz iodine lamp combined with narrow-band filters. The optical part of the detection system is based on classic principles of photometry with two doublets and a field diaphragm. A rotating polaroid polarizer is mounted at the entrance pupil. The scattered light is measured by a photomultiplier tube working in the analog mode. Two identical optical detection systems have been set up, one fixed to provide a reference so as to cancel any variation in the aerosol or in the illuminating source and one mounted on a rotating table moved by a stepping motor (Figure 1). The whole experiment is under computer control. For a given scattering angle, we take the average of typically 3 measurements, each one having been previously divided by the corresponding reference signal. Measurements are repeated for the set of selected filters (see below). Complementary measurements are performed so as to

253

E. Bussoletti et al. (eds.), Experiments on Cosmic Dust Analogues, 253–258.

normalize the incident light intensity for the various filters and for the two directions of polarization.

Figure 1.

1.2 Characteristics

The size range of dust particles allowed by the fluidized-bed generator depends upon the density of the dust but is typically 1 to 40 μm. The scattering volume has a diameter and an height of 5 mm and contains approximately 100 to 1000 particles; a 250 W quartz iodine lamp is used as illuminating source with an optical system giving a collimated beam having an internal angular dispersion of $\pm 2°$ and narrow band filters centered at 447, 550, 645, 706, 744 and 824 nm (typical bandpass $\sim$ 45 nm). The two detections systems have an angular aperture of 2° and are equipped with Hamamatsu R 378 PMT whose sensitivity extends over the interval 125–850 nm. Finally, the mechanical set-up has been optimized to allow a range of scattering angles as large as possible, 8 to 168°.

1.3 Main advantages

The presence of a large number of grains in the scattering volume directly average the measurements over the various parameters characterizing them (e.g., roughness). The flow being turbulent, the average further extends over the randow orientations of the grains. The scattering light is aslo sufficiently intense to allow spectral measurements in a wide spectral domain with a source of reasonable power. The system has the capability to directly measure the volume scattering function for a dust population having a selected size distribution. This function can be exactly determined independently using for instance a Coulter counter.

2   Results

The results presented here been obtained with commercially available powders of simple chemical composition and whose size distribution extends typically over a few microns.

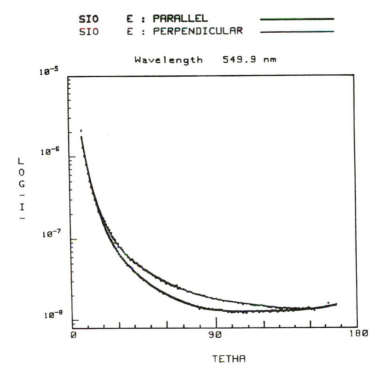

Figure 2: Scattering functions versus scattering angle for the two directions of polarization of silicon monoxide at $\lambda = 550$ nm. The curves show experimental data points and polynomial best fit representing the data.

Figure 2 display the scattering functions for the two directions of polarization of silicon monoxide SiO at $\lambda = 550$ nm. The experimental data points are given to appreciate the error in the measurements as well as polynomical best fits representing the data. The total scattering functions of SiO, SiO_2 and MgO powders at $\lambda = 550$ nm are regrouped in figure 3.

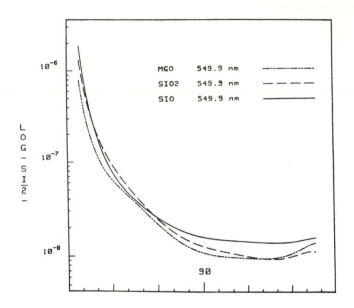

Figure 3: Total scattering functions of SiO, SiO₂, and MgO powders versus scattering angle at 550 nm.

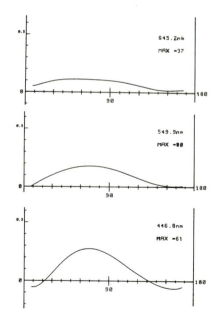

Figure 4: Polarization versus scattering angle for SiO at 447, 550 and 645 nm.

The polarization for the SiO sample powder at 3 wavelengths, $\lambda = 447$, 550 and 645 nm are shown in figure 4. These curves are polynomial fits to the data points. Figure 5 shows the polarization at the same 3 wavelengths far SiC.

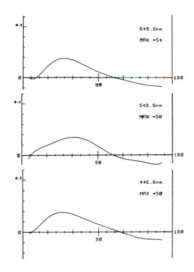

Figure 5: Polarization versus scattering angle for SiC at 447, 550 and 645 nm.

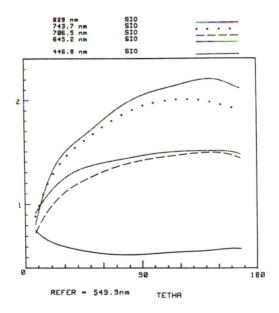

Figure 6: Color effect for SiO relative volume scattering functions at 447, 645, 706, 744 and 829 nm normalized to that at 550 nm versus scattering angle.

Another potential of these measurements is to study the color of the scattered light as function of the scattering angle. To illustrate this, the total scattering function at different wavelength is divided by that at $\lambda = 550$ nm used as a reference. The result for SiO is presented in figure 6 and indicates in fact an important color effect as a function of scattering angle.

MOLECULAR DIFFUSIONS IN ICES - IMPLICATIONS FOR THE COMPOSITION OF INTERSTELLAR GRAIN MANTLES AND COMET NUCLEI

B. Schmitt, R.J.A. Grim, J.M. Greenberg
Laboratory Astrophysics, Huygens Laboratory, Leiden University, Leiden, The Netherlands

1 Introduction

1.1 Composition of interstellar and cometary ices

It is known that the ice composition of interstellar grains varies from one molecular cloud to another. These differences already start with different cloud formation conditions, such as temperature, gas phase density, gas phase abundances and UV photoprocessing rates [d'Hendecourt et al., 1985 and 1986].

The shape of the 3 μm ice band can be regarded as a good indicator for the thermal evolution of interstellar ices [Van de Bult et al., 1985]. Furthermore, the CO/H_2O ratio in ice mantles may also be indicative for the thermal history. For instance, Grim and Greenberg (1987) summarizing recent observational data have found that the CO/H_2O ratio in interstellar grains varies between 0 and 26%. Unfortunately it is extremely difficult to deduce the distribution of grain composition in molecular clouds since the observed ratios are an average along the line of sight to the object. A direct investigation of the cometary ice composition is not possible yet. So far, all available data are inferred from gas phase production rates. For CO and CO_2 the information is limited to a few comets. CO production rates relative to H_2O range between less than 1% (comet Enke) to about 20% (comet West and Halley) and perhaps much more (comet Kohoutek) [Feldman and Brune, 1976; Opal and Carruthers, 1977; Festou, 1984; Feldman, 1986; Woods et al., 1986]. For CO_2, a value of 1.5% has been measured for Halley by the IKS experiment on board of the Vega spacecraft [Combes et al., 1986] but model calculations lead to upper limits as high as 25% for comets Bradfield and West [Feldman and Brune, 1976; Festou, 1984]. In addition CO_2 productions also seem to vary with time rapidly [Cochran et al., 1980; Feldman et al., 1986]. Other volatile parent molecules have been observed in Halley too, but only upper limits to the production rates could be deduced. The limits reported are 2% for CH_4 and for NH_3 [Allen et al., 1987], 1% for H_2CO [Combes et al., 1986] and 10% for N_2 [Eberhardt et al., 1986].

The large variability in interstellar grain composition and gas productions of comets raise questions on the origins of these differences: are they the result of formation conditions or evolutionary modifications? In either case there are three key parameters: the molecular composition, the temperature of the ice and the time evolution at any

259

E. Bussoletti et al. (eds.), Experiments on Cosmic Dust Analogues, 259–269.
© *1988 by Kluwer Academic Publishers.*

temperature. Besides the many parameters that control the grain mantle formation, once formed, the ice is also subjected to further processing, such as continued UV irradiation and thermal annealing. One may expect at a first thought that once a grain temperature is well above 17 K (the sublimation temperature of CO in the interstellar medium; Leger, 1983) the grains become depleted in CO. On the other hand from laboratory experiments it is now known that volatile molecules can remain trapped in ice well above their sublimation temperature [Bar-Nun et al., 1987; this proc]. Thus, one cannot regard the grain mantle composition as a simple mixture of volatile components, but one has to be aware of, and fully understand, the different trapping mechanisms in ice to unravel the history of a grain. For the comet formation the critical parameters are the time and the maximum temperature of the aggregation process. These two factors are intimately connected with the origin and the place of formation of comets. Secondly, the evolution experienced by a cometary nucleus since its entrance into the solar system depends on its thermal history and especially on the maximum temperature reached by each layer of the nucleus [Schmitt and Klinger, 1987]. If a temperature profile exists in the cometary nucleus, we may expect that at least a partial evaporation of the more volatile molecules present in comets occurs beneath the nuclear surface.

1.2   Ice mixture models

At present at least four different physical models have been proposed to describe the temperature dependence of the evolution of ice mixtures (see table 1).

Table 1: Comparison between our results and the different ice mixture models. The upper limits of the CO/H_2O ratio are the maximum values which have been experimentally measured at the corresponding temperatures. Lower values have been obtained depending on the initial amount of CO co-condensed with water.

MODEL	STRUCTURE	CO EVAPORATION controlled by	CO/H$_2$O RATIO at 80 K	170 K
DELSEMME HOUPIS	clathrate + pure CO	volatility of the clathrate	< 0.17	< 0.17
YAMAMOTO	pure H$_2$O ice pure CO ice	volatility of pure CO	0	0
BAR-NUN et al.	H$_2$O ice + clathrate + gas pockets	$\rightarrow$ 40 K: evap. CO 137 K: $I_a \rightarrow I_c$ 160 K: $I_c \rightarrow I_h$ 180 K: clathrate evaporation	< 3.2	< 0.45
OUR results	amorphous CO:H$_2$O ice mixture	$\rightarrow$ 45 K: evap. CO $\rightarrow$ 120 K: diffusion in amorphous ice 125 K: diff. in I$_c$	$\leq$ 0.15	0

First Delsemme and Miller (1970) proposed that ice mixtures are mainly composed of clathrate hydrates and that the evaporation of the gas trapped in the clathrate cages is controlled by the volatility of the H_2O lattice. This model implies a ratio of the volatile molecules, such as CO, CH_4 and N_2, relative to water, lower than 17% at all temperatures larger than 20 K. Secondly, a model considering macroscopic mixtures of pure solid ices (H_2O, CO, CO_2, ...) only has been developed by Yamamoto (1983–85). In this model, each ice component evaporates according to its vapor pressure in the pure solid phase. A model combining the clathrate model and the pure ice model has been developed afterwards by Mendis et al.(1985).

A completely different model is presented by Bar-Nun et al. (1985, 1987), who experimentally deduced that an ice mixture above 40 K is composed of pure H_2O ice, clathrate hydrate and gas pockets. Seven different temperature regimes and mechanisms of gas release have been argued. After the cubic ice crystallization, reported at 137 K, the amount of internally trapped gas can still be 1.25 times the amount of water ice. This ratio decreases to 0.45 after hexagonal ice recrystallization (160 K) and the final evaporation is reported to occur with a gas/H_2O ratio that can reach values as high as 1.

Finally, on the basis of experimental results for amorphous ices and clathrate hydrates, a tentative sketch of all the physical evolutions that a comet nucleus could have experienced has been drawn by Schmitt and Klinger (1987). One of the main conclusions reached is that CO clathrate does not probably exist at the surface of comets as a result of the very high gas pressure needed for its stabilization above 100 K and of the high decomposition rates of the clathrate structures.

All models lead to very different conclusions about the chemical differentiation process within the nucleus and the connections between the composition of the gas released into the coma, the initial comet composition and the formation conditions. In order to understand the basic behaviour of water rich ice mixtures we are currently performing experiments directly following the evolution of $CO:H_2O$, $CO_2:H_2O$ and $NH_3:H_2O$ ice mixtures versus time and temperature. The experiments are designed to measure the composition of the solid phase by means of infrared spectroscopy.

2 Experimental procedures

After preparation in a vacuum glass line, a gas mixture ($CO:H_2O$, $CO_2:H_2O$ or $NH_3:H_2O$) with the desired composition is deposited onto a cold (10 K) sample substrate (aluminum). The thickness and the condensation rate of the sample is evaluated by measuring the interference fringes with a He–Ne laser. Thicknesses range between 0.5 and 5 μm and condensation rates are always of the order of 5 μm per hour. The pressure in the cell (constantly pumped) is continuously recorded with an ionization gauge during the experiment. After deposition an infrared spectrum (4000–400 cm^{-1}) of the initial mixture is recorded at 10 K using a Fourier transform spectrometer with a resolution of 4 cm^{-1}.

The evolution of the ice mixtures is studied by performing several successive annealing steps at desired temperatures T_a and times t_a. Depending on the outcome (slow or rapid evolution), infrared spectra of the annealed samples were recorded at temperature T_a or after recooling to 10 K. Upon slow evolution of the ice at T_a, the sample composition did not change significantly during the 10 minutes scanning of an infrared spectrum. In such cases the IR spectra are taken from time to time during the annealing step at temperature T_a. If rapid sublimation occurs the sample is recooled to 10 K for recording the infrared spectrum, then it is heated again at T_a (or at $T > T_a$) for a next annealing step, and so on.

After each annealing, the number of molecules present in the sample is calculated by integrating over the vibrational bands of:

$$- H_2O \quad : O-H \text{ stretch } (3700-2800 \text{ cm}^{-1})$$
$$- CO \quad : C=O \text{ stretch } (2160-2120 \text{ cm}^{-1})$$
$$- CO_2 \quad : C-O \text{ stretch } (2380-2300 \text{ cm}^{-1})$$
$$- NH_3 \quad : \text{umbrella mode } (1200-1050 \text{ cm}^{-1})$$

The composition is expressed (in percent) as the ratios relative to water: CO/H_2O, CO_2/H_2O and NH_3/H_2O.

3   Experimental results and preliminary interpretation

3.1   Evolution of CO:H₂O ice mixtures

The main experimental results obtained on the evolution of CO:H₂O ice mixtures with increasing temperature are (see figures 1 to 4):

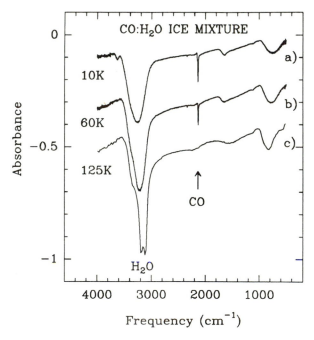

Figure 1: Successive infrared spectra at 10 K of a CO:H₂O = 1:3 mixture. a) initial sample deposited at 10 K. b) after several annealings from 20 to 60 K. c) after 3 hours at 125 K: CO has completely disappeared.

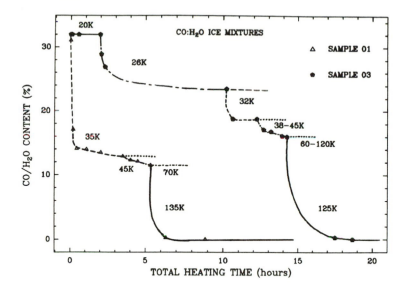

Figure 2: Evolution of the CO content (%) versus total heating time for two samples (initial CO/H₂O ratio = 32%). The temperatures noted along the curves are the successive annealing temperatures.

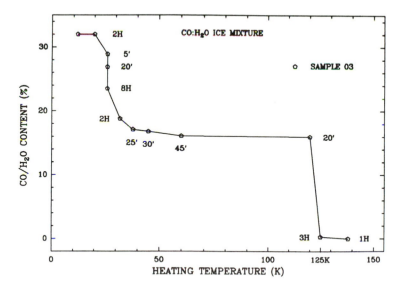

Figure 3: Evolution of the CO content (%) versus temperature for a CO:H₂O = 1:3 mixture. The time noted at each point is the maximum annealing time at the corresponding temperature (several points are shown at 26 K).

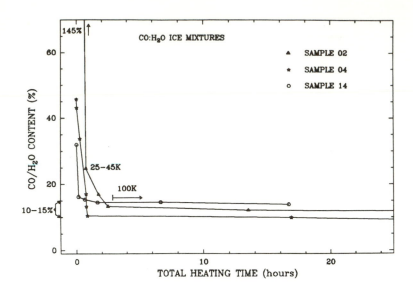

Figure 4: Evolution of the CO (%) versus total heating time for three samples with different initial CO/H₂O ratios (32%, 46% and 145%). The samples were first heated in the 25–45 K range where the fast decrease occurs until 10–15%. The temperature was then fixed at 100 K for about one day.

1. No CO evaporation occurs as long as the temperature is set below 25 K (Cell pressure $= 8 \cdot 10^{-8}$ mbar).

2. A fast decrease of the CO fraction trapped in the H_2O ice occurs between about 25 K and 45 K. In this range, CO diffusion strongly depends on both the CO fraction and the temperature, but the final CO/H₂O ratio at 45 K is always in the order of 10 to 15% independently of the initial CO fraction (Figure 4).

3. A slow decrease of the CO content (1% per day decrease of the CO/H₂O ratio, figure 4) is observed upon heating from 50 K to temperatures ranging between 95 K and 120 K. However, reliable measurements of slow CO diffusion on long time scales (several days) in the 50–120 K range are difficult to perform as a result of water vapour leaking into the system and forming an ice layer on top of the sample.

4. As soon as the temperature exceeds 125 K a fast and complete disappearance of CO occurs in the solid. This CO evaporation seems to be related to the H_2O ice crystallization.

From these results a tentative sketch of the possible physical mechanisms which may be responsibles of these different ranges of CO disappearance can be drawn (Figure 5):

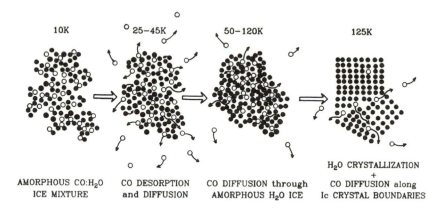

Figure 5: Artistic sketch for the possible physical mechanisms responsible for the CO:H_2O ice mixture evolution.

first, the fast decrease of the CO content occuring between 25 K and 45 K is probably the result of CO desorption and easy CO diffusion through a loose H_2O ice matrix. At the same time short scale reorganizations of the H_2O matrix probably orrurs. The final CO/H_2O ratio of 0.10 to 0.15 points out that this fast diffusion occurs until all the remaining CO molecules are fairly well isolated and trapped in an amorphous H_2O matrix. The further decrease occuring above 45 K and below the crystallization temperature of water ice (125–135 K), can be ascribed to a slow CO diffusion through amorphous H_2O ice. Finally, the rapid depletion of CO during cubic ice crystallization could be explained by a faster CO diffusion along the boundaries between cubic ice microcrystals.

3.2 Evolution of CO_2:H_2O ice mixtures

Only one experiment with a CO_2/H_2O ice mixture (initial CO_2/H_2O ratio of 22%) has been performed so far. This first experiment was mainly done in order to measure the CO_2 diffusion rate during the water ice crystallization. After a heating step at 60 K, where only a very small CO_2 evaporation has been observed, the sample was heated at 116 K. During this annealing step and after further heating at 127 K a very fast decrease of the CO_2/H_2O ratio has been observed until a value of 8% was reached. Then, with the temperature set at 127 K, the CO_2 content decreased much more slowly and nearly linearly with time (Figure 6).

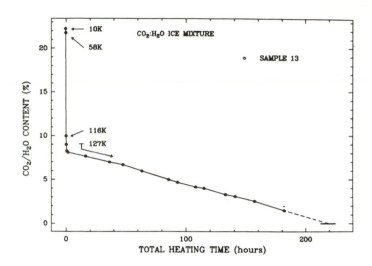

Figure 6: Evolution of CO_2 content (%) versus total heating time (initial CO_2/H_2O ratio = 22%). The sample was first heated 10 minutes at 58 K and 20 minutes at 116 K before studying the CO_2 diffusion during water ice crystallization at 127 K.

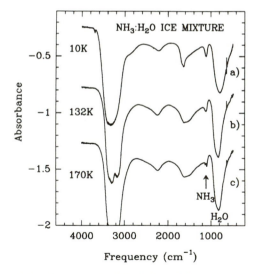

Figure 7: Successive infrared spectra of a $NH_3:H_2O = 1:7$ ice mixture recorded at T_a. a) initial sample deposited at 10 K. b) after 22 hours at 132 K. c) after 3 hours at 170 K.

The time needed to completely deplete the CO_2 in an ice sample about 5 μm thick is estimated to be 10 days at 127 K. The lower CO_2 diffusion rate, compared to CO_2, is due to both a lower volatility and a larger size of the CO_2 molecule.

3.3 Evolution of NH₃:H₂O ice mixtures

From several experiments with NH_3/H_2O ratios ranging from 5 to 25 % we observe that water-ammonia mixtures form crystalline ammonia hydrate at about 140 K. The peak location (1130 and 1100 cm⁻1) of the double structure of the NH_3 umbrella mode is in good agreement with that of pure ammonia monohydrate [Bertie and Morrison, 1980]. The crystallization temperature found for this phase is much higher than that found for pure NH_3 ice (about 70 K). Furthermore NH_3 molecules are not able to diffuse through water ice at temperature as high as 170 K and leave the substrate only when H_2O molecules evaporate (Figure 7). The strong hydrogen binding between NH_3 and H_2O molecules is certainly the main reason for this effect.

4 Comparison between ice mixture models

The main differences between our results, the other experimental results and the theoretical models published so far are well illustrated by the comparison of the CO/H_2O ratios at some astrophysically relevant temperatures: 80 K (evaporation temperature of H_2O ice in the interstellar medium) and 170 K (surface temperature of a comet nucleus at 2–3 A.U.)(Table 1). Our results disagree with the clathrate model of Delsemme and Miller (1970) because they assume that clathrate structures are stable and are able to trap CO molecules at high temperature and under vacuum conditions. Further the discrepancy between Yamamoto's model (1985) and our results below 120 K is not yet completely resolved on account of the lack of experimental results on CO diffusion at these low temperatures.

The first experimental results reported by Bar-Nun et al. (1985, 1987) differ substantially from our results. The main discrepancy is the amount of volatile gas (argon or CO) which can be trapped in the water ice at high temperature (T>80 K). Differences in condensation conditions (temperature, sample thickness, ...) and in the annealing procedure (continuous heating or constant temperature) should perhaps explain some of the differences. However, it seems difficult to understand how an equivalent volume of about 2 cm³ of solid argon can be trapped in one cubic centimeter of H_2O ice at temperatures as high as 140 K without allowing bulk gas pressure to build up inside the sample!

5 Implications for the evolution of the volatile cometary matter

We conclude from our results that all parts of comets that have experienced a temperature higher than about 35 K will have a CO/H_2O ratio at most equal to 15% and that all parts that have experienced a temperature higher than 125 K should have completely lost their CO and CO_2 molecules. Most likely, this conclusion is also true for other volatile, not hydrate forming, molecules present in comets such as CH_4, N_2 or H_2CO. On the other hand, NH_3 and H_2O molecules are expected to evaporate simultaneously at the comet surface. Consequently their ratio should be that of the interior of the nucleus. However, due to the complexity of the chemical differentiation processes of cometary nuclei, the large differences in CO and CO_2 productions observed from one comet to another cannot directly be assigned to differences in initial composition. An elaborated modelling including the thermal history and the gas diffusion through the comet porosity is necessary in order to understand how the gas production can be related to the initial composition. No matter what evolution has taken place, it is possible to assert that comets containing either CO or CO_2 have certainly been formed at temperatures lower than 125 K. This result is the first experimental proof of a maximum formation temperature for comets based on both CO and CO_2. The maximum comet formation temperature derived from S_2 as a parent molecule is consistent with this [Grim & Greenberg, 1987].

6   Implications for the composition of grain mantles

From the trapping of CO in H_2O ice at temperatures well above the sublimation temperature of pure CO we can expect that icy interstellar grain mantles may retain CO molecules at temperatures higher than 25 K. However, the presence of CO in grains cannot simply be used to derive the cloud conditions. Expecially, it is extremely difficult to draw any specific conclusions about the thermal histories of grains only from their measured CO/H_2O ratios. A more sophisticated grain modelling is needed to account for all phenomena.

7   Conclusions

CO diffusion through amorphous ice is a critical problem to be solved because the presence of CO in hot interstellar grains (T>25 K) and in comet nuclei directly depends on the capability of ice to retain CO molecules at a temperature higher than the sublimation temperature of pure CO. Accurate experimental determinations of CO depletion times of ice mantles in the range 20–80 K would provide useful basic input data for modelling the evolutions of grain composition linked with their thermal history. The upper limit for the formation temperature of a comet that we can deduce from the presence of CO in the nucleus corresponds to the temperature at which a complete CO depletion can occur over time scales in the order of comet aggregation times (10^6 years). An upper limit lower than 125 K for the formation temperature of a comet can possibly be deduced from CO diffusion studies below 125 K. For such a study ultra-high vacuum systems ($P< 10^{-8}$ mbar) are necessary in order to reduce the growing of the additional ice layer leading to a slowing down of the CO diffusion.

In addition a study of the influences of the condensation temperature and sample thickness is in progress. We hope that these experiments will help us to understand the origins of the discrepancies between the results reported by Bar-Nun and al. (1985, 1987) and ours.

References

Allen, M., Delitsky, M., Huntress, W., et al.: 1987, Astron. Astrophys., (in press).

Bar-Nun, A., Herman, G., Laufer D.: 1985, Icarus, **63**, 317.

Bar-Nun, A., Dror, J., Kochavi, E., Laufer, D.: 1987, Physical Review B**35**, 2427.

Bertie, J.E., Morrison, M.M.: 1980, J. Chem. Phys., **73**, 4832.

Cochran, A.L., Barker, E.S., Cochran, W.D.: 1980, Astron. J., **85**, 474.

Combes, M., Moroz, V., Crifo, J.F., Bibring, J.P. et al.: 1986, in "Proc. 20th ESLAB Symp. on the Exploration of Halley's Comet", ESA SP-250, 353.

Delsemme, A.H., Miller, D,C.: 1970, Planet. Space Sci., **18**, 717.

Feldman, P.D., Brune, W.H.: 1976, Astrophys. J., **209**, L45.

Feldman, P.D.: 1986, in "Proc. Asteroids, Comets, Meteors II", Uppsala 3–6 June 1985, eds. Lagerkvist et al., Uppsala Universiteit, 263.

Feldman, P.D., A'Hearn, M.F., Festou, M.C., McFadden, L.A., Weaver, H.A., Woods, T.N.: 1986, Nature, **324**, 433.

Festou, M.C.: 1984, Adv. Space Res., **4**, 165.

Grim, R.J.A., Greenberg, J.M.: 1987, Astron. Astrophys., **181**, 155.

Heberhardt, P., Krankowsky, D., Schulte, W. et al.: 1986, in "Proc. 20th ESLAB Symp. on the Exploration of Halley's Comet", ESA SP-250, 383.

d'Hendecourt, L.B., Allamandola, L.J., Greenberg, J.M.: 1985, Astron. Astrophys., **152**, 130.

d'Hendecourt, L.B., Allamandola, L.J., Grim, R.J.A., Greenberg, J.M.: 1986, Astron. Astrophys., **158**, 119.

Houpis, H.L.F., Ip, W.H., Mendis, D.A.: 1985, Astrophys. J., **295**, 654.

Léger, A.: 1983, Astron. Astrophys., **123**, 271.

Opal, C.B., Carruthers, G.R.: 1977, Astrophys. J., **211**, 294.

Schmitt, B., Klinger, J.: 1987, in Proc. "On the Diversity and Similarity of Comets" ESA SP-278 (in press).

Van de Bult, C.E.P.M., Greenberg, J.M., Whittet, D.C.B.: 1985, Mon. Not. R. astr. Soc., **214**, 289.

Woods, T.N., Feldman, P.D., Dymond, K.F., Sahnow, D.J.: 1986, Nature, **324**, 436.

Yamamoto, T., Nakagawa, N., Fukui, Y.: 1983, Astron. Astrophys. **122**, 171.

Yamamoto, T.: 1985, Astron. Astrophys., **142**, 31.

FORMATION OF PARTICULATES BY ION BOMBARDMENT OF CRYOGENIC ICE MIXTURES

T.J. Wdowiak, L.D. Brasher, E.L. Robinson, G.C. Flickinger, H.R. Setze
Physics Department University of Alabama, Birmingham, U.S.A.

1 Introduction

The irradiation of ices with charged particles is a technique utilized to study in the laboratory, molecular systems that may be of relevance to astrophysical questions (Moore and Donn 1982, Brown et al. 1982, Pirronello et al. 1982, Haring et al. 1983, Foti et al. 1984, de Vries et al. 1984, Johnson et al. 1984, and Lanzerotti et al. 1987). We have adapted a Van de Graaff accelerator capable of accelerating various ions under a potential of up to 400 kV, for such experiments by constructing a target chamber that incorporates a closed cycle cryostat, quadrupole mass spectrometer, and optical windows for spectroscopy and observation (Wdowiak et al. 1985). Utilizing the model for the ς Ophiuchi cloud calculated by Black and Dalgarno (1977) and an observational determination of the interstellar argon abundance (Simpson et al. 1986), we have utilized a mixture of gases (Table 1) that reflects cosmic molecular abundances in experiments where 200 Torr liters of the mixture is frozen as a $\sim$750 μm matrix on a 2.5 cm diameter sapphire disk at 10 K.

Table 1.

SPECIES	MELTING POINT	PARTS
Ar	84 K	170
CO	74 K	170
H_2O^*	273 K	25
N_2	63 K	20
CH_4^{**}	91 K	15

* for OH; ** for CH

The ice matrix is then iradiated with H_2^+ ions accelerated at 350 kV (175 keV/proton). The 2-3 μ ampere, 6 mm diameter beam irradiates the target for a time typically on the order of one hour. Species eroded from the target travel directly into the ionizer of a

271

E. Bussoletti et al. (eds.), Experiments on Cosmic Dust Analogues, 271–279.

quadrupole mass spectrometer. Figure 1, shows the irradiated dark area on the matrix prior to warm up.

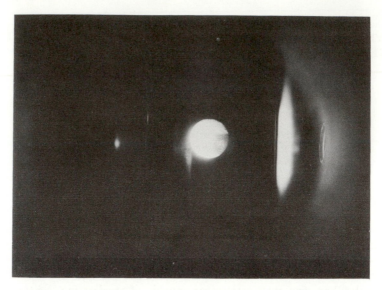

Figure 1: A matrix composed in parts of 170 Ar/170 CO/25 H_2O/20 N_2/15 CH_4 at 10 K after H_2^+ irradiation at 175 keV/proton showing a dark area on the right resulting from irradiation.

Figure 2: Optical microscope photograph of a typical residue after warmup under vacuum. The field of view is approximately 275 μm.

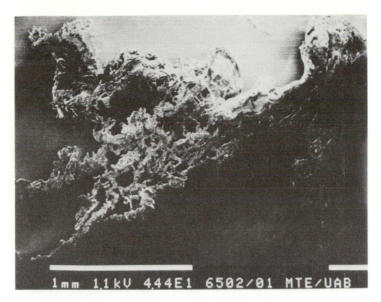

Figure 3a: SEM photograph of a residue prepared from a gas mixture in which Ar was absent. The image was made with 1.1 keV electrons and the white bars at the bottom indicate the scale.

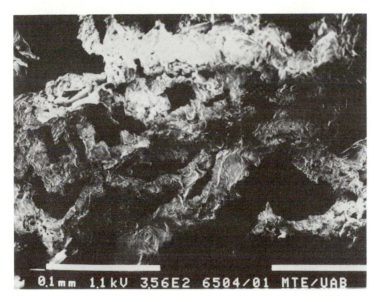

Figure 3b: For caption see figure 3a.

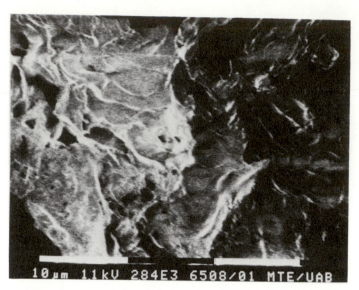

Figure 3c: For caption see figure 3a.

Figure 3d: For caption see figure 3a.

A starcell ion pump maintains chamber pressures at $P<10^{-6}$ Torr during all phases of the experiment except during rapid sublimation of the ices. The accelerator is pumped with a LN_2 trapped oil diffusion pump followed by a large ion pump on the beam line to reduce contaminents. The effectiveness of reduction of contamination from vacuum pump oil was demonstrated in a control experiment by irradiating a pure argon matrix and finding no residue on the sapphire upon warmup.

2   Experimental results

A residue was noted visually upon sublimation under vacuum in the very first accelerator run attempted. Examination with an optical microscope revealed particles and filaments of brown matter (Figure 2). A complex three dimensional morphology was seen at all magnifications used. The material was extremely fragile as exhibited by an accidental blowing away of a portion of one sample by the experimenters breath during microscopy. Scanning electron microscopy (SEM) was accomplished on an uncoated specimen by utilizing a 1.1 keV electron beam without significant charging of the sample or sapphire substrate (Figure 3). That particular sample was prepared with a gas mixture that did not include argon as a constituent in an experiment to determine the effect of its presence on particulate formation. The result was that argon apparently does not determine whether or not a particulate residue is formed.

Material from several runs was incorporated into a 100 mg KBr/0.5 cm^2 pellet, and the 2 μm to 25 μm infrared absorption spectrum was obtained using a Mattson Polaris FTIR spectrometer.

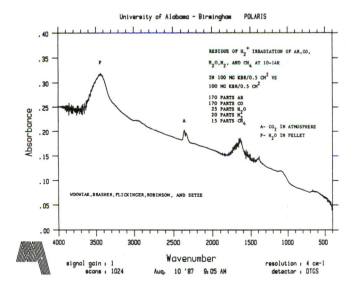

Figure 4: FTIR absorbance spectrum of a residue prepared from a frozen gas mixture in parts of 170 Ar/170 CO/25 H$_2$O/20 N$_2$/15 CH$_4$ irradiated with H$_2^+$ ions having an energy of 175 keV/proton. The material was pressed into a 100 mg KBr pellet and run against a 100 mg KBr pellet background.

The spectrum (Figure 4) exhibits aliphatic CH stretch at 2940 cm^{-1} (3.4 μm), a broad feature at 1700 cm^{-1} (5.9 μm), a strong 1640 cm^{-1} (6.1 μm) band, 1400 cm^{-1} (7.1 μm), and 1100 cm^{-1} (9.1 μm) bands.

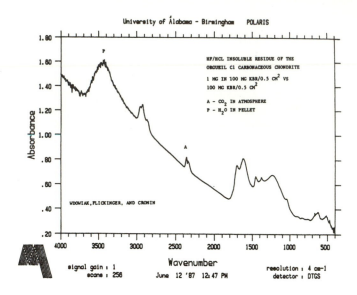

Figure 5: FTIR absorbance spectrum of the HF/HCL insoluble residue of the Orgueil C1 carbonaceous chondrite. 1 mg of material was pressed into a 100 mg KBr pellet and run against a 100 mg KBr pellet background.

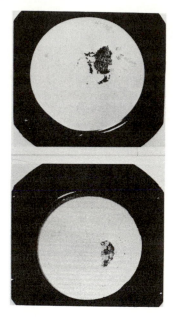

Figure 6: Typical residues on 2.5 cm diameter sapphire substrates.

In addition a broad plateau seems to exist in the 1800 cm^{-1} to 1000 cm^{-1} range, and another broad feature in the 700 cm^{-1} to 500 cm^{-1} region. Many of these features are also evident in the FTIR KBr pellet spectrum of the acid insoluble residue of the Orgueil meteorite (Figure 5) (Wdowiak et al. 1987).

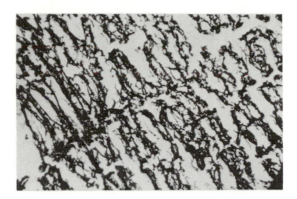

Figure 7: Large scale structure of a residue. The field of view is approximately 1 mm.

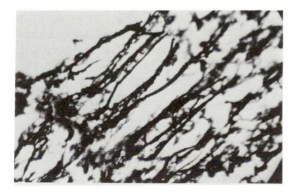

Figure 8: Filaments that exist in certain areas of the residue. The optical microscope field of view is approximately 275 μm.

3   Conclusions

The results of the experiments provoke two questions - (1) Why do particulates result from these experiments, where none have been reported in others, and (2) are there astrophysical implications? In terms of (1) it should be noted that the ion energies (175 keV/proton) involved are an order of magnitude lower than the typical energies utilized by others (see the reference list at the beginning of part 1.). Foti et al. (1984) have reported the formation of micrometer films synthesized by irradiation of CH_4 ice at 4 K with 1.5 MeV proton. The targets used had surface densities of $\approx 10^{19} - 10^{20}$ C atoms cm^{-2} while the targets of the experiments reported here have a surface density of $\sim 6 \cdot 10^{20}$ C atoms cm^{-2}. The lower energies and thicker targets may be responsible in that the ions would not penetrate deep into the ice. The filamentary nature of much of the residue could be a consequence of the ion tracks in the ice. Experiments involving 200 keV Xe ions impinging on molybdenum-nickel films at LN_2 temperature have resulted in fractal structures (Liu et al. 1987). Perhaps the complex structure of the residues of our experiments results from an initial fractal structure in the irradiated ice matrix that becomes distorted during sublimation on warmup.

Mass spectrometers onboard the Giotto and Vega space crafts revealed hydrocarbon particles composed of C, H, O, and N as significant fraction of the dust emitted by Comet Halley (Kissel et al. 1986a, Kissel et al. 1986b). Data from the Giotto heavy-ion analyzer (RPA2-PICCA) has been utilized to argue that the inner coma of Comet Halley contains polymerized formaldehyde/polyoxymethylene (Mitchell et al. 1987, Huebner 1987) and that the material may have originated by cosmic ray bombardment of pre-cometary ices. The experiments described here appear to simulate that process or a similar one involving hydrogen and helium ions accelerated during the sun's T Tauri stage.

It is planned to carry onboard the Comet Rendezvous/Asteroid Flyby (CRAF' space craft a SEM for microscopic examination of cometary particles. Because of the nonconducting nature expected for the "CHON"-like material, consideration needs to be given to the problems of imaging such material. Our experience in obtaining SEM images of what may be a laboratory analog, indicates that with proper technique "CHON" particles can be studied with the SEM without having to resort to caoating techniques that might alter the sample and complicate the sampling process on board the spacecraft.

This research was supported by NASA grant NAGW-749, and we thank Scott Brande and Scott Walck for assistance with the microscopy.

References

Black, J,H, Dalgarno, A.: 1977, Ap. J. Suppl., **34**, 405.

Brown, W.L., Lanzerotti, L.J., Johnson, R.E.: 1982a, Science, **218**, 525.

Foti, G., Calcagno, L., Sheng, K.L., Strazzulla, G.: 1984, Nature, **310**, 126.

Haring, R.A., Haring, A., Klein, F.S., Kummel, A.C., de Vries, A.E.: 1983, Nucl. Inst. and Meth., **211**, 529.

Huebner, W.F.: 1987, Science, **237**, 628.

Johnson, R.E., Lanzerotti, L.J., Brown, W.L.: 1984, Adv. Space Res., **4**, 41.

Kissel, J., Brownlee, D.E., Buchler, K., Clarck, B.C., Fechtig, H., Grun, E., Hornung, K., Igenbergs, E.B., Jessberger, E.K., Krueger, F.R., Kuczera, H., McDonnell, J.A.M., Morfill, G.M., Rahe, J., Schwehm, G.H., Sekanina, Z., Utterback, N.G., Volk, H.J., Zook, H.A.: 1986a, Nature, **321**, 336.

Kissel, J., Sagdeev, R.Z., Bertaux, J.L., Angarov, V.N., Audouze, J., Blamont, J.E., Buchler, K., Evlanov, E.N., Fechtig, H., Fomenkova, M.N., von Hoerner, H., Inogamov, N.A., Khromov, V.N., Knabe, W., Krueger, F.R., Langevin, Y., Leonas, V.B., Levasseur-Regourd, A.C., Managadze, G.G., Podkolzin, S.N., Shapiro, V.D., Tabaldyev, S.R., Zubkov, B.V.: 1986b, Nature, **321**, 280.

Lanzerotti, L.J., Brown, W.L., Marcantonio, K.J.: Ap. J., **313**, 910.

Liu, B.X., Huang, L.J., Tao, K., Shang, C.H., Li, H.D.: 1987, Phys. Rev. Let. **59**, 745.

Mitchell, D.L., Lin, R.P., Anderson, K.A., Carlson, C.W., Curtis, D.W., Korth, A., Reme, H., Sauvaud, J.A., d'Uston, C., Mendes, D.A.: 1987, Science, **237**, 626.

Moore, M.H., Donn, B.: 1982, Ap. J., **257**, L47.

Pirronello, V., Brown, W.L., Lanzerotti, L.J., Marcantonio, K.J., Simmons, E.H.: 1982, Ap. J., **262**, 636.

Simpson, J.P., Bregman, J.D., Dinerstein, H.L., Lester, D.F., Rank, D.M., Witteborn, F.C.: 1986, Bull. Am. Astron. Soc., **18**, 1022.

de Vries, A.E., Pedrys, R., Haring, R.A., Haring, A., Saris, F.W.: 1984, Nature, **311**, 39.

Wdowiak, T.J., Wills, E.L., Robinson, E.L., Bales, G.S.: 1985, Nucl. Instr. and Meth., B10/11, 735.

Wdowiak, T.J., Flickinger, G.C., Cronin, J.R.: 1987, (this volume).

SOME CONSEQUENCES OF FROZEN GAS IRRADIATION BY ENERGETIC IONS

V. Pirronello[1,2], D. Averna[3]
[1]*Dipartimento di Fisica, Università della Calabria, Cosenza, Italy*
[2]*Osservatorio Astrofisico di Catania, Catania, Italy*
[3]*Istituto di Astronomia, Università di Catania, Catania, Italy*

1 Introduction

The interaction of fast particles with icy mantles on grains in dense interstellar clouds can be relevant for the formation of some molecules (Pirronello et al., 1982; Pirronello et al., this volume). The simplest from the chemical point of view, but probably one of the most difficult to be formed H_2, is among them (Pirronello, 1987). Molecular hydrogen is of particular importance because can be considered the precursor of all the other molecules observed in these clouds (Duley and Williams, 1984) but cannot be formed in gas phase, because of the problem of getting rid of the formation energy. The problem is overcome if H_2 is formed on the surface of a grain. In this case, in fact, the grain will absorbe the energy excess.

Calculation relative to this process have been performed by many authors and among them: Hollenbach and Salpeter (1971), Watson and Salpeter (1972), Smoluchowski (1983).

2 The formation of H_2 according to Smoluchowski calculations

In order to form hydrogen molecules on grains starting with H atoms, several conditions have to be fulfilled: two H atoms have to collide with a grain and stick on its surface (i.e. be adsorbed), furthermore they need to be mobile to get the chance to encounter each other. If this happens, within their residence time (defined as the interval of time they spend on the average before evaporating) on the surface of the grain, there is a finite probability that they will recombine forming a H_2 molecule.

On a more quantitative ground one can define the average collision time t_c between a H atom and a grain of radius a as

$$t_c = \left(v_H n_H \pi a^2\right)^{-1} \tag{1}$$

where

E. Bussoletti et al. (eds.), Experiments on Cosmic Dust Analogues, 281–286.
© *1988 by Kluwer Academic Publishers.*

v_H = average velocity of a H atom
n_H = number of Ha atoms per unit volume

and the adsorption time as

$$t_{ad} = t_c/S \qquad (2)$$

where $S=$ sticking coefficient for H atoms on the grain surface. Any adsorbed atom will remain, according to a semiclassical estimate, on the surface before evaporating for an average time

$$t_r \simeq \nu^{-1} \exp(E_b/KT) \qquad (3)$$

where

ν = vibrational frequency of the adsorbed atom in the potential well
E_b = binding energy
T = grain temperature
K = Boltzmann constant

A necessary condition for the formation of H is that two atoms remain adsorbed together on the same grain for long enough to meet each other while migrating from one site to another; this implies that it must be

$$t_r > t_{ad} \qquad (4)$$

If such a requirement is fulfilled, mobility has to be retained by at least one of the adsorbed hydrogen atoms. Mobility will consist in jumping from one adsorption site to another. The time t_s required to perform one of such jumps is strongly dependent on the grain temperature and below a critical value $T \sim 30$ K thermal hopping becomes absolutely inefficient and only quantum mechanical tunneling provide enough mobility. The actual time t_{en} required for H atoms to encounter each other is a multiple of t

$$t_{en} \sim K t_j \qquad (5)$$

where $K = N^2$, if N is the number of sites and a pure a random walk applies for hydrogen motion (Hollenbach and Salpeter, 1971). Still a necessary condition is

$$t_r > t_{en} \qquad (6)$$

When conditions (4) and (6) are satisfied the two atoms, if they are mobile enough to encounter each other, have a probability p to form a H_2 molecule, which is then presumably released in the gas phase, because the 4.5 eV formation energy released to the grain, which has at 10 K a very low thermal capacity, can raise its temperature (overall, if it is tiny enough, or only locally) to let the molecule just formed evaporate.

As it is evident one of the critical points concerns the mobility of atoms on the surface of grains. The problem of mobility has to be treated quantum mechanically. In the case of the periodic structure of a crystalline solid, the wave packet describing the atom spreads quickly, assuring the required mobility and giving an almost equal probability of finding it in any suitable adsorption site. In these conditions the probability of encounter between two adsorbed H atoms becames close to unity. In the case in which the structure of the solid is amorphous, as it is for ice mantles the mobility of H atoms has been studied by Smoluchowski (1983). In this case because of the lack of periodicity in the distribution of molecules also the adsorption sites are distributed randomly and are characterized by a continuous distribution of depths. Due to this non periodic spacing, the wave packet describing the adsorbed atom becames quickly localized in one of the binding sites, as it was first argued by Anderson (1958). In order to treat the localization of the wave packet, Smoluchowski used Mott's (1969) arguments. In this notation the wave function of the adatom decreases in all directions as $\exp[-r]$ and the resulting tunneling frequency is

$$p \sim \nu \exp[-4\alpha(N_e \pi KT)^{-1/2}] \tag{7}$$

where

$$\nu = 10^{12} \div 10^{13} s^{-1}$$
$$\alpha = \text{reciprocal distance at which the amplitude of the wave function}$$
$$\text{of the atom residing on a site drops to its } e \text{ value}$$
$$N_e = \text{number of sites}$$

with values of N_e that can be obtained from figure 5 of Smoluchowski's 1983 paper.

The use of equation (7), with $\alpha^{-1} \sim 10$ Å or 15 Å, shows that a hydrogen atom initially adsorbed in a low energy binding site would quickly migrate to a deeper site in its vicinity, remaining almost immobile in there. Then at low temperatures H atoms on amorphous grains will form H_2 molecules only if they are adsorbed two by two within a distance comparable to the size at which their structural ordering holds, i.e. about 10 Å (Konnert and Karle, 1973). Following these results on adatom mobility in the quantum mechanical regime, Smoluchowski evaluated the production rate of H_2 in interstellar conditions as a function of the grain temperature T_g, the density of atomic n_H and molecular n_{H_2} hydrogen in the gas phase; in fact he also considered the presence of H_2 which would compete with H atoms in occupying adsorption sites on grains. Such a competition is naturally stronger in molecular clouds (which we are interested in), where the ratio $n_{H_2}/n_H \sim 500$. In this case he evaluated the rate of formation of H_2 through the equation

$$R = Z/2(S_1 v_H)^2 n_H n_g \pi a^2 \sigma(1 - S_2 n_{H_2} v_{H_2} \sigma t_2) t_1 \tag{8}$$

where

$$Z = \text{average number of sites surrounding a particular adsorption site}$$
$$S_1 = \text{sticking coefficient for a H atom}$$
$$S_2 = \text{sticking coefficient for a } H_2 \text{ molecule}$$
$$n_H = \text{number of H atoms per unit volume}$$
$$n_{H_2} = \text{number of H molecules per unit volume}$$
$$v_H = \text{average velocity for a H atom}$$
$$v_{H_2} = \text{average velocity for a } H_2 \text{ molecule}$$
$$n_g = \text{grains density in the clouds}$$
$$a = \text{average radius of the grain}$$
$$\sigma = \text{average size of an adsorption site}$$
$$t_1 = \text{the average lifetime of incident H atoms}$$
$$t_2 = \text{the average lifetime of incident H molecules}$$

A general feature of his results is that they are orders of magnitude lower than those obtained considering crystalline ice (e.g. Hollenbach and Salpeter, 1971), especially at the lowest temperatures ($T < 15$ K), and remain at least three orders of magnitude smaller than the others even in the temperature range where they reach the maximum.

These low rates may render other mechanisms more relevant.

3 H_2 production by cosmic rays

We will try to evaluate the production of H_2 as a direct release in gas phase from ice bombarded by cosmic rays and to compare them with the calculation of Smoluchowski which, among those that consider the recombination of adsorbed H atoms on grains, treat the mobility of adatoms in the most rigorous quantum mechanical way. In order to estimate these rates we have experimental results on the release H_2 by ices bombarded by fast ions obtained by us (Brown et al., 1982) with the equipment and the techniques described by Pirronello et al. (this volume).

A Monte Carlo simulation of the process of interaction of cosmic rays with grain mantles at various depths in the core of a spherical molecular cloud has been performed and the resulting production of H has been obtained by means of a computer code built for this purpose (Pirronello and Averna, 1987). Models of Boland and de Jong (1984) have been used to describe interstellar cloud like L134; grains having either graphite or silicate core with a dirty ice mantle made of H_2O, CH_4, NH_3 and CO and with radius between 400 on 4000 Å have been taken into account. Several depths in the clouds have been considered; the simulation has been continued until 40.000 ions (H^+, He^+) had hit each grain of the type (silicate or graphite) and of the size chosen. We have preferred to estimate conservatively the production of H_2, considering only H and He ion bombardment of mantles. Ion energies have been generated pseudo-randomly, using the well known technique of building the cumulative distribution function of a probability density function that mimics the Morfill et al. (1976) energy spectrum of cosmic rays. Each cosmic ray has been injected inside the cloud with a pseudo random direction and the energy lost by the ion interacting with the gas, inside the cloud before impinging on the grain, has then been subtracted from the initial energy. Our results, in comparison with those of Smoluchowski, are shown in figure 1. In this three-dimensional plot the production rate of diatomic hydrogen per cubic centimeter per second $R_{H_2}(T, d)$ is shown versus the grain temperature T and the depth in the core of the cloud. Unbroken lines represent Smoluchowski temperature dependent results; dashed lines show our results, which are not temperature dependent till 30 K.

As it is evident, the Smoluchowski rates are higher then ours, and hence recombination of H atoms even on amorphous surfaces is dominant over production of H_2 molecules by cosmic rays bombardment only in a very restricted interval of temperature.

When the cloud temperature is inside this range, H_2 is mainly formed according to Smoluchowski calculations, and this occurs only for $A_v > 2$ magnitudes. *In the core of the cloud, where the temperature of grains is close to 10 K, bombardment by cosmic rays is by several orders of magnitude the dominant process into enriching the environment of molecular hydrogen.*

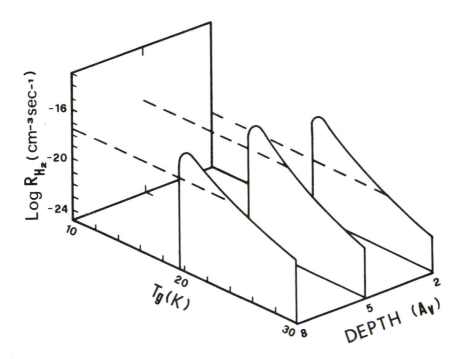

Figure 1: Production rate of molecular hydrogen per cubic centimeter per second versus temperatures and depth in the cloud. Rates due to cosmic rays bombardment of grain mantles (our results) as well as rates due to recombinations of H atoms on grains (Smoluchowski results) are shown.

Acknowledgements

Helpfull discussions with R. Smoluchowski, B. Donn, J.M. Greenberg are gratefully acknowledged. Thanks are also due to Mr. A. Calí for drawing the figures.

References

Anderson, P.W.: 1958, Phys. Rev. **109**, 1492.

Boland, W., de Jong, T.: 1984, Astron. Astrophys. **134**, 87.

Brown, W.L., Augustiniak, W.M., Simmons, E., Marcantonio, K.J., Lanzerotti, L.J., Johnson, R.E., Boring, J.W., Reimann, C.T., Foti, G., Pirronello, V.: 1982, Nucl. Inst. Meth. **198**, 1.

Duley, W.W., Williams, D.A.: 1984, "Interstellar Chemistry", Academic Press, London.

Hollenbach, D.J., Salpeter, E.E.: 1971, Astrophys. J. **163**, 155.

Konnect, J.K., Karle, J.: 1973, Acta Cryst.. Sect. A **29**, 702.

Morfiil, G.E., Volk, H.J., Lee, M.A.: 1976, J.G.R. **181**, 5841.

Mott, N.F.: 1969, Phyl. Mag. **19**, 835.

Pirronello, V., Brown, W.L., Lanzerotti, L.J., Marcantonio, K.J., Simmons, E.: 1982, Astrophys. J., **262**, 636.

Pirronello, V.: 1987, in Proc. Summer School on "Genesis and Propagation of Cosmic Rays", M.M. Shapiro ed., Reidel Publ. Co., Dordrecht, in press.

Pirronello, V. Averna, D.: 1987, Astron. Astrophys. (accepted).

Watson, W.D., Salpeter, E.E.: 1972, Astrophys. J., **174**, 321.

HD AND CO RELEASE DURING MeV ION IRRADIATION OF H$_2$O/CD$_4$ FROZEN MIXTURES AT 9 K

V. Pironello[1,2], W.L. Brown[3], L.J. Lanzerotti[3], C.G. MacLennan[3]

[1]*Dipartimento di Fisica, Università della Calabria, Cosenza, Italy*
[2]*Osservatorio Astrofisico di Catania, Catania, Italy*
[3]*AT&T Bell Laboratories, Murray Hill, New Jersey, U.S.A.*

1   Introduction

Chemical effects induced in ice mixtures by energetic (keV-MeV) ions in laboratory experiments (Brown et al., 1982; Pironello et al., 1982, Roessler, 1986) can be of fundamental importance in understanding the continuous processing suffered by grain mantles bombarded by cosmic rays inside molecular clouds, where UV radiation is strongly suppressed. The interplay of gas phase (Duley and Williams, 1984) and surface reactions on grains (d'Hendecourt et al., 1985) together with the formation of molecules in the bulk of the ice layer, induced by the release of ion energy, will allow to develop the comprehension of the chemical history of these clouds. An example of this last process is given by the production of molecular hydrogen in the core of dense clouds by cosmic rays (Pironello and Averna, 1987). These type of experimental studies have often been referred as laboratory "simulations" of what takes place in space. However the word simulation is a dangerous one; in fact it can be interpreted as the effort to reproduce "now" in the laboratory what occurs in nature. This is certainly the final goal, but it is still untimely to attempt to simulate in its full complexity the behaviour of nature. The real risk is, in fact, to succeed into reproducing what is observed, for instance, in interstellar space without gaining any deep insight of the processes that compete among themselves to give rise to what is observed because no unique solution to the problem exists.

We do believe that before reaching the point in which we can attempt to fully simulate the interaction of energetic particles with complex mixtures of frozen gases we have to proceed in steps trying to obtain the production rates per impinging ion of molecules that are formed. We think that the best way is to start bombarding homogeneus ice layers or at most binary mixtures, as we already did few years ago investigating the production and release of H$_2$ and O$_2$ by water ice (Brown et al., 1982) and of H$_2$CO by H$_2$O + CO$_2$ targets bombarded with 1.5 MeV He$^+$ at T $\geq$ 200 K (Pironello et al., 1982). Even using binary mixtures problems to overcome do exist, not only because a rich network of chemical reactions is already involved with the production and destruction of several molecules when they are irradiated, but also because the preparation of the mixture by slow deposition from the gas phase on the cold finger has to be done carefully. Two main problems may in fact happen:

287

E. Bussoletti et al. (eds.), Experiments on Cosmic Dust Analogues, 287–291.

a) the deposited mixture may not have the same compisition it had in the gas phase in the vessel in which it was prepared;

b) a significant segregation effect may rise among the different components of the mixture.

Both of them are probably due to the different propensity of different species to be adsorbed on the walls of the vessel, where the gas phase mixture is prepared and on the walls of the pipe that is used to deposit the mixture on the cold finger. While the first type of problems can be overcome in a relatively easy way with a careful quantitative analysis of the species released by mass spectrometric tecniques or by "in situ" infrared spectroscopy, the second type of problems are more subtle and cannot be easily monitored with the mentioned tecniques. At AT&T Bell Laboratories in Murray Hill (New Jersey) we succeeded into detecting these problems and then avoiding them by means of the so-called Rutherford backscattering technique (RBS). Such a technique, in fact, does not only give a quantitative measure of the relative amounts of species in the deposited mixture but also through the shape of the peaks either confirms or deny the occurrence of a non uniform mixing (segregation) of species. When such an effect is detected another film has to be accreted on the cold finger.

2 Experimental procedure and results

According to the line of thought we just told in the introduction, we prepared in gas phase several binary mixtures of H_2O and $^{13}CD_4$. They have been then deposited on a finger cooled at 8 K by a continuous flow of liquid helium, that was then released in open air, and irradiated by 1.5 MeV helium ions produced by the van de Graaff accelerator in Murray Hill. The ion dose rate has always been mantained in the range where a linear relation holds between the effect (molecular sputtering) and the flux of impinging ions in order to measure single particle effects, an approach that is mandatory for applications to astrophysical environments where particle fluxes are always very small with respect to laboratory ones. The deposition on the cold surface has been done at a very slow rate to produce an amorphous ice layer (Rice, 1975) because the "dirty" ice mantle on dust grains in interstellar clouds is almost certainly amorphous (Leger et al., 1979). Isotopically labelled compounds have been used mainly to monitor the origin of atoms forming daughter molecules. RBS has been performed on the deposited frozen layer to check the composition and uniformity of the mixture before irradiating it with fast particles, while a UTI residual gas analyzer (RGA) has been used to monitor the release in the gas phase of molecules from the target. This RGA was controlled by computer, ten masses have been monitored; the measuring time of 150 msec has been chosen for each species and 50 msec were necessary to set the RGA on the next mass to be measured for a total duty cicle of 2 seconds. In this note some preliminary results on the chemical modifications induced by fast ion bombardment are presented. Several species are produced, some of them are released in gas phase during bombardment while other species remain trapped in the icy layer due to their low mobility. To the first group H_2, HD, D_2, H_2O, CD_4, CO belong, to the second one H_2CO, D_2CO, CO_2. Here we should like to show the dose dependence in the ejection of two sinthetized species HD (see figure 1) and CO (see figure 2) from a roughly equal mixture of the mentioned components deposited in a layer $5.7 \cdot 10^{17}$ cm^{-2} thick.

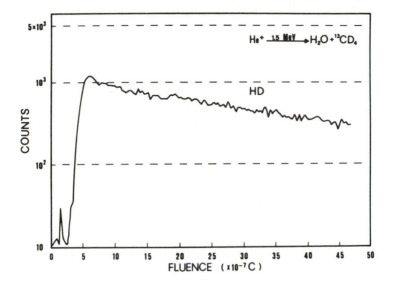

Figure 1: Release of HD versus He$^+$ dose.

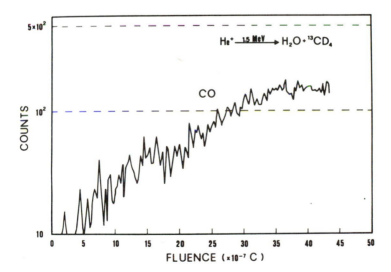

Figure 2: Release of CO versus He$^+$ dose.

The choice of growing such thin films is not only due to the attept of "simulating" thin mantles on grains but, according to our philosophy, is mainly connected with the will of performing *clean* experiments having the least number of parameters varying during their working out. In this samples, in which the ion looses only a very minor fraction of its initial energy, all the production yields of species can be considered as obtained at the energy of the impinging ion; in thick samples, on the contrary, the ion energy loss changes with the depth inside the layer and the yields that are then obtained are only average results often far from the values relative to the initial energy of the impinging ion. In the two figures the number of counts obtained from the RGA during irradiation is plotted versus the helium fluence. The releases of the two species show, as it should have been expected, a quite different behaviour.

HD, as also H_2 and D_2, are released after a short delay which is probably due either to the formation of percolation paths or to the necessary build up of a partial pressure of the formed species untill the overcoming of a suitable threshold value occurs. During this "delay time" some spikes in the release of molecules occur; they are probably due to deuterated molecular hydrogen that is synthesized in the most external part of the icy film. A steady state ejection of HD is not reached within the total dose of projectiles used in this case and this huge transient effect may have its own relevance for interstellar applications.

CO is released in the gas phase at a much lower rate than HD. The long time scale (at constant dose rate) of CO to reach the maximum yield is probably due to a low production rate but also to its low mobility in the frozen matrix; CO is in fact one of the species that also remains trapped in the layer, after being formed, untill it is induced the total sublimation of the ice: a fact that will be shown elsewhere. In the light of this other result the release of CO in the gas phase shown in this short note has to be clearly regarded only as a portion of the production yield of such a specie during ion irradiation and by no means as the total one.

Because the molecular yield of the RGA is not calibrated we used as an external calibration RBS measurements of the components of the mixture before irradiation; after this measurements we let sublime the icy layer warming up the cold finger to the room temperature. The ratio between the number of H_2O molecules deposited on the cold finger (measured with RBS) and the number of counts from RGA during sublimation in our experimental configuration was $w = 5.5 \cdot 10^{11}$. Using this conversion value one can obtain molecular yields "Y" (number of molecules formed and released per impinging ion). For instance for HD we have roughly at the peak (see figure 1)

$$Y_{HD}(1.5 \text{ MeV}) = 8.9 \cdot 10^2$$

and at the plateau in the CO release (see figure 2) it is

$$Y_{CO}(1.5 \text{ MeV}) = 25.$$

3   Conclusions

In this short note we have shown some preliminary results on the production and the release in the gas phase of HD and CO from a fairly well mixed frozen layer of roughly equal amounts H_2O and $^{13}CD_4$ bombarded with helium ions of 1.5 MeV kinetic energy. The dose dependence in the release is shown; its behaviour is due to the simultaneously competing processes of formation, release in the gas phase and trapping inside the frozen layer of the formed species. The production yield at the maximum for the two species is also given.

References

Brown, W.L., Augustiniak, W.M., Simmons, E., Marcantonio, K.J., Lanzerotti, L.J., Johnson, R.E., Boring, J.W., Reimann, C.T., Foti, G., Pirronello, V.: 1982, Nucl. Instr. & Methods, **198**, 1.

d'Hendecourt, L.B., Allamandola, L.J. Greenberg, J.M.: 1985, Astron. Astrophys. **152**, 130.

Duley W.W., Williams, D.A.: 1984, "Interstellar Chemistry", Academic Press, London.

Leger, A., Klein, J., de Cheveigne, S., Guinet, G., Defourmeon, D., Belin, M.: 1979, Astron. Astrophys. **79**, 256.

Pirronello, V., Brown, W.L., Lanzerotti, L.J., Marcantonio, K.J., Simmons, E.: 1982, Astrophys. J. **262**, 636.

Pirronello, V., Averna, D.: 1987, Astron. Astrophys. (in press)

Rice S.A.J.: 1975, Top Curr. Chem **60**, 109.

Roessler K.: 1986, Radiation Effects **99**, 21.

STUDIES OF ABSORPTION FEATURES IN THE 3 μm SPECTRA OF HIGHLY OBSCURED STELLAR AND PROTOSTELLAR OBJECTS

R.G. Smith[1], K. Sellgren[2], A. Tokunaga[2]
[1]Max-Planck-Institute für Physik und Astrophysik, Institut für extraterrestrische Physik, Heidelberg, F.R.G.
[2]Institute for Astronomy, University of Hawaii, Honolulu, U.S.A.

1   Introduction

The results reported here represent a part of a more extensive investigation on the absorption features in the 2.8–3.8 μm spectra of highly obscured objects. Our aim is to obtain high S/N spectra of bright sources with known 3 μm absorption features so that these features may be intercompared. Since the 3.08 μm water ice feature dominates the spectra in our sample, we therefore have a means of studying this feature over a range of conditions. By comparison with available laboratory data, we can estimate a grain temperature for the molecular clouds in which these sources are embedded, although the accuracy of this estimate is affected by the dependance of the shape of the 3.08 μm feature on the temperature history of grains. It is particularly important that we build up a very good understanding of the 3.08 μm water ice feature so that we can then turn our attention to weaker features in the spectra. For example, an absorption feature near 2.96 μm was originally attributed to N-H stretching vibration in ammonia ice by Knacke et al. (1982). However, recent observations by Knacke and McCorkle (1987) have cast doubt on this identification and they have proposed scattering effects to explain this feature in the Becklin-Neugebauer object (BN). The reason that ammonia ice was attractive is because a mixture of water and ammonia ice can produce ammonia hydrate, which exhibits a broad absorption in the 3.4 μm region, very similar in shape to what is actually seen in the spectra of many molecular cloud sources (e.g. Willner et al. 1982). However, many workers (e.g. Duley and Williams 1983) have shown that absorptions near 3.4 μm can also arise from the C-H stretching vibration. Since there is a great deal of evidence to suggest the presence of carbon compounds in the interstellar medium (ISM), this explanation for the 3.4 μm absorption is also quite plausible. At present however, available observational data are insufficient to specifically identify the source of this feature. Thus, while the identification of the strong absorption feature near 3.08 μm is quite secure, this is clearly not the case for weaker absorption features. One of the first steps towards an identification is the gathering of high quality spectra of a variety of sources and a detailed study of these spectra.

E. Bussoletti et al. (eds.), Experiments on Cosmic Dust Analogues, 293–297.
© 1988 by Kluwer Academic Publishers.

2   Results

We take only a few sources from our sample to illustrate our results and the information potential of these spectra. The spectra were obtained with the cooled-grating array spectrometer (CGAS) on the NASA infrared telescope facility (IRTF). Spectra were obtained at low-resolution grating installed in the CGAS, giving resolution of 140 at 3 μm. In all spectra shown here, in the wavelength range 2.82–3.39 μm spectra have been double sampled by moving the grating the distance equivalent to half a resolution element. This allows us to discuss, with some confidence, the presence or absence of narrow absorption features in spectra. In addition, the CGAS has a substantial advantage over single detector CVF systems because it contains an array of 32 integrating InSb detectors in the focal plane. This results not only in increased observing efficiency but also in a significant reduction in the point-to-point scatter in a spectrum.

2.1   OH231.8+4.2 (OH0739–14)

We now turn to the objects themselves. To provide some contrast for the molecular cloud souces, we first consider a late-type, high mass-loss star surrounded by a thick circumstellar shell, OH231.8+4.2, which has a spectrum characteristic of pure water ice absorption (Smith et al. 1987).

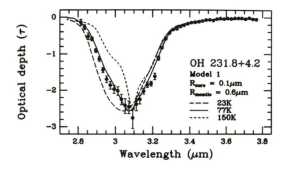

Figure 1: Spectrum of the OH/IR source OH231.8+4.2 compared with the absorption spectrum of small silicate grains coated with water ice.

As shown in figure 1, a simple model involving grains with silicate cores and mantles of 77 K amorphous water ice provides a very good fit to the observed spectrum. Unfortunately, with published optical constants available for only three temperatures, 23 K, 77 K, and 193 K (Kitta and Krätschmer 1983, Léger et al. 1983, and Bergren et al. 1978, respectively), it is difficult to accurately estimate the grain temperature. However, the fit with the 77 K model is sufficiently good that we conclude that the present grain temperature must be close to 80 K. This conclusion is helped by the nature of the source: OH231.8+4.2 is a high mass-loss source and so grains are cooling as they move away from the central star. This additional factor is necessary to tie down the grain temperature because detailed studies of amorphous water ice (Hagen et al. 1983) have shown that the half-width or wavelength of maximum absorption we see can arise either as a result of ice formed at roughly 80 K, or from ice formed at a lower temperature which has subsequently cooled. Also apparent in the OH231.8+4.2 spectrum are weak absorptions near 2.95 and 3.2 μm. Since the 2.95 μm feature appears to be very broad, we cannot accept an identification with ammonia ice unless some way can be found to broaden the fundamental ammonia ice absorption feature (i.e. the N-H stretch at 2.96 μm) under astrophysical conditions. However, features near both 2.95 and 3.2 μm appear in crystalline ice. Thus we suggest that the ice absorption in this souce can probably be explained by considering ice formed at a range of temperatures where the observed spectrum, while being dominated by ice formed near 80 K, also shows effects due to ice formed at higher temperatures, now covered by the 80 K ice.

2.2 The Becklin-Neugebauer object (BN)

The Becklin-Neugebauer object (BN) spectrum (Figure 2) shows some surprising similarities to the spectrum of OH231.8+4.2, with weak features near 2.95 and 3.2 μm.

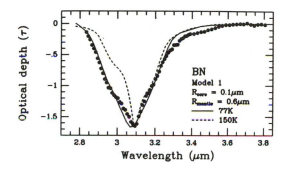

Figure 2: Spectrum of the Becklin-Neugebauer object compared with the absorption spectrum of small silicate grains coated with water ice.

There are however noticeable differences in the wavelength of maximum absorption and in presence of a long wavelength wing (or 3.4 μm feature), extending to about 3.5 μm. For comparison, we show the 77 K ice model superimposed on the BN spectrum although clearly the fit is not very good. The long wavelength wing has previously been explained in terms of a feature arising in an ammonia+water ice mixture; however, as was the case for OH231.8+4.2, the absence of any narrow absorption feature near 2.96 μm which is fundamental to ammonia ice, suggests that we must look to other molecules to explain the long wavelength wing. The features near 2.96 and 3.2 μm are again suggestive of a mixture of ices at least partly composed of higher formation temperature (crystalline) ice. The substantially different formation conditions, compared to OH231.8+4.2, make it difficult to assign a grain temperature with any certainty; however, the width and shape of the water ice absorption is sufficiently similar to OH231.8+4.2, neglecting for the moment the 3.4 μm absorption, that the ices in the two objects may be very similar. In particular, since the optical depth in OH231.8+4.2 is so high (i.e. 2.5), the band may be saturated near 3.08 μm, in which case the absorption features become even more alike. We thus suggest that grain temperatures >80 K probably existed for both objects at the time the ice was formed. The appearance of the absorption feature can of course be modified by subsequent cooling of grains. Unfortunately there is not sufficient quantitative information in the literature for us to judge the extent of this effect.

2.3   AFGL 989

We now consider GL 989, which shows a much broader absorption feature and stronger long wavelength wing (3.4 μm feature). The basic 3.08 μm feature can be adequately explained by a lower temperature water ice model (i.e. 23 K), as shown in figure 3.

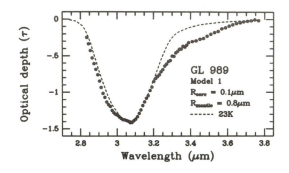

Figure 3: Spectrum of GL 989 compared with the calculated absorption of small silicate grains coated with amorphous water ice at a temperature of 23 K.

Interestingly, while the 2.95 μm absorption feature seems to be present, although substantially broadened, no sign is seen of the 3.2 μm feature. This is consistent with water ice at a temperature lower than that seen on OH231.8+4.2 and BN since the observations of Fink and Sill (1982) have shown that the 2.95 μm feature is present in water ice at >90 K while the 3.2 μm feature does not appear until >135 K. Once again, as was the case for BN and OH231.8+4.2, we see evidence for existence of ice coated grains at a range of temperatures. In the case of GL 989, the absorption is dominated by low temperature grains, near 20 K, but the sub-features are characteristic of grains that have been as hot as about 130 K (i.e. <135 K, where the 3.2 μm feature appears).

3   Conclusions

We draw the following preliminary conclusions from an intercomparison of three of the spectra taken from our study of 3 μm absorption features in highly obscured objects.
(1) Based on an examination of published spectra, we cannot find any reasonable grounds for the identification of the 3.4 μm absorption feature with a mixture of ammonia and water ices. In all cases, laboratory absorption spectra resulting from this mixture (i.e. ammonia hydrate) show a sharp absorption near 2.96 μm due to the fundamental N-H

stretching vibration. This is not seen in spectra of any of the astronomical objects presented here.

(2) In the studied objects, the absorption feature near 3.08 μm, plus weaker absorptions on both short and long wavelength sides of this feature can be qualitatively explained by the presence of water ice on grains with a large range of temperatures. A quantitative result is made more difficult here by the lack of published data on amorphous ice, particularly on the effects seen when such ice is cooled, as must undoubtedly happen in many astrophysical situations.

References

Bergren, M.S., Schuh, D., Rice, S.A.: 1978, J. Chem. Phys. **69**, 3477.

Duley, W.W., Williams, D.A.: 1983, M.N.R.A.S. **205**, 67.

Fink, U., Sill, G.T.: 1982, in "Comets", L.L. Wilkening ed., (Tucson, University of Arizona Press), p.164.

Hagen, W., Tielens, A.G.G.M., Greenberg, J.M.: 1981, J. Chem. Phys. **56**, 367.

Kitta, K., Krätschmer, W.: 1983, Astr. Ap. **122**, 105.

Knacke, R.F., McCorkle, S., Puetter, R.C., Erickson, E.F., Krätschmer, W.: 1982, Ap. J. **260**, 141.

Knacke, R.F., McCorkle, S.: 1987, personal communication.

Léger A., Gauthier, S., Defourneau, D., Rouan, D.: 1983, Astr. Ap. **117**, 164.

Smith, R.G., Sellgren, K., Tokunaga, A.T.: 1987, Ap. J., submitted.

Willner, S.P., Gillett, F.C., Herter, T.L., Jones, B., Krassner, J., Merrill, K.M., Pipher, J.L., Puetter, R.C., Rudy, R.J., Russell, R.W., Soifer, B.T.: 1982, Ap. J. **253**, 174.

AN IONIC LOOK AT INTERSTELLAR GRAIN MANTLES

R.J.A. Grim, J. Mayo Greenberg
Laboratory Astrophysics, Huygens Laboratory,
Leiden University, Leiden, The Netherlands

1 Photoprocessing of interstellar grain mantles

Our knowledge of the interstellar medium and its components is still limited to the tip of an iceberg. However, with the coming of new and better observing facilities the amount of information has tremendously increased over the last decade and more and more people jump in. Observations (towards diffuse clouds) in the ultraviolet have revealed at least three populations of interstellar dust. Large grains $(+/-0.15\ \mu m)$ are responsible for the blocking of the visual star light. Small $(0.01\ \mu m)$ carbonaceous particles are required to explain the strong 216 nm hump. The exact nature of those small particle is still unknown and many explanations have been and are still being put forward. However, in the Leiden Laboratory strong indications have been found that they result from unsaturated carbon chain molecules like polyacetelyne. Finally, a third dust component consists of small $(0.01\ \mu m)$ silicate grains providing the far ultraviolet part of the extinction curve. How these grains evolve physically and chemically is a complicated story. In this contribution we want to focus on one aspect only: the chemical composition of the mantles that have accreted on large interstellar grains in molecular clouds.

One of the best ways to study the chemical composition of these mantles is by infrared spectroscopy since the 2.5 to 25 μm region can be regarded as a fingerprint of the molecular composition of the grain mantle. In this spectral region the stretching, bending and rocking motions of molecules are found. Since 1976 several molecules have been identified (Table 1) to be present in the mantle. In the laboratory, where grain mantles are simulated by accreting simple gases on cold substrates, matrix isolation spectroscopy has proven to be a very powerful technique and now provides one of the basic elements in the systematic study of the identification of the remaining unknown features.

Infrared observations towards protostellar sources as W33A or Mon R2-IRS2 (Lacy et al., 1984; Geballe et al., 1985; Tielens et al., 1985) have made it clear that the ice mantles must have been subjected to a degree of photoprocessing capable of changing the mantle composition significantly. But also thermal processing (annealing) of the cold interstellar grains becomes possible when the grain environment is heated by a protostellar source or when the grains travel closer to the edges of the cloud. Evidence

E. Bussoletti et al. (eds.), Experiments on Cosmic Dust Analogues, 299–311.
© *1988 by Kluwer Academic Publishers.*

for mantle annealing has been found in the infrared spectra towards e.g., W33A, Mon R2-IRS2 and HL Tau.

Table 1: List of molecules found in dust mantles by means of infrared absoption spectroscopy. Tentative identifications are marked by a question mark.

molecule	λ (μm)	ν (cm^{-1})	object
NH$_3$(?)	2.97	367	associated with 3.08 μm H$_2$O bands
H$_2$O	3.08	3247	several protostellar sources (W33A, BN, ect.), OH/IR stars and field stars (Elias 3,16)
-CH$_2$, -CH$_3$	3.38, 3.42	2959, 2924	Galactic centre sources (Sgr.W), OH 01-477
CH$_3$	3.53	2833	W33A
H$_2$S	3.95	2532	W33A
"XCN"	4.617	2166	W33A, NCG 7538/IRS 9(?), Mon R2/IRS 2(?), GL 961(?)
CO	4.675	2139	W33A, NCG 7538/IRS 2, NCG 7538/IRS 9, W3/IRS 5, Elias 1, Elias 16, GL 961, GL 989, GL 2136, OMC 2/IRS 3, NCG 2024/IRS 2
"XCS"	4.9	2041	W33A
?	5.5	1818	NCG 7538/IRS 9
H$_2$O	6.0	1667	several protostellar sources, see 6.8 μm absorption
"6.8 μm"	6.85	1460	W33A, NCG 7538/IRS 9, GL 2136, W3/IRS 5 GL 961, NCG 2024/IRS 2, Mon R2/IRS 2
?	7.6	1315	NCG 7538/IRS 9

The high extinction towards W33A provides us the richest infrared spectrum showing absorption by H$_2$O, CO, H$_2$S, OCS, CH$_3$OH and several unknown absorbers of which the 4.62 μm feature, historically referred as "XCN", and the 6.8 μm feature are the strongest (see table 1). Both absorption have been reproduced in numerous experiments in our laboratory. Principally these experiments consisted of the deposition of simple molecules such as H$_2$O, CO, NH$_3$ and CH$_4$ on a cold (12 K) aluminum sample substrate situated in a pressure chamber (P< 10^{-7} mbar) and a subsequent ultraviolet irradiation by a hydrogen discharge lamp. Next, the sample could be annealed if this was required by the experiment.

Although many candidates for the 4.62 and 6.8 μm features have been suggested not one of them seems to satisfy all restrictions. Therefore we studied the infrared spectra again, but now by using isotopic labelling of the starting products, a method well known in matrix isolation spectroscopy in order to identify unknown absorbers. The mayor advantage of this method is that substitution of an isotope in a molecule leads to generally well defined frequency shifts. For instance, in the simple diatomic case the frequency of a vibrational mode depends on the reduced mass and varies with $\mu^{-1/2}$. For a triatomic molecule XYZ the frequency shifts of the symmetric and antisymmetric vibration are coupled but are still simple to analyze. For larger molecules the relations become complex but laboratory studies have yielded a lot of data on isotopic substitution. The theory of isotopic substitution is well explained by Pinchas and Laulicht (1971) who also provide a good data base of the experimental data available.

Using this powerful technique we have come to some interesting results. First, we have identified the ion OCN$^-$ as giving rise to the 4.62 μm absorption (Grim and Greenberg, 1987a) and, next, the fundamental ionic group $-NH_3^+$ as giving rise to the 6.8 μm absorption (Grim and Greenberg, 1987b). The possibility that ionic species can also contribute to the infrared spectra of interstellar grains has escaped anyones attention sofar and we are thus forced to change our view of the grain mantle composition rather drastically.

2 The 4.62 μm "XCN" absorption

The first observations of this feature in W33A were reported by Soifer et al. (1979). The feature was unresolved due to the low resolution of the CVF spectrum. It was Lacy et al. (1984), see figure 1, and later on Larson et al. (1985) and Geballe et al. (1985) who provided high resolution spectra by which the two distinct bands could be resolved.

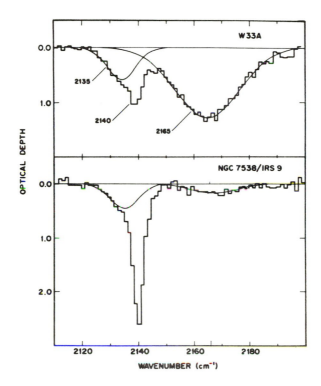

Figure 1: The infrared absorption spectra of the 2100–2200 cm^{-1} region of W33A and NCG 7538/IRS 9 showing the CO (2140 cm^{-1}) bands. The presence of an "XCN" absorption in NCG 7538/IRS 9 still remains questionable. The spectra are reproduced from Lacy et al. (1984).

Several possible candidated have been proposed to account for the 4.62 μm (2166 cm^{-1}) absorption including methylisocyanide (CH$_3$-NC), cyanogen (NC-CN) and pyruvoisoni-trile (CH$_3$-CO-NC). A literature search has added candidates such as carbodiimide

(HN=C=NH), isocyanamide (NH$_2$-NC), isodiazomethane (HC=N=NH) and, finally, OCN$^-$ (Grim and Greenberg, 1987a). Several of these molecules could be rejected immediately using simple arguments and leaving CH$_3$NC and, finally, OCN$^-$ as the most plausible candidates for XCN. Both possibilities were tested using isotopic substitution with ^{13}C and ^{15}N in CO/NH$_3$ mixtures.

2.1 Laboratory experiments

The CO/NH$_3$ mixtures were irradiated for 4 hr with the hydrogen lamp and subsequently heated to 80 K to eliminate CO which dominates the XCN feature at 12 K. Figure 2 shows the infrared spectra of the irradiated CO/NH$_3$, ^{13}CO/NH$_3$ and CO/^{15}NH$_3$ mixtures after heating to 80 K.

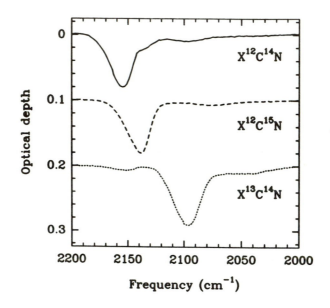

Figure 2: The infrared absorption spectra (2200–2000 cm^{-1}) of the mixtures: ^{12}CO/^{14}NH$_3$ = 1/1 (upper), ^{12}CO/^{15}NH$_3$=1/1 (middle) and ^{13}CO/^{14}NH$_3$=1/1 (lower) after 4 hr ultraviolet photolysis and subsequent warmup to 80 K.

In table 2 the observed frequencies and frequency shifts are listed and compared with the peak positions and shifts of isotopic CH$_3$NC and OCN$^-$.

Table 2: Infrared absorption frequencies of OCN^- and CH_3NC compared with the observed XCN features.

Species	$\nu(^{12}C^{14}N)$ (cm^{-1})	$\nu(^{12}C^{15}N)$ (cm^{-1})	$\nu(^{13}C^{14}N)$ (cm^{-1})	$\Delta(^{15}N)^e$ (cm^{-1})	$\Delta(^{13}N)^f$ (cm^{-1})
XCN (W33A)^a	2165				
CH_3NC^b	2165		2124		41
OCN^- (KI)^b	2155.8	2138.7	2099.0	17	56
OCN^- (KBr)^b	2169.6	2152.5	2112.8	17	57
OCN^- (KCl)^b	2181.8	2124.7			57
OCN^- (NaCl)^b	2211.2		2153.7		57
KNCO^c	2165		2112		53
NaNCO^c	2226		2160		66
NH_4NCO^c	2217				
CO/NH_3=1/1^d	2157	2139	2100	18	57

^{a}XCN in protostellar source W33A.
^{b}Matrix isolated OCN^-, the matrix material is placed between brackets.
^{c}Cyanate salts.
^{d}XCN in laboratory dirty ice analogue.
$^e\Delta(^{15}N) = \nu(^{12}C^{14}N) - \nu(^{12}C^{15}N)$.
$^f\Delta(^{13}C) = \nu(^{12}C^{14}N) - \nu(^{13}C^{14}N)$.

The 41 cm^{-1} frequency shift of CH_3NC is incompatible with the observed shift of 57 cm^{-1}. On the other hand the frequency shifts for OCN^- agree excellently with the observed ones. The peak position of OCN^- in the CO/NH_3 ices is shifted by 8 cm^{-1} to lower frequencies compared with OCN^- in W33A (2166 cm^{-1}; table 2). However, by simply adding H_2O one is able to account for the shift in the interstellar peak position (Table 3).

Table 3: OCN^- band positions in various photolyzed dirty ices.

Composition	Frequency (cm^{-1})
CO/NH_3=1/1	2157
$H_2O/CO/NH_3$=1/1/1	2160
$H_2O/CO/NH_3$=2/1/1	2163
$H_2O/CO/NH_3$=3/1/1	2166
$H_2O/CO/NH_3$=5/2/1	2166
$H_2O/CO/CH_4/NH_3$=6/2/1/1	2167

2.2   OCN⁻ in interstellar ices

The laboratory experiments have demonstrated that one may account for the interstellar position and shape of the 4.62 μm absorption in W33A (Figure 3).

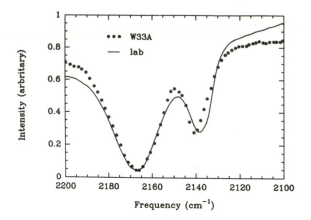

Figure 3: The spectrum of W33A compared with a photolyzed $H_2O/CO/CH_4/NH_3=$ 6/2/1/1 mixture heated to 110 K. Similar comparison have been shown in Lacy et al. (1984) and Geballe et al. (1985). The optical depths have been scaled arbitrary.

Furthermore, we can calculate the minimum integrated absorbance value of the 2166 cm⁻¹ band of OCN⁻. This value, $2 \cdot 10^{-17}$ cm molecule⁻¹, is in good agreement with earlier reported intensities for this band (Schettino and Hisatsune, 1970). Combining the largest value of $8.5 \cdot 10^{-17}$ cm molecule⁻¹ for OCN⁻ (measured in KI; Schettino and Hisatsune, 1970) and Lacy's data, we calculate the OCN⁻ column densities towards W33A and NCG 7538/IRS 9 as $4.5 \cdot 10^{17}$ *and* $1.4 \cdot 10^{17}$ cm⁻², respectively. Both values lead to OCN⁻ concentrations relative to H_2O of 1%.

3   The 6.8 μm absorption

Numerous observations of this features towards HII regions and protostellar objects (see figure 4) have been reported. Willner et al. (1982), Tielens and Allamandola (1987) and Grim and Greenberg (1987b) have summarized all observations and candidates. Among them are carbonates, silicates and saturated hydrocarbons (particularly simple alcohols). Carbonates absorb strongly near 6.8 μm but their strong funadmental at 25 μm has never been observed. Silicates, especielly the hydrated ones which are rich in Mg, could also be responsible for the 6.8 μm band (Hecht et al., 1986) although there is no correlation with the 10 μm absorption bands (Willner et al., 1982). Finally, there are problems with the methyl (-CH$_3$) and metylene (-CH$_2$) bending vibration explanation since one would generally expect a much higher 3.4 μm absorption feature than observed and, secondly, there is no correlation between the 3.4 and 6.8 μm absorption feutures in various protostars (Willner et al., 1982).

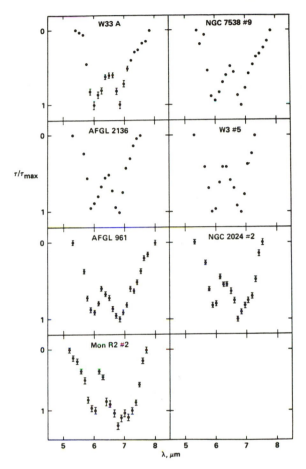

Figure 4: The 5–8 μm spectra of various protostars showing the 6.0 and 6.8 μm absorption bands. Note the differences in the half widths of the various 6.8 μm bands demonstrating the different environments of the dust mantles (Tielens and Allamandola, 1987). The figure is reproduced from Tielens and Allamandola (1987).

Tielens et al. (1984) proposed CH_3OH to account for the 6.8 μm absorption in W33A which implies a concentration relative to H_2O of 50%. The maximum CH_3OH concentration in grain mantles towards W33A as deduced by Bass et al. (1987) is only 5%. Again, in order to discriminate between the possible candidates we have performed a series of isotopic substituted experiments.

3.1   Laboratory experiments

In table 4 the experimental conditions and results for a number of experiments are summarized while figure 5 shows the infrared spectra of the $NH_3/CO/O_2=2/2/1$ experiment.

Table 4: Absorption frequencies and identifications for the 2000–1200 cm^{-1} region of various isotopically labelled photolyzed dirty ice mixtures. Total photolysis time is 4 hr. Resolution is 4 cm^{-1}. As a result of the broad long wavelength wing in the H_2O/NH_3 experiment absorption features below 1600 cm^{-1} are difficult to observe.

H_2O/NH_3 (1/1)	CO/NH_3 (1/1)	NH_3/O_2 (1/1)	$CO/NH_3/O_2$ (2/2/1)	$^{13}CO/NH_3/O_2$ (2/2/1)	$CO/^{15}NH_3/O_2$ (2/2/1/)	Assignment
–	–	1233	1233	1233	1207	NO_2^-
–	–	1337	1341	1333	1302	NO_3^-
–	1391	1385	1389	1387	1389	NH_4^+
1500	1499	1497	1499	1497	1497	NH_3^+
1620	1634	11626	1632	1628	1626	NH_3
–	–	1875	1879	1877	1844	NO

A key element in the results is that the 6.8 μm carrier obviously does not contain any carbon atom (se the NH_3O_2 experiment; figure 6).

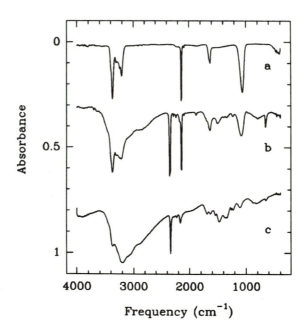

Figure 5: Infrared spectra (4000–500 cm^{-1}; 2.5–20 μm) of $NH_3/CO/O_2= 2/2/1$ after deposition (a), after 4 hrs UV photolysis (b) and after heating to 140 K (c). For convenience of presentation the depths of the CO and CO_2 absorption in (a) and (b) are cut arbitrarily.

Further, in samples where NH_3 is absent the 6.8 μm absorption is not observed at all (N_2O_2, $H_2O/O_2/N_2$, CO/N_2, CO/CH_4) or is present very weakly ($CH_4/CO/N_2$, H_2O/CO). All this information rules out carbonates and saturated hydrocarbons, and particularly CH_3OH, as possible candidates and points into the direction of a molecule that is strongly related to NH_3.

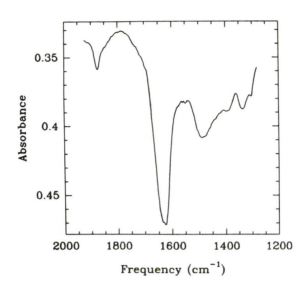

Figure 6: The infrared spectrum from 2000 to 1200 cm^{-1} of $NH_3/O_2 = 1/1$ after 4 hr UV photolysis at 10 K clearly showing the absorption at 1500 cm^{-1} due to the 6.8 μm carrier.

This is supported by the fact that Willner et al. (1982) have found a strong correlation between the 6.8 and 2.92 μm bands, of which the latter is characteristic for N-H stretching vibrations. In table 5 we have listed all possible candidates, their peak positions and isotope frequency shifts for comparison with the laboratory observed features. Only the bending vibration of the $-NH_3^+$ fundamental group absorb close enough to 1500 cm^{-1} to be considered as the most likely candidate. This is supported by the isotopic substitution experiments (Tables 4 and 5).

Annealing of the samples tend to shift all frequencies observed in the 1800–1000 cm^{-1} region to lower wavenumbers indicating physical and chemical changes in the sample material. Figure 7a shows the infrared spectrum of an $NH_3/O_2/N_2 = 2/1/1$ mixture after 4 hr UV photolysis and subsequent heating to 240 K. The spectrum is in good agreement with that of crystalline NH_4NO_3 at 230 K (Figure 7b).

Table 5: Possible 6.8 μm candidates nitrogen and their ^{14}N and ^{15}N absorption frequencies.

molecule	^{14}N (cm^{-1})	^{15}N (cm^{-1})	assignment
$N_2O_2^-$	1031	1015	(ν_5)
HONO	1265	1261	O-H bending (ν_4)
NO_2^-	1270	1243	anti symm. N-O stretching (ν_3)
N_2O	1285	1265	N^+-O^- (ν_1)
NO_2^-	1318	1306	symmetric N-O stretching (ν_1)
	1618	1580	anti symm. N-O stretching (ν_3)
NO_3^-	1390	–	(ν_3)
NH_4^+	1404	1399	N-H deformation (ν_4)
NH_3^+	1500*		symmetric NH_3^+ deformation
	1531	1525	symmetric NH_3^+ deformation
NH_3	1628	1625	symmetric NH_3 deformation
NO	1876	1843	N-O stretching

(*)-NH_3^+ measured in carboxylic acids.

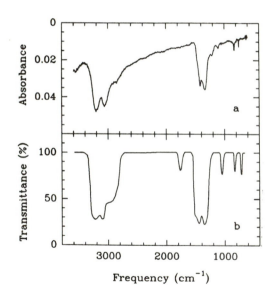

Figure 7: The infrared spectrum after heating to 240 K of a photolyzed $NH_3/O_2/N_2=$ 2/1/1 mixture compared with the spectrum of crystalline NH_4NO_3 at 230 K.

The integrated absorbance value for the 1500 cm^{-1} absorption of -NH_3^+ needed to be calculated from the experiments because no data could be found in literature. By assuming nitrogen conservation a minimum value can be estimated. The nitrogen conservation for the $NH_3/O_2=1/1$ experiment is tabulated in table 6. Assuming that the available nitrogen atoms are in NH_3^+ we calculate in combination with the observed integrated absorbances a typical value of $2 \cdot 10^{-17}$ cm molecule^{-1}. Due to the errors in the quoted literature values and in the choice of an appropriate baseline for the 6.8 μm feature in our spectra the error bar for this value can easily be 100%.

Table 6: Nitrogen conservation for the $NH_3/O_2=1/1$ experiment. Not included are NH_4^+ and molecules which do not absorb in the infrared, e.g. N_2, or are too weak to be observed, e.g. NH_2. The number of molecules are in units of 10^{18} cm^{-2}.

Molecule	Σ_{NH_3}	Σ_{N_2O}	Σ_{NO}	$\Sigma_{NO_2^-}$	$\Sigma_{NO_3^-}$	Σ_N	$\Sigma_{NH_3^+}$
deposition	2.01	–	–	–	–	2.01	–
1 hr UV	1.70	0.006	0.16	0.006	0.006	1.88	0.13
3 hr UV	1.56	0.006	0.15	0.007	0.008	1.73	0.28

3.2 NH_3^+ in interstellar ices

Using the integrated absorbance value estimated above we can derive the number of - NH_3^+ absorbers in interstellar grain mantles towards W33A. The derived column density, $0.7 \cdot 10^{19}$ cm^{-2}, is 15% of the H_2O column density.

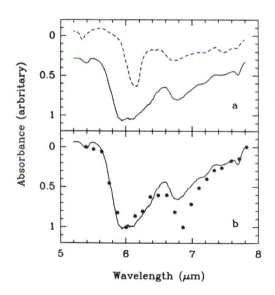

Figure 8: The 5–8 μm infrared absorption spectrum of $H_2O/CO/O_2/NH_3/CH_4/N_2=$ $1/1/1/0.3/0.3/0.03$ after 24 hr UV and subsequent heating to 130 K (solid line) compared with: a) The equivalent spectrum of $NH_3/O_2= 1/1$ after 4 hr UV photolysis at 10 K (dashed line) showing that the 6.8 μm absorption feature is produced without carbon. b) The spectrum of W33A (dots) obtained by Tielens and Allamandola (1987).

This seems quite high but the number should be considered with some caution since the errors in the calculation are at least a factor of 2. Fits of some laboartory spectra to observations of the 6.8 μm feature are shown in figures 8 and 9.

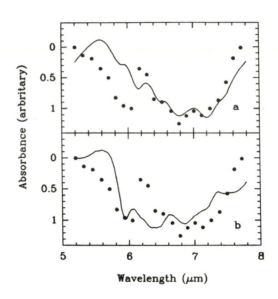

Figure 9: Comparison of the 5–8 μm absorption of Mon R2/IRS2 with: a) $NH_3/CO_2/O_2=$ 1/1/1 after 4 hr UV photolysis and subsequent heating to 100 K (solid line) and with: b) $H_2O/CO/NH_3= 4/2/1$ after 6 hr UV photolysis and subsequent heating to 220 K (solid line). Laboratory spectra have been reduced to 16 cm^{-1} resolution.

4   Future grain mantle modelling

We have identified in our laboratory the following ions: OCN^-, NH_3^+, NH_4, CN^-, NO_2, and NO_3, of which only the first two have been found in interstellar grain mantles so far. However, the presence of OCN^- and -NH_3^+ in grain mantles has great implications for our understanding of interstellar grain chemistry. Although, evidently, the photon energies are not high enough to ionize the molecules present, in the course of the photoprocessing (recombination, excitation) proton and/or electron exchange reactions can also occur, leading to the formation of the ions. As shown by our experiments the presence of NH_3 as an electron donor appears to be crucial. Ions other than OCN^- and NH_3^+ might be present in grains too, but especially CN^- will be difficult to detect because it has a low integrated absorbance value. The ions that might be of astrophysical relevance are listed in table 7 together with their peak positions and absorbance values.

Table 7: Absorption frequencies and integrated absorbance values of ions of astrophysical interest in unannealed ices.

Ion	ν (cm^{-1})	λ (μm)	Integrated absorbance value (cm molecule^{-1})
NO_2^-	1233	8.11	$1.0 \cdot 10^{-16}$
NO_3^-	1340	7.46	$1.4 \cdot 10^{-16}$
NH_4^+	1390	7.19	?
NH_3^-	1500	6.66	$> 2 \cdot 10^{-17}$
NCS^-	2052	4.87	?
CN^-	2090	4.78	$1 \cdot 10^{-19}$
OCN^-	2165	4.62	$6 \cdot 10^{-17}$

Figures 3, 8 and 9 are good examples in what way we can reproduce interstellar absorption spectra in the laboratory. However, one must realize that the laboratory analog can not be more than an approximation of the real physical and chemical state of the interstellar garin mantles, because the interstellar spectra are always an average grain mantle composition which can be, along the line of sight, variously affected by different temperature and photolysis conditions. This makes an interpretation of the interstellar grain mantle composition from interstellar and laboratory spectra certainly not straightforward.

In the future more sophisticated experimental methods have to be developed to study the chemical processing and annealing in dust mantles. But even with its limitations, we feel that the present method of experimental study will be fruitful for many years to come, particularly when much more detailed spectroscopic information becomes available with ISO.

References

Baas, F., Grim, R.J.A., Geballe, T.R., Greenberg, J.M.: 1987, in preparation.

Geballe, T.R., Baas, F., Greenberg, J.M., Schutte, W.: 1985, Astr. Ap. **146**, L6.

Grim, R.J.A., Greenberg, J.M.: 1987a, accepted for publication in Ap. J. Letters.

Grim, R.J.A., Greenberg, J.M.: 1987b, submitted to Ap. J. Letters.

Hecht, J.H., Russell, R.W., Stephens, J.R., Grieve, P.R.: 1986, Ap. J. **309**, 90.

Lacy, J.H., Baas, F., Allamandola, L.J., Persson, S.E., McGregor, P.J., Lonsdale, C.J., Geballe, T.R., van de Bult, C.E.P.M.: 1984, Ap. J. **276**, 533.

Larson, H.P., Davis, D.S., Black, J.H., Fink, U.: 1985, Ap. J. **299**, 873.

Pinchas, S., Laulicht, I.: 1971, "Infrared Spectra of Labelled Compounds", Academic Press, London.

Schettino, V., Hisatsune, I.C.: 1970, J. Chem. Phys., **52**, 9.

Soifer, B.T., Puetter, R.C., Russell, R.W., Willner, S.P., Harvey, P.M., Gillett, F.C.: 1979, Ap. J. Letters **232**, L53.

Tielens, A.G.G.M., Allamandola, L.J.: 1987, in "Physical Processes in Interstellar Clouds", eds. Morfill, G., Scoles, M.

Willner, S.P., Gillett, F.C., Herter, T.L., Jones, B., Krassner, J., Merrill, K.M., Pipher, J.L., Puetter, R.C., Rudy, R.J., Russell, R.W., Soifer, B.T.: 1982, Ap. J. **253**, 174.

INFRARED PROPERTIES OF INTERSTELLAR GRAINS DERIVED FROM IRAS OBSERVATIONS

G. Chlewicki, R.J. Laureijs, F.O. Clark, P.R. Wesselius
*Space Research Laboratory and Kapteyn Astronomical Institute,
Groningen, The Netherlands*

1   IRAS observations of diffuse interstellar clouds

We have analyzed a sample of more than 10 isolated interstellar clouds, which have been observed by IRAS in all four bands (12, 25, 60 and 100 μm). The clouds range from very diffuse with no measurable extinction to dense objects with more than 4^m of blue extinction. The analysis of the distribution of energy emitted by each cloud over the four IRAS bands provides several results from which information on the general properties of grains in the infrared can be derived (Figure 1):

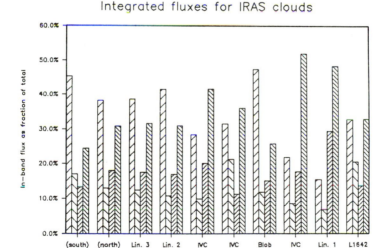

Figure 1: Energy distribution between the four IRAS bands for a selected set of diffuse interstellar clouds. The bars show the flux integrated over the apparent area of the cloud, as percentage of the total energy received by IRAS.

E. Bussoletti et al. (eds.), Experiments on Cosmic Dust Analogues, 313–319.
© *1988 by Kluwer Academic Publishers.*

– **total energy emitted at 100 μm** typically corresponds to a value of $I_{100}/A_B = 7.5$MJy sr^{-1}mag^{-1}($7.5 \cdot 10^{-17}$erg cm^{-2}s^{-1}Hz^{-1}sr^{-1}mag^{-1});

– **the radio of intensities at 60 and 100 μm** is high in most diffuse clouds (0.2–0.23) corresponding to colour temperatures of ~26 K (with a λ^{-1} emissivity law); in several clouds, the radio is much lower (<0.15);

– **the energy contained in the 12 and 25 μm bands** (covering the interval from 8 μm to 30 μm) is approximately the same as the emission in the 60 and 100 μm bands (45 μm < λ < 120 μm); there is considerable variability in the amount of short wavelength emission from cloud to cloud;

– **energy distribution between short wavelength bands** - typically the total energy emitted in the 12 μm band (7.8–14 μm) is 3 times higher than at 25 μm (15–30 μm); as for all other flux ratios, there are large differences between individual clouds.

2   IR brightness distribution in individual clouds

The IR surface brightness distribution in individual clouds can be used to study the dependence of emission from grains on the energy density in the radiation field and its spectral distribution. This analysis is only possible for several clouds with exceptionally regular morphology, such as G299–16 (Figure 2).

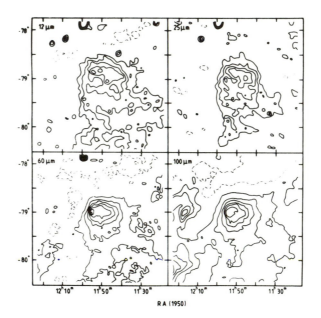

Figure 2: Surface brightness distribution in the four IRAS bands for an isolated interstellar cloud (G229–16 in Chamaeleon). Note the difference in morphology between far IR (60 and 100 μm) and mid IR bands (12 and 25 μm). The increrased 12 and 25 μm emission occurs on the side facing the galactic plane.

Several features in surface brightness distribution have direct implications for grain models:

- I_{60}/I_{100} **rises with increasing opacity**: the ratio of fluxes emitted at 60 and 100 μm is nearly independent of radiation field if the energy density is close to that of diffuse galactic radiation in the solar vicinity; in a few particularly regular clouds there is clear evidence that I_{60}/I_{100} increases towards the more opaque, central part of the cloud (Figure 3a);

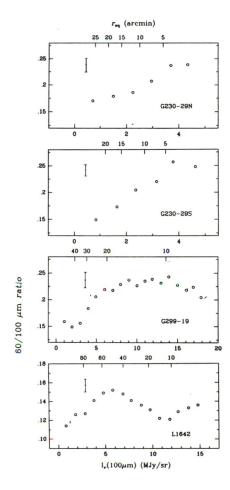

Figure 3a: IRAS flux ratios as a function of offset from cloud centre (defined by 100 μm brightness contours), and the 100 μm intensity (I_{100} is proportional to extinction almost throughout the range shown in the diagrams; the value of I_{100}/A_B is typically $\sim$7.5 MJy/sr/mag).

I_{60}/I_{100}: the apparent levelling off seen in L1642 and G229–16 may be due to the complex morphology of the central part of each cloud.

- **the morphology of IR emission** depends on the wavelength: the appearance of clouds tends to be similar at 60 and 100 μm, but is very different in the short wavelength bands (12 and 25 μm), where the emission seems to be associated with the UV component of the interstellar radiation field;

- **the energy emitted at short IR wavelengths declines with respect to far IR** emission as the opacity increases (Figure 3b);

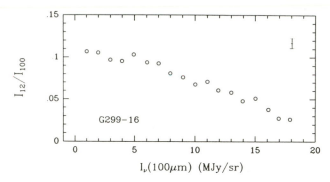

Figure 3b: IRAS flux ratios as a function of offset from cloud centre (defined by 100 μm brightness contours), and the 100 μm intensity (I_{100} is proportional to extinction almost throughout the range shown in the diagrams; the value of I_{100}/A_B is typically $\sim$7.5 MJy/sr/mag).

I_{12}/I_{100} for G229-16: this is the only cloud among the four shown in figure 3a with sufficiently regular morphology at 12 μm to allow derivation of meaningful values for the 12/100 μm flux ratio.

- **the radio of I_{100} to A_B remains constant up to $A_B \simeq 1^m$**; within this range of opacities there is no indication that infrared emission levels off as a result of lower grain temperatures inside the cloud.

3   Grain model

The observed properties and theoretical considerations (emission mechanism) divide the range covered by IRAS into two regions:

- **Far infrared (60 and 100 μm).** The explanation of IRAS observations in this range does not require a substantial modification of existing grain models. In particular, it is not necessary to resort to temperature fluctuations as an explanation for the strong 60 μm emission. Although the 60/100 μm colour temperature observed by IRAS is much higher than equilibrium temperatures of dielectric grains (15–20 K), it is well within the expected range for conductors. Both the strength of 60 μm emission and the rise in the 60/100 μm ratio towards higher opacities can be explained as a simple consequence of a bimodal distribution of grain temperatures (Figures 4 and 5).

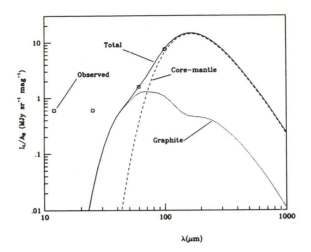

Figure 4: Predicted and observed infrared energy distribution for diffuse medium grains. The results shown in the diagram are for a simplified model which omits the contributions of small silicate grains responsible for the far UV rise in extinction as well as the "cool" graphite grains (Hong and Greenberg, Astron. Astrophys. **88**, 194). The short wavelength emission is attributed to aromatic molecules (see figure 6).

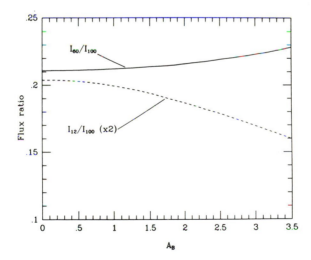

Figure 5: Predicted flux ratios in clouds as a function of apparent opacity along the line of sight for the grain model shown in figure 4. The simplified model is sufficient to reproduce the observed trend in I_{60}/I_{100} (rising towards higher opacities) and the opposite behaviour of I_{12}/I_{100}. The predicted decline in relative strength of the 12 μm band increases significately if the dielectric grains are assumed to absorb more in the near IR.

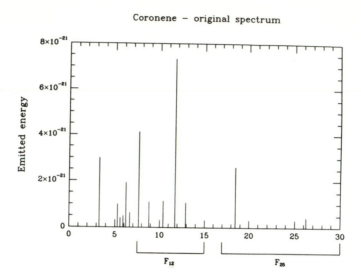

Figure 6a: Vibrational fluorescence from coronene excited by the diffuse interstellar radiation field. The bars represent the integrated emission (per atom) in each line. The calculations are based on the tabulation of fundamental vibrations of coronene and the IR absorption spectrum from Cyvin et al. (Z. Naturforsch., 1359, 37a).
Emission expected for an unmodified absorption spectrum of coronene.

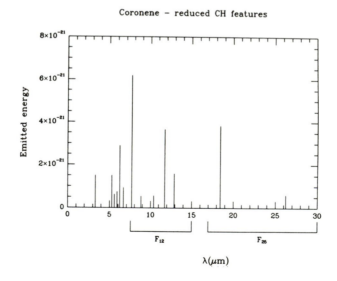

Figure 6b: Predicted emission with C-H features in the 12 μm range reduced by a factor of 3 to match the radio of C-C to C-H features observed in astronomical spectra.

The model shown in figure 4 uses core-mantle grains (silicate core, organic refractory mantle) as "cool" dielectrics with a small population of "hot" grains added as an explanation for the 60 μm emission. We have assumed that the photoprocessed molecular ice adopted for the mantle absorbs strongly in the visual and has an effective energy gap of $\sim$ 1 eV. High absorptivity in the visual and in the near IR is required by IRAS observations at 100 μm, which show a slow decline of grain temperature with opacity in clouds (the absorption at $\sim$ 1 μm should in fact be significantly stronger than in the model shown in figures 4 and 5). The model does not depend on the specific nature of the "warm" grains. We suggest that the most likely identification for these particles is small (in the "classical" sense, i.e., a radius of $\sim$ 0.01 μm) graphite grains. The rather low temperatures (T < 20 K) currently calculated for graphite particles result from the presence of a weak axial component of conductivity, which dominates the cooling of the grains (for strong absorbers, the IR cross-section is inversely proportional to the imaginary part of the dielectric constant). It seems likely that imperfect stacking in interstellar graphite grains may suppress the axial conductivity and increase the particle temperatures to $\sim$35–40 K. Only 20% of the abundance of graphite implied by the strength of the 2200 Å hump is required to explain the observed 60/100 μm flux ratio.

Mid infrared (12 and 25 μm). We have attributed the emission in the 12 μm band to aromatic molecules (all calculations have been carried out for coronene). The spectrum emitted by coronene under diffuse medium conditions has been calculated using the method developed by Léger and Puget (Astron. Astrophys. **137**, L5) of estimating the effective specific heat per carbon atom. The results show that aromatic molecules can give rise to all the observed emission not only 12 μm but also at 25 μm. The strenght of the 12 μm band can be matched with only $\sim$ 10% of cosmic carbon in aromatic molecules. The total depletion of carbon in the model is $\sim$ 60%.

References

Léger, A., Puget, J.L.: 1984, Astron. Astrophys. **137**, L5.

Hong, S.S., Greenberg, J.M.: 1980, Astron. Astrophys. **88**, 194.

INTERSTELLAR CIRCULAR POLARIZATION AND THE DIELECTRIC NATURE

G. Chlewicki[1], J.M. Greenberg[2]
[1]Space Research Laboratory and Kapteyn Astronomical Institute,
Groningen, The Netherlands
[2]Laboratory Astrophysics, University of Leiden, The Netherlands

1 Circular polarization criterion

Measurements of interstellar circular polarization are regarded as one of the essential constraints on theoretical models of interstellar grains. Observations accumulated since 1972 indicate that in all lines of sight, the interstellar circular polarization changes sign at a wavelength, λ_c, which is always close to the peak wavelength of linear polarization, λ_{max}. Mie theory calculations carried out by Martin (1972) apparently indicated that the relationship between linear and circular polarization depended on the optical properties of the grains: the observed coincidence of λ_c and λ_{max} could only be matched if the grains were transparent dielectrics (the imaginary part of the index of refraction in the visual should not be higher than 0.1). Martin's criterion has been widely used to evaluate the explanations of interstellar polarization in various grain models. A complementary constraint can be derived from the measurements of the visual albedo in reflection nebulae, which yield a typical value of 0.6, and therefore imply that at least one grain population at visible wavelengths must be significantly absorptive. Differences in the application of these criteria have led to a long controversy about the validity of specific grain models, which we have attempted to resolve with a new set of Mie theory calculations.

Martin's criterion has been challenged already in 1975 by Shapiro, who showed that a strongly absorbing conductor (magnetite with room temperature optical constants) can reproduce the observed coincidence of λ_c with λ_{max}. Shapiro has also developed a general argument based on Kramers-Kronig relations to show that the relationship between the circular and linear polarization should be independent of the optical constants of the grains. This elegant argument has been ignored in most of the recent studies of interstellar polarization and the discussion of grain models continued to be based of Martin's original criterion. We have caried out extensive Mie theory calculations to test the validity of Shapiro's argument by means of numerical experiments and we are able to confirm its universal applicability. The Kramers-Kronig relations also lead to an explanation of the origin of the spurius "circular polarization criterion".

321

E. Bussoletti et al. (eds.), Experiments on Cosmic Dust Analogues, 321–327.

2   Origin of interstellar polarization

Linear polarization of starlight passing through the interstellar medium results from preferential alignment of non-spherical interstellar grains induced by the magnetic field (the Davis-Greenstein alignment mechanism; all calculations shown in the figures are for partial spinning alignment). The projection of the galactic magnetic field on the plane of the sky defines the orientation of the electric vector with the lowest extinction efficiency for an ensemble of partly or totally aligned grains (the efficiency is highest for the perpendicular orientation). The difference between extinction efficiencies for these two orientation of the electric vector measures the linear dichroism of the interstellar medium and defines the wavelength dependence of the linear polarization: $p \propto Re\{Q_1 - Q_2\}$ (Q_2 corresponds to the projected direction of the magnetic field). Circular polarization arises from a more complex mechanism, which is related to the linear birefringence of the interstellar medium (the dependence of the phase of the electromagnetic wave scattered by the particle on the polarization of the incident radiation). The phase lag sustained by the scattered wave can be formally included in the definition of the extinction cross-section of the particle as its imaginary part. Circular polarization, which occurs only in the lines of sight in which the direction of the magnetic field varies, is then proportional to the product $Re\{Q_1 - Q_2\} \times Im\{Q_1 - Q_2\}$.

The effects of interstellar grains on the propagation of starlight through the interstellar medium can also be described by defining an index of refraction which measured the global optical properties of the interstellar medium. It can be shown that this index of refraction is related to the properties of individual grains through the formulae:

$$\tilde{m}' - 1 = \frac{1}{2k} N Im\{C_{ext}\}$$

$$\tilde{m}'' = \frac{1}{2k} N Re\{C_{ext}\}$$

with N representing the volume density of scattering particles, and $k = 2\pi/\lambda$. The polarizing properties of the medium can be described by the differential index of rerfraction $\tilde{m}_{diff} = \tilde{m}_1 - \tilde{m}_2$ (for the two previously defined orientations of the electric vector). The imaginary part of $\tilde{m}_{diff}$ represents the linear dichroism of the interstellar medium; the birefringence is characterized by the real part, and the observed circular polarization is determined by their product.

3   Mie theory results

We have carried out Mie theory calculations for infinite cylinders of either homogeneous or core-mantle structure (silicate cores, organic refractory mantles). The material adopted for grain mantles was the refractory component of photoprocessed molecular ice (organic refractory), for which we have "synthesized" a set wavelength dependent optical constants on the basis of the general physical and chemical properties of the material.

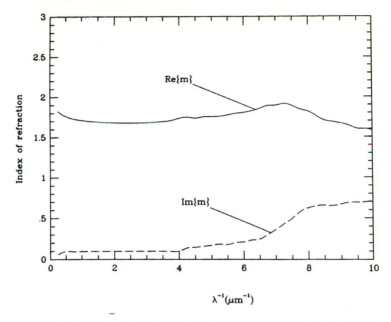

Figure 1a: The index of refraction of organic refractory with moderate visual absorptivity $(m'' \simeq 0.1)$.

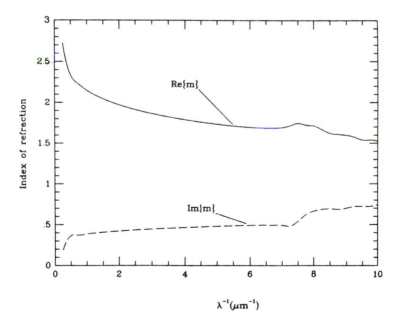

Figure 1b: The same as in figure 1a for a more strongly absorbing material $(m'' \simeq 0.4$ at 5500 Å). Note the strong rise in m' towards infrared wavelengths.

We allowed the imaginary part of the dielectric constant at visible wavelengths to be modified arbitrarily so that the effects of varying visual absorptivity on the calculated polarization could be studied. The real part of ϵ was always derived from the Kramers-Kronig relations. The same material was also used as a "generic" dielectric in calculations involving homogeneous grains.

Examples of the optical constants (index of refraction $m = m' - im''$) of the "synthetic" organic refractory have been shown in figure 1 for moderate and high values of visual absorptivity. The resulting linear and circular polarization has been shown in figures 2a and 2b. The calculations used in figure 2b show that even for a value of m'' as high as 0.4 in the visual, i.e. 4 times higher than Martin's upper limit, the coincidence between λ_c and λ_{max} is preserved (!). We have continued the calculations with increasing values of m'' up to $m'' = 1$ with no change in the calculated ratio of λ_c and λ_{max}. These results cannot be reconciled with the existence of an upper limit on the allowed value of m'' in the visual.

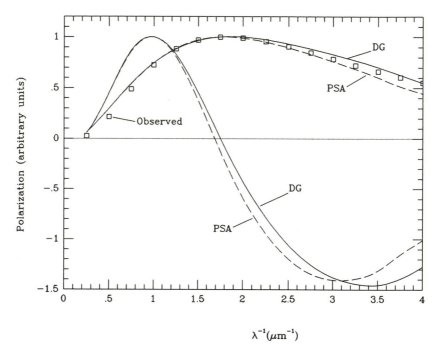

Figure 2a: Linear and circular polarization derived from Mie theory calculations for core-mantle infinite cylinders with moderately absorbing mantles (optical constants as in figure 1a). Solid line: partial Davis-Greenstein alignment (DG); dashed line: perfect spinning alignment (PSA). The average particle size is 0.09 μm. The observed linear polarization (squares) is from Wilking et al. (A.J., 87, 695), adjusted for $(\lambda_{max})^{-1} = 1.75$ μm^{-1}.

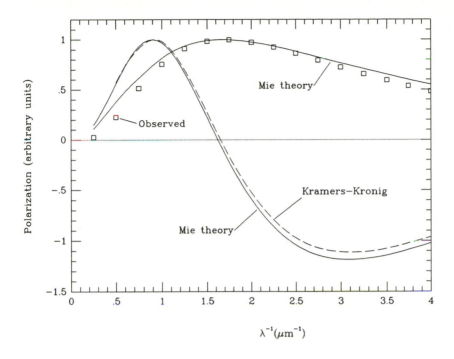

$$\lambda^{-1}(\mu m^{-1})$$

Figure 2b: Linear and circular polarization for homogeneous infinite cylinders of organic refractory with high visual absorptivity (index of refraction as in figure 1b). Only DG results have been shown. The size distribution has been adjusted to match the observed wavelength dependence of linear polarization. The Kramers-Kronig curve shows the circular polarization obtained by explicitly applying the Kramers-Kronig relations to the calculated linear polarization. The deviation from directly calculated circular polarization is due only to the limited accuracy of the Kramers-Kronig integration.

4 Kramers-Kronig relations and interstellar polarization

Our Mie theory results are no more than an illustration of the general relationship pointed out by Shapiro. The differential index of refraction of the interstellar medium has to satisfy and Kramers-Kronig relations, in particular:

$$\tilde{m}'_{diff}(\omega) = \frac{2}{\pi} \int_0^\infty \frac{\omega' \, \tilde{m}''_{diff}(\omega')}{\omega'^2 - \omega^2} d\omega'$$

The consequences of this equation have been illustrated in figure 3, which shows the real and imaginary part of $\tilde{m}_{diff}$ derived from the same Mie theory cross-sections that have been presented in a different form in figure 2b. The observed shape of the linear polarization curve corresponds to an isolated "resonance" in the imaginary part of $\tilde{m}_{diff}$.

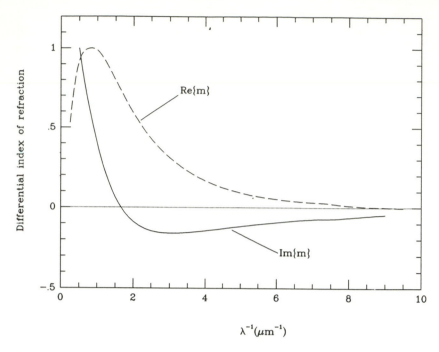

Figure 3: Real and imaginary part of the differential index of refraction of the interstellar medium. The results are derived from the same Mie theory calculations as in figure 2b.

It is a well-known property of any dispersive medium that in the vicinity of an isolated resonance, the real part of the responce function must have a region of anomalous dispersion, in which it decreases with frequency. Anomalous dispersion forces the real part of $\tilde{m}_{diff}$ to change sign at a wavelength which must be close to the peak of the resonance; this change of sign is observed as a cross-over in the interstellar circular polarization (because of its proportionally to $Re\{\tilde{m}_{diff}\}$). Therefore, the Kramers-Kronig relations requirer that λ_c be close to λ_{max}; the nearly exact coincidence observed in the interstellar medium is due to the particular shape of the linear polarization peak. This result cannot depend on the optical constants of grains.

5   The arrow of time

The Kramers-Kronig relations can also explain Martin's incorrect interpretation of the relationship between linear and circular polarization. Figure 1b shows that if the imaginary part of the index of refraction of the particle remains high in the visible, the real part of the index must rise towards longer wavelengths. Martin's calculations were based on the assumption -sanctioned by a long tradition- that both the real and imaginary part of the index of refraction were constant. If m'' is low in the visible, this is approximately true (see figure 1a), but the same assumption drastically violates the Kramers-Kronig relations if m'' is as high as in our figure 1b. Such a violation of causality is inevitably carried through all the steps in Mie theory calculations and, in particular, results in wrong predictions for circular polarization. Therefore, the deviations from the observed value of λ_c that Martin derived for high values of m'' measured not so much the effects of grain properties as simply the extent to which the Kramers-Kronig relations were violated. This is an interesting example of the practical effects of the Titchmarsh theorem which states the equivalence between the Kramers-Kronig relations and the preservation of causality in the time-dependent response of a physical system. A violation of the Kramers-Kronig relations effectively allows electromagnetic signals to propagate back in time, which in Mie theory must lead to incorrect predictions for the phase of the scattered wave. As a result, the predicted values of extinction and linear polarization, which depend only on the amplitude of the scattered wave, remain qualitatively correct; calculated quantities such as circular polarization, which involve the phase of the wave, are entirely wrong.

6   Implications for grain models

We believe that in spite of the absence of any upper limit on the absorptivity of polarizing grains, it is unlikely that polarization is due to conductors (the observed polarization of the 9.7 μm feature provides the strongest argument). Instead, the removal of the "circular polarization criterion" implies that the polarizing grains could contain any dielectric, including an important class of materials, such as organic refractory and hydrogenated (or hydrogen-free) amorphous carbon, which absorb strongly in the visual. Any criticism of existing grain models based on the "circular polarization criterion" is obviously invalid.

References

Martin, P.G.: 1972, M.N.R.A.S., **159**, 179.

Shapiro, P.R.: 1975, Ap.J., **201**, 151.

SIZE DISTRIBUTION OF COMETARY DUST, METEOROIDS AND ZODIACAL GRAINS

M. Fulle
ISAS - International School for Advanced Studies, Trieste, Italy

1 Introduction

It is commonly accepted there is a strong relation among dust released from comets, meteoroids and dust in the zodiacal cloud. In this paper we investigate this relation starting from size distribution analysis of cometary dust grains. Size distribution of dust grains from comets is strongly time-dependent. A detailed analysis of this quantity was possible only with analysis of the dust tails of bright comets, or with the space missions to comet Halley. In the former case, the most successfull approach was the inverse approach (Fulle, 1987a) to the Finson-Probstein method (Finson and Probstein, 1968), which allowed to obtain the time-dependence of the size distribution of dust grains over periods of weeks for the comets Bennett 1970II (Fulle, 1987a), Arend-Roland 1957III, Seki-Lines 1962III (Fulle, 1987b), and Halley (Fulle et al., 1987a, b). The dust mass loss rates resulted regularly dependent on the Sun-Comet distance, whereas the number loss rates, ruled by smaller grains, resulted dominated by outbursts. These results were confirmed by space probes to comet Halley, which reported strong variations in time of the size distribution, probably due to fragmentation processes which are very active in the jets characterizing the inner coma (Zarnecki et al., 1986). To compare the size distribution of cometary grains to the same quantity of the zodiacal dust, we have to use averaged quantities. We assume the mean size distribution averaged over long times (weeks or months) given by the Finson-Probstein method as the mean distribution of cometary dust grains. This choice is supported by the strong correlation shown by the time-averaged size distributions obtained for the four comets already considered (Fulle et al., 1987a, b). The introduction of this approximation will only allow us to obtain qualitative indications about the constraints which are necessary to link cometary and interplanetary dust size distributions.

2 The time-averaged dust size distribution g of comets

Results of the inverse approach (Fulle, 1987a) of the Finson-Probstein method (Finson and Probstein, 1968) allowed to obtain the power index k of the time-averaged size distribution of cometary dust grains $g(\rho_d d)$ from the Finson-Probstein modified distribution

329

E. Bussoletti et al. (eds.), Experiments on Cosmic Dust Analogues, 329–336.
© 1988 by Kluwer Academic Publishers.

$f(\beta)$:

$$f(\beta) \quad \propto \quad (\rho_d d)^4 g(\rho_d d) \tag{1}$$

$$g(\rho_d d) \quad \propto \quad (\rho_d d)^{-k} \tag{2}$$

$$(\rho_d d) \quad = \quad C_{pr} Q_{pr} \beta^{-1} \tag{3}$$

where ρ_d is the bulk density of the dust grain of diameter d, $C_{pr} = 1.19 \cdot 10^{-4}$ g cm^{-2}, $Q_{pr} \approx 1$ for $d > 10$ μm (Burns et al., 1979) and β is the ratio between the solar radiation pressure force and the solar gravitational force. The value of k depends on the actual $(\rho_d d)$-interval considered. Fulle et al., (1987a, b) obtained the value of k which are reported in table 1 from the analysis of comets Bennett 1970II (Fulle, 1987a), Arend-Roland 1957III, Seki-Lines 1962III (Fulle, 1987b) and Halley (Fulle et al., 1987a, b).

Table 1: Power indexes of cometary size distributions $(\rho_d d)$: value at which size distribution is considered (g cm^{-2}). k_O: Power index observed. k_C: Power index given by expression (5). Comet: comet to which the distribution power index is related.

$(\rho_d d)$	k_O	k_C	Comet
10^{-1}	4.7	4.7	C/1962III
10^{-2}	4.0	4.0	C/1957III
$5 \cdot 10^{-3}$	3.9	3.8	C/Halley
$5 \cdot 10^{-3}$	3.7	3.8	C/1970II
$2 \cdot 10^{-3}$	3.5	3.5	C/Halley
10^{-3}	3.3	3.3	C/1957III
$2 \cdot 10^{-4}$	3.0	2.8	C/1970II

We do not consider the data about the dust size distribution of comet Halley given by space missions for the subfemtogram grains, because we do not know the bulk density values of these particles (Sekanina, 1986). We note that the power indexes reported in table 1 are fitted very well by the Jambor's size distribution (Kerker, 1969; Jambor, 1973) plotted in figure 1:

$$g(\rho_d d) = \frac{\exp\left\{-ln^2[(\rho_d d)/(\rho_d d)_m]/2\sigma^2\right\}}{\sqrt{2\pi}\sigma(\rho_d d)_m \exp[\sigma^2/2]} \tag{4}$$

with the value of the two free parameters $(\rho_d d)_m = 2 \cdot 10^{-8}$ g cm^{-2} and $\sigma = 1.8$. The logarithmic slopes given by (4):

$$d[ln\ g(\rho_d d)]/d[ln(\rho_d d)] = \sigma^2 ln[(\rho_d d)/(\rho_d d)_m] \tag{5}$$

are compared with the observed power indexes k in table 1 to show this satisfactory agreement. In the following section we will consider this distribution as the mean size distribution of dust grains from comets.

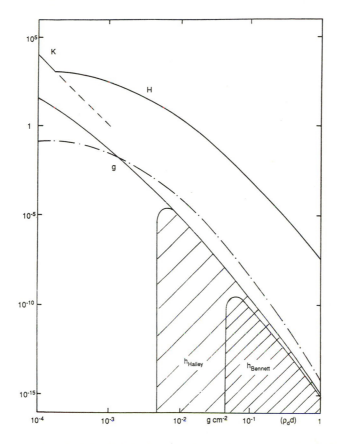

Figure 1: *g*: cometary dust size distributions (continuous line, from dust tail analysis (normalized); dashed and dotted line, from the linking to the size distribution of interplanetary dust). *h*: meteoroid size distribution for comet Halley (single dashed area) and Bennett (double dashed area). *H*: distribution of the bounded population of the zodiacal cloud. *K*: distribution of the hyperbolic population of the zodiacal cloud.

3   Size distribution *h* of meteoroids from comets

Fulle (1987c) showed that dust grains which are injected into a bounded orbit by a comet with perihelion q and eccentricy e must satisfy the following non-dimensional inequalities:

$$2\sqrt{(1 - \xi)\chi}\,\beta^{1/6} - \chi\,\beta^{1/3} - \beta + \xi > 0 \tag{6}$$

$$\sin(\varphi - \Psi) > \varsigma = \frac{\chi\,\beta^{1/3} + \beta - \xi}{2\sqrt{(1 - \xi)\chi}\,\beta^{1/6}} > -1 \tag{7}$$

where $\chi = rv^2(t)/2GM$, $\xi = (1-e)r/2q$, G is the gravitational constant, M is the solar mass, $\pi/2 - \Psi$ is the direction of the velocity vector of the comet nucleus. The dust grain is ejected in the direction φ (φ corotates with the comet true anomaly, $\varphi = 0$ at the subsolar point) with the velocity $v(t)$ at $\beta = 1$ and at the Sun-Comet distance r. Let the inequality (6) be satisfied for $\beta < \beta_1$ is related to $(\rho_d d)_1$ by expression (3), and the inequality (7) for $\beta > \beta_2$ where β_2 is related to $(\rho_d d)_2$. Fulle (1987b, c) showed that the dust mass of meteoroids is almost insensitive to the anisotropies of dust ejection from the inner coma. Therefore the size distribution $h(\rho_d d)$ of the meteoroids injected into bounded orbits by a comet with a dust size distribution $g(\rho_d d)$ can be expressed as follow:

$$h(\rho_d d) = 0 \qquad \text{for} \quad (\rho_d d) < (\rho_d d)_1 \tag{8}$$
$$h(\rho_d d) = g(\rho_d d)(1 - \varsigma)/2 \quad \text{for} \quad (\rho_d d)_1 < (\rho_d d) < (\rho_d d)_2$$
$$h(\rho_d d) = g(\rho_d d) \qquad \text{for} \quad (\rho_d d) > (\rho_d d)_2$$

The distributions of $h(\rho_d d)$ of comets Bennett (Fulle, 1987c) and Halley (Fulle et al., 1987c) are plotted in figure 1. The distribution of comet Bennet is characterized by the value $(\rho_d d)_1$ only. In fact, since the orbit of comet Bennett is quasi-parabolic, we have $\beta_2 = 0$. This means that ejection velocity can inject into open orbits also the largest grains. On the contrary, in the case of comet Halley the value $(\rho_d d)_1$ is very close to $(\rho_d d)_2$. This means that injection of meteoroids into bounded orbits is dominated by the orbital eccentricity of the parent comet and is insensitive to the ejection velocity. In other words, a grain larger than $(\rho_d d)_2$ is injected into a bounded orbit whatever is its ejection velocity. Since $\chi \ll 1$, for $e < 0.95$ we have $(\rho_d d)_1 \approx (\rho_d d)_2$. Moreover, the largest amount of dust and meteoroids is produced near perihelion. Therefore a suitable approximation of size distribution of meteoroids from short-period comets is the following:

$$h(\rho_d d) = 0 \qquad \text{for} \quad (\rho_d d) < 2C_{pr}(1-e)^{-1} \tag{9}$$
$$h(\rho_d d) = g(\rho_d d) \quad \text{for} \quad (\rho_d d) > 2C_{pr}(1-e)^{-1}$$

4   Size distributions H and K of zodiacal dust

We consider the results of LeSergeant and Lamy (1978, 1980), who found a combination of two populations of zodiacal grains, characterized by two distributions, $H(\rho_d d)$ and $K(\rho_d d)$, which overlap each other. When we consider a dust bulk density of ≈ 2 g cm^{-3} (Delsemme, 1982), $H(\rho_d d)$ results well fitted by expression (4) with parameters $(\rho_d d)_m = 1.2 \cdot 10^{-4}$ g cm^{-2}, $\sigma = 1.3$, whereas $K(\rho_d d)$ by expression (2) with $k = 3.85$. The distributions H and K are plotted in figure 1.

The contribution of $H(\rho_d d)$ to the total size distribution of zodiacal dust vanishes for $(\rho_d d) < 2 \cdot 10^{-4}$ g cm^{-2}. If we consider a grain released at zero velocity from a planetary body at perihelion, we have that it can be injected into a bounded orbit if $\beta < (1-e)/2$ (Burns et al., 1979), where e is the eccentricity of the parent body orbit. Therefore it is very unlikely to have dust with size $(\rho_d d) < 2.4 \cdot 10^{-4}$ g cm^{-2} in bounded orbits. This value is very close to the turning point between H and K distributions. Therefore, as pointed by LeSergeant and Lamy, we can consider that H distribution characterizes the population of meteoroids in bounded orbits, whereas K distribution characterizes dust grains injected into ·hyperbolic orbits by fragmentation processes in the inner Solar System (Leinert et al., 1983; Grün et al., 1985).

5 Relation between H and h distributions

In this section we analyse the relation between the H distribution of the zodiacal cloud and the h distribution of meteoroids from short-period comets, in order to obtain the characteristics of the original cometary distribution g which are required by the condition that the H population is replenished directly from the dust released by short-periods comets. Since we are interested in qualitative indications about time-averaged size distribution, we will assume that short-period comets produce dust grains characterized by a time-independent size distribution. We consider the total number of grains released from a comet in function of its orbital eccentricity $N(e)$:

$$N(e) \propto <\dot{N}(e)> = T^{-1} \int_0^T \dot{N}_q(e)(q/r)^2 \, dt = \dot{N}_q(e)(1-e)^{3/2}(1+e)^{-1/2} \qquad (10)$$

where T is the orbital period of the comet. In (10) we have assumed an inversely square dependence of the dust number production rate $\dot{N}$ on the Sun-Comet distance, following the results of Ney (1982) and Fulle (1987b) about dust mass loss rates. We remember that, since we consider a time-independent size distribution, the ratio between dust number and mass loss rates is time-independent. We consider the H population, which is composed of N_0 dust grains, as the sum of the h populations of cometary meteoroids composed of $N(e)$ dust grains:

$$N_0 G(\rho_d d) d(\rho_d d) = g(\rho_d d) \int_0^{E(\rho_d d)} N(e) \, de d(\rho_d d) \qquad (11)$$

$$E(\rho_d d) = 1 - 2C_{pr}(\rho_d d)^{-1}$$

$$G(\rho_d d) = H(\rho_d d) - K(\rho_d d)$$

$$K(\rho_d d) = H(2C_{pr})[2C_{pr}/(\rho_d d)]^k$$

where E is the highest eccentricity which contributes to the H distribution in the range between $(\rho_d d)$ and $(\rho_d d) + d(\rho_d d)$. When we invert this integral equation, we obtain:

$$\frac{N(e)}{N_0} = \left. \frac{\frac{dG}{d(\rho_d d)} - \frac{G}{g}\frac{dg}{d(\rho_d d)}}{g\frac{dE}{d(\rho_d d)}} \right|_{(\rho_d d)=2C_{pr}(1-e)^{-1}} \qquad (12)$$

We substitute (4), (9) and (10) into (11) and (12), obtaining:

$$\dot{M}(e) \propto \dot{N}_q(e) \propto (1+e)^{1/2}(1-e)^{-5/2}\{(s_G/\sigma_G^2 - s_H/\sigma_H^2)\exp[s_G^2/2\sigma_G^2 - s_H^2/2\sigma_H^2] +$$
$$+(1-e)^k(k - s_G/\sigma_G^2)\exp[s_G^2/2\sigma_G^2 - s_0^2/2\sigma_H^2]\} \qquad (13)$$

$$s_{G,H} = \ln[2C_{pr}/(\rho_d d)_{mG,H}] - \ln(1-e)$$

$$s_0 = \ln[2C_{pr}/(\rho_d d)_{mH}] \qquad (14)$$

where $(\rho_d d)_{mG}$ and σ_G are related to $g(\rho_d d)$, while $(\rho_d d)_{mH}$ and σ_H to $H(\rho_d d)$.

6   Discussion

The results given by expression (13) are plotted in figure 2 following the parameters of the size distributions of cometary and interplanetary dust discussed in previous sections. We note a large disagreement with respect to the actual dust mass loss rates obtained by Newburn and Spinrad (1985) with spectrophotometric methods, and by Sekanina and Schuster (1978), Hanner et al. (1985) and Fulle et al. (1987a, b) with the Finson-Probstein method of dust tail analysis. In particular, comets with orbits of low eccentricity ($e < 0.7$) produce too much dust with respect to that required from comets with orbits of moderate eccentricity ($e > 0.7$).

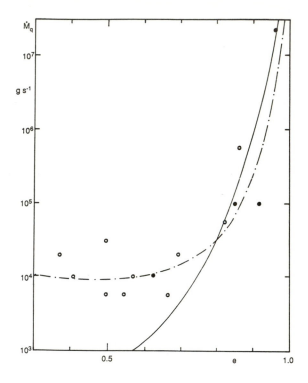

Figure 2: Observed dust mass loss rates from short-period comets (full dots, from dust tail analysis; unfilled dots, from spectrophotometric analysis) and theoretic dust mass loss rate required by relation (13) between cometary and interplanetary dust size distributions; continuous line: cometary distribution of parameters $(\rho_d d)_m = 2 \cdot 10^{-8}$ g cm^{-2} and $\sigma = 1.8$ from dust tail analysis is assumed; dashed and dotted line: best fitting obtained with a size distribution of parameters $(\rho_d d)_m = 1.2 \cdot 10^{-4}$ g cm^{-2} and $\sigma = 1.25$.

A sufficient agreement with the observations is obtained with a distribution which is much flatter (Figure 1, Figure 2) than that observed for long-period or parabolic comets. This leads us to two alternative conclusions. If we assume that short-period comets eject dust with a size distribution close to that of long-period comets, we must conclude not only that fragmentation cannot produce the observed shape of the zodiacal dust distribution, but also that small grains must be converted into large grains, since the required size distribution of grains given by the source of the zodiacal dust must have a prominent lack of small grains with respect to that from comets. It is difficult to conceive efficient accretion processes among meteoroids during their motion in space. The second hypothesis is that the size distribution of dust from short-period comets effectively is much flatter than that characterizing long-period comets, even flatter than that plotted in figure 1 in order to take into account the probable conversion of large grains into smaller ones due to collision. This large difference is hardly explicable. We could suppose that the largest amount of small grains produced by long-period comets is due to fragmentation processes very active in the jets (Zarnecki et al., 1986) observed in the inner coma (Keller et al., 1986) of comet Halley. Short-period comets of the Jupiter family have not jets, and this could result in a lack of fragmentation of the original large grains. In other words, the much flatter size distribution would be the true distribution of cometary dust grains. This scenario could also explain why short-period comets do not show dust tails, which are composed of small grains, whereas they show anti-tails (Sekanina and Schuster, 1978), which are composed of large grains. Moreover, a much flatter distribution could increase the estimate of dust mass loss rate, of a factor $\exp[7(\sigma_2^2 - \sigma_1^2)/2](\rho_d d)_{m2}/(\rho_d d)_{m1} \approx 15$ for the assumed values of the parameters in expression (4). This could decrease the difference between the necessary mass to replenish the zodiacal cloud and the mass effectively produced by short-period comets. On the contrary, if future space mission should confirm that the short-period comets of the Jupiter family produce dust with a size distribution close to that observed in the case of comet Halley or long-period comets, we should conclude that short-period comets cannot be considered the main source of interplanetary dust, both for the lack of mass, and for the incompatibility between h and H size distributions. In this case the main candidate would be given by long-period comets, which produce a sufficient amount of mass of meteoroids (Fulle, 1987c). Fragmentation could be the only way to obtain grains smaller than 0.4 mm from this source of dust, that is to obtain the zodiacal H distribution from the original h distribution.

References

Burns, J.A., Lamy, P.L., Soter, S.: 1979, Icarus **40**, 1.

Delsemme, A.H.: 1982, in "Comets", L.L. Wilkening ed., (University of Arizona Press, Tucson) p. 85.

Finson, M.L., Probstein, R.F.: 1968, Astrophys. J. **154**, 327, 354.

Fulle, M.: 1987a, Astron. Astrophys. **171**, 327.

Fulle, M.: 1987b, Astron. Astrophys., in press.

Fulle, M.: 1987c, Astron. Astrophys., in press.

Fulle, M., Barbieri, C., Cremonese, G.: 1987a, Proc. ESLAB Symposium on the Diversity and Similarity of Comets, Bruxelles, ESA SP-278, in press.

Fulle, M., Barbieri, C., Cremonese, G.: 1987b, Astron. Astrophys. (submitted).

Fulle, M., Barbieri, C., Cremonese, G.: 1987c, Astron. J. (submitted).

Grün, E., Zook, H.A., Fechtig, H., Giese, R.H.: 1985, Icarus **62**, 244.

Hanner, M.S., Knacke, R., Sakanina, Z., Tokunaga, A.T.: 1985, Astron. Astrophys. **152**, 177.

Jambor, B.J.: 1973, Astrophys. J. **185**, 727.

Keller, H.U., Arpigny, C., Barbieri, C., Bonnett, R.M., Cazes, S., Coradini, M., Cosmovici, C.B., Delamere, W.A., Huebner, W.F., Hughes, D.W., Jamar, C., Malaise, D., Reitsema, H.J., Schmidt, H.U., Schmidt, W.H.K., Seige, P., Whipple, F.L., Wilhelm, K.: 1986, Nature **321**, 320.

Kerker, M.: 1969, in "The Scattering of Light", (Academic Press), ch.7.

Leinert, C., Roser, S., Buitrago, J.: 1983, Astron. Astrophys. **118**, 345.

LeSergeant, L.B., Lamy, P.L.: 1978, Nature **276**, 800.

LeSergeant, L.B., Lamy, P.L.: 1980, Icarus **43**, 350.

Newburn, R.L.Jr., Spinrad, H.: 1985, Astron. J. **90**, 2591.

Ney, E.P.: 1982, in "Comets", L.L. Wilkening ed. (University of Arizona Press, Tucson), p. 323.

Sekanina, Z.: 1986, Proc. 20th ESLAB Symposium on the Exploration of Halley's comet, ESA SP-250, vol.II, p.131.

Sekanina, Z., Schuster, H.E.: 1978, Astron. Astrophys. **68**, 429.

Zarnecki, J.C., Alexander, W.M., Burton, W.M., Hanner, M.S.: 1986, Proc. 20th ESLAB Symposium on the Exploration of Halley's comet, ESA SP-250, vol.II, p.185.

A ZODIACAL DUST EMISSION MODEL

P. Temi[1], P. de Bernardis[1], S. Masi[1], G. Moreno[1], A. Salama[2]
[1]*Dipartimento di Fisica, Università "La Sapienza", Roma, Italy*
[2]*Max-Planck-Institute für Astronomie, Heidelberg, F.R.G.*

1 Introduction

Several experiments have been devoted to the study of the zodiacal dust cloud emission in the infrared (ZIP, IRAS, ARGO etc.), but the agreement between theory and experimental results is poor: only semi-empirical models fit in satisfactory way the data (Price et al., 1985, Salama et al., 1987) while the standard model (Roser and Staude 1978) predicts fluxes about 20 times smaller than observed (Murdock and Price, 1985). Here we describe a deterministic model which, starting from known optical properties of the grains and using the expected interplanetary dust density and size distributions, predicts the infrared zodiacal emission. In order to reduce the number of free parameters we assume the simplest hypothesis of spherical graphite and silicate grains. Comparison with all available data shows an agreement within a factor 3 in the range of wavelengths between 5 and 200 μm and, for solar elongations, between 2° and 180°.

2 The model

Our model is based on the following hypothesis:
a) optical properties of grains: we used the values of the complex refractive index $n(\lambda)=m(\lambda)+i\,k(\lambda)$ quoted by Draine (1985) for graphite and astronomical silicates. Using Mie formulas we calculated the scattering phase function $F(\lambda, s, \theta)$ and the absorption efficiency $Q_{abs}(s, \lambda)$: λ is the wavelength (in the range 0.14 μm to 300 μm) and s is the grain size (in the range 0.01 μm to 100 μm). The calculated $Q_{abs}(s,\lambda)$ is shown in figure 1.
b) density and size distributions: we assumed that the number of grains in the ecliptic plane per unit volume and unit size interval is

$$f(s,r) = n(s)w(r) \tag{1}$$

where

$$n(s) = n_0 s^{-\xi} \tag{2}$$

$$w(r) = r^{-\nu} \tag{3}$$

337

E. Bussoletti et al. (eds.), Experiments on Cosmic Dust Analogues, 337–344.

(r being heliocentric distance in A.U.).

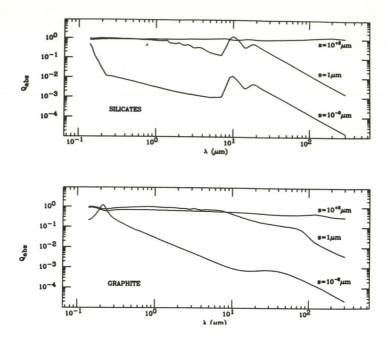

Figure 1: Absorption efficiency $Q_{abs}(\lambda, s)$ for silicates and graphite, calculated for three different particle sizes (s) using the Mie formulas.

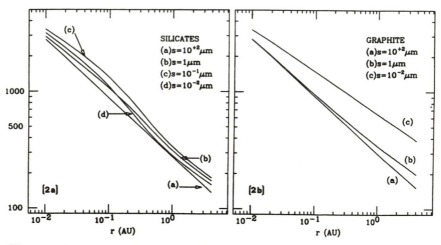

Figure 2: Grains temperatures as a function of eliocentric distance r and particle size s calculated for silicates and graphite.

The constants n_0 and ξ have been obtained by Grun et al. (1985) using "in-situ" observations of meteoroid flux. The value of ν has been taken as a free parameter. Assuming balance between absorbed and re-emitted radiation we calculated grain temperatures as functions of distance r and size s (Figure 2). We computed emission coefficients j_{th} and j_{sca} (i.e. the energy emitted and scattered per unit volume at distance r and at wawelength λ):

$$j_{sca}(\lambda, x(r)) = I_0(\lambda)\omega(x) \int \pi s^2 Q_{sca}(\lambda, s) F(\lambda, s, \theta) f(s, x)\, ds \qquad (4)$$

$$j_{th}(\lambda, x(r)) = \int \pi s^2 Q_{abs}(\lambda, s) f(s, x) BB(T(s, x), \lambda)\, ds \qquad (5)$$

where x is the distance measured along the line of sigh; BB is the Planck function; I_0 is the mean solar intensity; ω is the solid angle subtended by the sun at x. In order to compare model predictions with observations, we integrated the emission coefficients along the line of sight, obtaining the expected brightness as a function of the elongation ε:

$$I(\varepsilon, \lambda) = \int [j_{sca}(\lambda, x) + j_{th}(\lambda, x)]\, dx \qquad (6)$$

3 Comparison with observations

The available infrared data have been collected by the experiments listed below:

reference	wavelength	elongation	ecliptic latitude
Price et al., 1980 rocket (AFGL)	11; 20 μm	30°–80° 150°–195°	$-40°/\ 40°$
Murdock & Price 1985 rocket (ZIP)	2–30 μm	22°–180°	$-60°/\ 90°$
Hauser et al., 1984 satellite (IRAS)	12; 25 μm 60; 100 μm	68°–103°	$-90°/\ 90°$
Salama et al., 1987 balloon (ARGO)	11; 19; 50 μm 108; 225 μm	10°–90°	ecliptic plane

Data taken in ecliptic plane are plotted vs elongation in figure 3 and vs wavelength in figure 4. As a first step we determined the value of ν (Equation 3), by fitting with equation 6 the IR fluxes at 11 μm of AFGL and ZIP experiments as a function of elongation. The shape of the theoretical curve for both silicate and graphite was very similar to the trend of the data. However, absolute values are a factor K=2.6 lower for silicates and a factor 3.2 lower for graphite. This can be due to an underestimate of the density of interplanetary dust. Using the appropriate value for K we obtained a good fit to the data for silicate grains with $\nu=0.9$ and for graphite grains with $\nu=1$. These values are not far from $\nu=1.3$ derived from optical measurements (Leinert et al., 1981) and very close to the value (1.0) theoretically predicted for a steady equilibrium state with Poynting-Robertson drag. Finally we fitted with equation 6 the observed spectral data between 5 and 100 μm at elongations of 45°, 90°, 180°. The best fit curves corresponding to different values of ν are almost identical (Figures 4a, b, c). Values of K obtained in this way are in close agreement with values obtained before for $\varepsilon = 90°$ and 180° while values at 45° are somewhat lower (K=1.8).

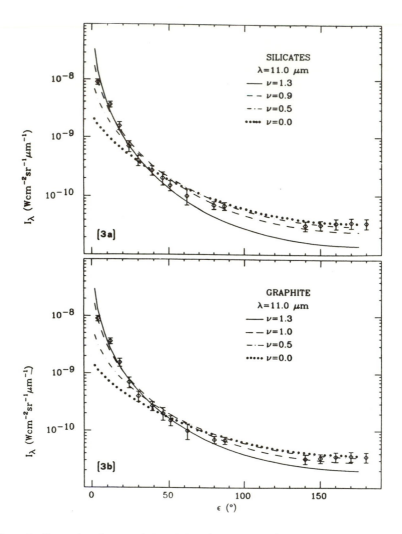

Figure 3: Comparison between infrared data (at λ=11.0 μm) taken in the ecliptic plane and theoretical infrared flux as a function of elongation for four values of ν; each of these curves was chosen minimizing the deviation from the data.

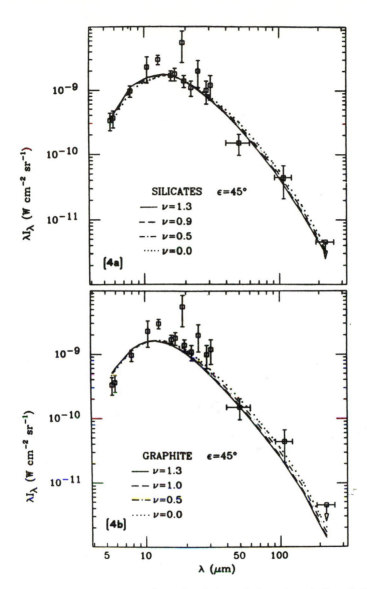

Figure 4a, b: Comparison between observed and theoretical spectrum in the ecliptic plane at three different elongations. Theoretical curves were obtained for graphite and silicates with the same normalization procedure used in figure 3.

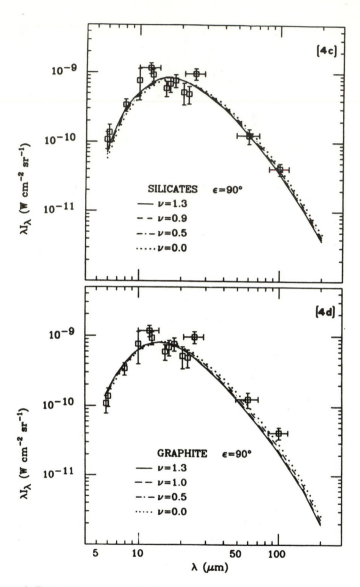

Figure 4c, d: For caption see figure 4a, b.

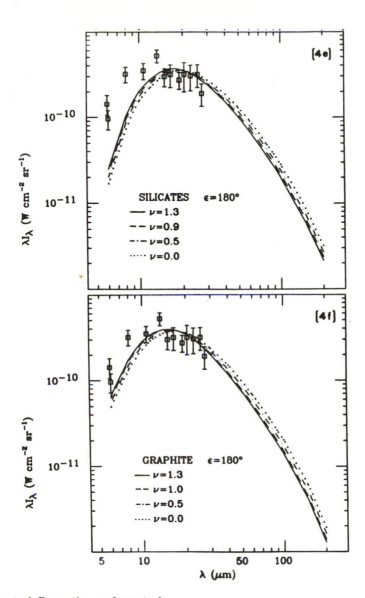

Figure 4e, f: For caption see figure 4a, b.

4 Conclusions

We conclude that our model fits satisfactory all available data of IR zodiacal emission under the assumption that the density of interplanetary dust is about three times higher that the one obtained from direct measurements. Observations may be accounted by silicate as well as by graphite grains. In both cases, the radial trend of grain density is described by equation 3, with $\nu \sim 1$. We stress the fact that our analysis does not exclude a much complex composition and structure of interplanetary grains as proposed by current theories (see for a review Mukay, 1986) and indicated by recent observations of cometary dust (Kissell et al., 1986). We have just proved that present infrared observations can be fitted by simple assumptions on shape and chemistry of the grains.

References

Draine, B.T.: 1985, Ap. J. Suppl. **57**, 586.

Grün, E., Zook, H.A., Fechtig, H., Giese, R.H.: 1985, Icarus **62**, 244.

Hauser, M.G., Gillett, F.C., Low, F.J., Gautier, T.N., Beichman, C.A., Neugebauer, G., Aumann, H.H., Baud, B., Boggess, N., Emerson, P., Houck, J.R., Soifer, B.T., Walker, R.G.: 1984, Ap. J. Lett. **L15**.

Kissel, J., Brownlee, D.E., Buchler, K., Clark, B.C., Fechtig, H., Grün, E., Hornung, K., Igenberg, E.B., Jessberger, E.K., Krueger, F.R., Kuczera, H., McDonnell, J.A.M., Morfill, G.M., Rahe, J., Schwehm, G.H., Sekanina, Z., Utterback, N.G., Volk, H.J., Zook, H.A.: 1986, Nature **321**, 336.

Leinert, C., Richter, I., Pitz, E., Planck, B.: 1981, Astron. Astrophys. **103**, 177.

Mukay, T.: 1986, in "Proceedings of the International School of Physics E. Fermi", (Varenna), Course 101,.

Murdock, T.L., Price, S.D.: 1985, Astron. J. **90**, 375.

Price, S.D., Murdock, T.L., Marcotte, L.R.: 1980, Astron J. **85**, 765.

Roser, S., Staude, H.J.: 1978, Astron. Astrophys. **67**, 381.

Salama, A., Andreani, P., Dall'Oglio, G., de Bernardis, P., Masi, S., Melchiorri, B., Melchiorri, F., Moreno, G., Nisini, B., Shivanandan, K.: 1987, Astron. J. **92**, 467.

REPORT OF THE WORKING GROUP ON CARBON

D.R. Huffman
Department of Physics, University of Arizona, Tucson, Arizona, U.S.A.

1 Introduction

It was clear to members of the group that carbon is important in the interpretation of several spectral regions of cosmic dust: the UV region, primarily because of the 2200 Å interstellar band, the 3 to 15 μm IR region, because of various unidentified bands in this region (including those at 3.3, 6.2, 7.7, 8.8 and 11.3 μm), and the far infrared region because of the necessity of fitting emissivity observations in this region. Carbon may also be important in the visible for explaining diffuse interstellar bands (4430 Å, etc.) following tentative suggestions for explanations based on clusters of carbon atoms in linear chains or in polycyclic carbon molecules. It has became more and more clear, however, that carbon is an extremely complicated and variable material. The first part of this summary expresses our concern for evaluating and describing the nature of carbon used in experiments. Following this, three significant problems are summarized and in the final section a list of recommended work that is needed is given.

2 Dealing with the complexity of carbon

It is now obvious that there are many kinds of carbon. Terms such as graphite, amorphous carbon and disordered carbon have been used very loosely in the past, both in private discussions and in the published literature. Part of the reason is simply the complex nature of carbon. In attempting to focus on some of the semantic and classification problems the group listed the following types of carbon:

1. **Disordered graphitic carbon** ranging over a continuous scale of structural disorder from completely disordered (amorphous?), if there is such, to single crystal graphite, which there is.

2. **Graphitic as appared to diamond-like** (as tetrahedrally bonded).

3. **Cyclic and acyclic carbon molecules**, with and without attached hydrogen and other atoms.

4. **Polycyclic aromatic hydrocarbons** (PAH).

5. **Quenched carbonaceous composite** (QCC).

E. Bussoletti et al. (eds.), Experiments on Cosmic Dust Analogues, 345–347.
© *1988 by Kluwer Academic Publishers.*

6. C_{60}: the proposed 60 atom cluster of carbon in a soccer ball geometry

This list is not meant to be exhaustive, but rather it illustrates complexities of carbon materials and the particular interests of the participants. It appears that insufficient effort has been made in the past to quantify the degree of disorder. After much discussion the group concluded that x-ray and electron diffraction as commonly used are not able to distinguish adequately short range order variations that are commonly encountered in making small particles of carbon. The group recommends the use of Raman spectroscopy and high resolution transmission electron microscopy for this purpose. Several studies have shown that Raman spectrum of carbon is a sensitive measure of the degree of present disorder. Whether or not these techniques can be employed to quantify the degree of disorder, it is strongly urged that experimentalists who make carbon particles should give complete details of productions conditions, sufficient that another experimenter at any time in the future could have the opportunity to prepare similar particles. In this way, even if inappropiate terminology is used it would not matter. It might also be possible in future to bring newly developed experimental techniques to bear on the kinds of carbon particles previously studied.

3   Some significant problems

In the experimental study of carbon for purposes of understanding cosmic dust the group identified three problem areas (in addition to the problem of quantifying degree of disorder). These are (1) how to produce the desired particles, (2) how to isolate the particles in an unclustered way and (3) what optical constants to use. Point (3) emphasizes the use of optical constants in conjunction with a suitable small particle theory, such as Mie theory or ellipsoid calculations. This has the great advantage over direct measurements of providing a very easy way of changing parameters such as size distribution. One simply enters a different set of parameters into the computer. Problems arise, however, when optical constants have not been measured accurately for the solid of concern in the wavelength range of interest. A valuable addition to the field would be more wide optical constants for carbon in various states of order. It is worth pointing out, however, that optical constants derived from anything other than homogeneous bulk samples with well prepared surfaces may be substantially in error. Production of particles may successfully be done by often-used methods for somewhat disordered, sub-micron particles. Such techniques are condensation of carbon vapor in inert gas and decomposition of hydrocarbon gases, for example. Some thoughts were shared on the possibility of producing particles like de-hydrogenated PAH's (or molecular-size pieces of graphite flakes) by starting with crystalline graphite and disrupting it. Although the idea is appealing in principle, no one has been able to suggest how to accomplish it in practice. A more serious potential problem in direct optical measurements may be the almost-unavoidable clustering of particles which significantly affects extinction (for example) in both UV and IR regions. Suggested possibilities for doing experiments on unclustered particles are the following:

1. Spectroscopy of particles in the gas phase, using continuous particle production and conventional spectrometry, or using rapid-time spectroscopyc methods.
2. Isolating particles in low temperature matrices, such as solid or rare gases.

3. Isolation in other solid matrices such as a stock of fused polyethylene sheets (for far IR), or a silica gel with appropriate spaces (for visible and perhaps near UV).

4. Possible production in cluster beams and/or trapping in ion traps, in the case of molecular clusters of carbon.

A further promising experimental technique, both for producing and maintaining isolation in the particulate sample, is to study carbon particles "in situ" in flames in which they are formed. This age-old scientific problem seems to have considerable relevance in these days of strong interest in carbon in connection with hydrocarbon species. It has been suggested, for example, that both PAH and C_{60} are detectable in suitable flames. In order to encourage further interest in the optical physics and the chemistry of flames, we have listed a few key references on the subject at the end of this report.

4 Further work recommended

Other suggested work that does not fit neatly into the above paragraphs is suggested below:

1. Spectroscopic studies of PAH molecules in IR, visible, and UV. It was felt by most members of the group that insufficient spectral data constitutes a seriouos draw back to the PAH in interstellar dust hypotesis. In addition, it is desirable to have absorption and emission spectra at high temperatures for PAH's, for de-hydrogenated members of the group, and for ionized ones.

2. Studies of temperature dependence of optical properties including both optical constants and direct small particle measurements.

3. A recommendation was made that important data of use to others should be included in tabular form, rather than as tiny graphs alone.

4. Studies of the development of optical properties in the transition region between molecular and small particle solids (which can be described by bulk optical constants). This has become particularly pertinent in the discussion of large molecular clusters of carbon and of polycyclic aromatic hydrocarbons.

A final suggestion relates to our interaction with other scientists. The experimental solution to the problems of cosmic dust demands the very best of modern experimental techniques and equipment. This can be extremely expensive. The group feels that it is incumbent upon all of us to try to enlist the very best talents and facilities from all the physical sciences in order to bring their techniques to bear on the fascinating problems of interstellar and interplanetary dust.

REPORT OF THE WORKING GROUP ON SILICATES

J. Nuth, W. Krätschmer, B. Kneißel, J. Stephens, T. Tanabé

There are at least two ways to understand the nature of grains in the universe. In one approach, one can try to understand the *life history* of typical particles. The fate of a grain may e.g. be as follows: vapours of refractory compounds are synthesized in the depths of stars and ejected into circumstellar regions by processes which are not yet well understood. In the circumstellar shell, refractory vapors nucleate into grains which continue to accrete additional material and may coagulate into larger aggregates. While still in the shell, grains can be annealed at temperatures on the order of 1000 K for timescales of up to a year. Both, within the circumstellar outflow and the interstellar medium (ISM), interaction of the grain with energetic particles could produce considerable change in the composition and structure of the particle. In ISM shock processes (including sputtering, heating and grain-grain collisions) can again alter or even destroy typical grains. Atoms from destroyed grains can recondense on grain cores which survived the shock, probably as SiH_4, $Fe(CO)_x$, $Al(OH)$, MgH along with the more abundant molecules such as CO, H_2O, NH_3. UV or cosmic ray processing of these ices could lead to a refractory metal oxide or organo-metallic residue over silicate or metal-oxide grain cores. At some point in a grain's life it evolves part of a collapsing molecular cloud which eventually evolves into a star. Of the grains in this collapsing cloud, a small percentage survive to from solid bodies such as planets, asteroids or comets; in a few of these cases the grains are sufficiently robust or the processing is sufficiently gentle that particles survive this processing intact and retain clues to the environments in which they previously existed. In such cases, careful study of meteoritic particles can reveal much about the grains in circumstellar and interstellar environments which can be obtained in no other way. An alternate and equally profitable approach to the question of the composition, crystal structure, grain morphology and degree of aggregation of cosmic grains is to deduce these quantities from the optical properties of grains. This implies to measure the optical constants for a wide range of particle compositions and degrees of crystallinity, understand completely the effects of particle shape and degree of aggregation on the optical properties of the particle and then proceed to make careful observations of relevant astronomical regions. Measurements of albedo, extinction, phase function, emission and degree of polarization at wavelengths ranging from the far UV to the far IR (and maybe into the radio) would then be used to derive the properties of the particles of interest. In actuality,

E. Bussoletti et al. (eds.), Experiments on Cosmic Dust Analogues, 349–352.
© 1988 by Kluwer Academic Publishers.

both approachers are useful, and indeed, both approachers must be pursued in order to streamline the amount of experimental data required to use either approach. Ideally, both approaches should yield the same result when applied to a particular astronomical object. There are two additional points which are important if experimental studies of either optical properties or physical processes are to be useful. First, the effect of the experimental apparatus (e.g. the method of particle production) on the experimental results must be completely undestood. Second, experimental results must be presented in such a way as to present a clear challenge to observers. In first instance the effect of the experimental system can be minimized by recommending that all studies of complex silicates include whenever possible a study of the pure silica system. Results of this *control* study should be reported comprehensively over as wide a range in the experimental parameters and measuring capability as possible and should be included where practical in the initial published description of the experimental apparatus. The second point can be accomplished if the results of specific experiments which have potentially observable consequences are presented in much a way that a clearly defined observational test of the results is obvious. Clear predictions of observable differences in specific astrophysical environments should be sufficient justification for an observer to invest time in a study to refute or confirm the applicability of specific experimental results. In what follows the working group has recommended specific studies which need to be performed in order to understand both the processes which effect the properties of cosmic dust (A) and the interpretation of astronomical observations of solid particles which may have been subjected to such processing (B).

A   "Cradle to grave" processing of cosmic dust

1. Careful studies of vapor phase nucleation of refractory grains are needed in order to develop a theoretical understanding of this very complex process. Under this broad heading we include both matrix isolation and cluster beam type experiments designed to study the properties of small refractory clusters, experiments to determine the chemical stabilities and reaction paths of the more important clusters, experiments to determine T, P conditions under which grain formation can occur as a function of temperature, stochastic-chemical dynamic studies of the nucleation process, and studies of the composition, crystal structure and morphology of particles condensed in a wide variety of environments.

2. Experiments are needed on the effects of specific metamorphic influences on the physical properties of freshly condensed grains. These experiments would specifically study vacuum thermal annealing, hydrous alteration, chemical sputtering yields and radiation induced changes in both the crystal structure and the chemical composition of the grains. In this regard it is important to try to study as many of these effects as possible on very small particles which have very large surface to volume ratios, small heat capacities and very small interaction volumes.

3. Experiments are needed to determine specific fragmentation/destruction probabilities in grain-grain collisions over a range of velocities comparable to those found in the interstellar medium. Current experimental velocities possible today are at least an order of magnitude to low to allow the full spectrum of those experiments to be carried

out, but preliminary experiments of lower velocity collisions (up to $\sim$ 20 km/s) can be carried out at present time. These experiments are necessary to determine the destruction efficiency of grains in the interstellar medium.

4. Experiments are needed which address the problem of the consequences of grain destruction in the interstellar medium. Specifically, what is the composition, structure and morphology of recondensed refractory material and what is the mechanism by which this recondensation process takes place? Are the grains produced in circumstellar environments observably different from grains formed in the interstellar medium? At present time, no work has been published which addresses this problem.

5. It would be very helpful if experimental results which hope to address the life cycle of silicate grains could make specific predictions which relate to meteoritic and cometary grains. These are very complicated and, in some cases, very highly processed materials, but they are the only cosmic grains available for study in our laboratories. These resources could serve as valuable tools to test our experimental results and we should make as much use of these resources as is practical.

B Observational properties of solid particles

1. Careful measurements are needed of the optical constants (in the bulk) of a variety of solids of *cosmic abundance*. In particular optical constants for a wide range of cation/silicate ratios must be tabulated over a spectral range as wide as possible, from the far UV to the far IR. If possible an attempt should be made to determine absolute rather than relative efficiencies and particular attention should be paid to the ratio of specific peak strengths to the strength of the continuum.

2. Similar measurements should be made for a variety of structural types of materials discussed in B.1 including as much of the range as possible from completely amorphous through perfectly crystalline meterials.

3. It is necessary to measure the optical properties of a large range of grain shapes in order to understand extinction, scattering and polarization properties of materials.

4. It is necessary to measure or, if possible, to calculate, the optical properties of a variety of complex composite microcrystalline small grains to see if there are diagnostic effects which might be observable indicators of the morphology.

5. We highly recommend the measurement of albedo, phase function and degree of polarization of a cloud of well characterized *unclustered* particles and careful studies of changes in the properties as these particles coagulate.

6. We feel that the measurement of single scattering effects from $\sim 10^3$ particle clusters of various morphology would possibly provide clues by which such aggregates might be detected.

7. Experiments similar in nature to above B.3-B.6 should also be carried out with coated grains where ice, refractory organic, icy inorganic (?) and refractory inorganic coatings of various thicknesses would be used.

8. Microware analog experiments are an extremely important means of determining the effects of varying grain shapes, coatings and aggregate geometries and these techniques should be exploited fully in order to develop an experimental data base sufficient to

serve as the basis upon which a thorough theoretical understanding of scattering processes can be based.

9. It will be very important to determine at what aggregate size *collective effects* become detectable and wavelengths at which such effects might be expected to occur. Both theoretical and experimental work is needed in this area.

10. Attempts to measure optical properties of naturally occuring interplanetary dust particles and to relate these properties both to cosmic dust observations and to theoretically predicted properties of composite particles will serve as a useful test of our ability to deal with real particles which occur in the universe.

11. Attempts to match specific observations with detailed grain models should be made whenever possible. Some examples of relevant recent questions include the variability of the strength of the 10 μm cometary silicate feature, or the explanation of what particular properties are actually characteristic of a *dirty silicate*.

12. The question of the effects of grain charging on the observed spectrum, degree of coagulation and probability of nucleation of refractory grains is at present open and experiments to address these questions could be very useful.

Finally we must realize that understanding cosmic dust will be an iterative process which will rely on experiments to give us clues as to the expected grain properties, measurements of the optical properties of relevant standard materials and careful comparison with observational data. We have barely begun to tachle the problem at the present time and considerable progress in this area can be expected in the future.

REPORT OF THE WORKING GROUP ON ICE AND RELATED TOPICS

A. Bar-Nun, B. Donn, R.J.A. Grim, F.R. Krueger, V. Pirronello, B. Schmitt, G. Strazzulla

A   Overview

Ice mantles are accreted on refractory cores (mainly silicates) in interstellar grains (ISG). Their identification has been possible by comparison of astronomical infrared absorption spectra with those known from laboratory spectra (e.g. the 3.1 μm feature of water). Grains are believed to be the sites where molecules (e.g. H2) can be formed and under certain circumstances returned to the gas phase. Again, ice mantles are subjected to erosion processes as sublimation (in regions where the temperature is high enough), photo (?) and ion sputtering, grain-grain collisions. Erosion processes are of fundamental importance in many contexts as e.g. in explaining why some material is left over in dense clouds in the gas phase instead to completely condense on cold grains. The ultimate fate of ice mantles (together with their cores) is to partecipate to the collapse of the cloud to which they belong to form, at least in the case of the Sun, a planetary system. Here different scenarios are possible: at one extreme there are researchers who believe that ISG are completely destroyed before planetary objects are formed, on the other extreme it is belived that grains, included ice mantles, have never experienced high (> 150 K) temperatures and that they preserve their status e.g. in comets where now we can go to search for them. During all their life ice grains suffer a continuum irradiation by external agents, as photons and ions, inducing not only erosion phenomena but also chemical alteration. These include formation of new organic materials due to the formation of radicals and their reactions. The presence of all the above sketched phenomena had, have and will have to be supported by laboratory work. This working group discussed, in some cases with great technical detail, some of those laboratory works. We report here some of these items.

B   IR spectroscopy of laboratory and interstellar ices

Infrared spectroscopy is a very powerful tool in studying laboratory ice samples in situ. During an experiment the composition of ice may change significantly as a result of high energetic processing, such as UV radiation or proton bombardment. In laboratory

E. Bussoletti et al. (eds.), Experiments on Cosmic Dust Analogues, 353–357.
© *1988 by Kluwer Academic Publishers.*

studies one normally starts off with simple molecules like H_2O, CO, NH_3, CH_4, N_2, O_2. In special cases molecules such as H_2CO, CH_3OH, or CO_2 can be added.

C   IR spectra of photolyzed ices

It is commonly known that processing of ices by UV irradiation or proton bombardment creates a wide variety of new species. However, they are dependent heavily on the conditions of the experiments, e.g. initial ice composition, irradiation dose and temperature. Generally speaking what happens upon UV irradiation is as follow: the energetic UV photon is able to break the vibrational bonds in a molecule. If no cage-recombination immediately occurs, a radical is formed which can be stored at low temperature for a long time. Subsequently, the photolyzed sample can be annealed slowly or rapidly. In the first case radicals are allowed to diffuse and recombine forming larger aggregates, either molecules or new radicals. This process in principle, continues until all radicals are recombined. During fast heating the energy released upon recombination may trigger an explosion. That is, the released energy locally the sample heats allowing further radical diffusion and recombination. In this way a chain reaction is set and may lead to an explosive evaporation of the sample. New molecules formed in this way can be stable to high temperature, in which the left residue is now called organic refractory material. A wide variety has been observed in different laboratories.

   Until recently nobody has looked seriously into the possibility that ionized species may also contribute to the IR spectra of these ices. In a recent study (see Grim and Greenberg, these proceedings) it appeared that OCN^- and $I_3(NH_3^+)$ are respectively responsible for the 4.62 and 6.8 μm absorption feature. It also appeared that the NH_3 is a critical molecule in the formation of these and other ions. Although photon energies are not high enough to ionize the starting products, surprisingly we observe these ions already 5 minutes after UV photolysis. Their formation is not yet fully understood and certainly needs further investigation.

   Our present understanding of all aspect of photolyzed ice mixtures is far from complete. All what we can say so far is that we have only a rough feeling of the average grain mantle composition towards several objects, of which W33A is the best studied (Grim and Greenberg, these proceedings). At this time we can point out some interesting projects to be studied in more detail:

- What is the formation mechanism of ions in ices?
- How is CH_3OH formed?
- Is there CO_2 in ISG mantles?
- Can there be e.g. salts mixed with residues?
- What are the radical concentrations in photolyzed/bombarded samples (ESR spectra)?
- How does the roughness of the surface, possibly fluffy (as found by Bar-Nun), affect all processes both in space and in lab?

D   Effects of ion irradiation

As far as the particle irradiation of ice is concerned both physical and chemical effects are to be investigated. Two approaches have been and can be used. One is to bombard relatively complex mixtures of frozen gas that may mimic the composition of either grain mantles in dense clouds or cometary material, the other one is to use simple targets made either of single species or at most binary mixtures in order to obtain rates or both physical and chemical effect per impinging particle. It seems that mainly in this second case energy of projectiles has to be chosen in such a way that their range in the solid is much larger than the thickness of the bombarded layer in order not to have a drastic change in the energy of the ion while it is penetrating. This procedure will prevent from getting data nominally obtained at a well defined energy of the ion but really obtained at several different energy values mixed together and therefore of difficult interpretation.

          Another important question that has been raised has to do with the checks that have to be performed, "in situ", on the already deposited mixture in order to be sure that during the process of deposition from the gas phase no mayor changes occur in the relative amounts of the mixture components and, not less important, that segregation effects also do not occur. Suggested methods could be either backscattering (see Pirronello et al., these proceedings) or IR spectroscopy; but as far as segregation is concerned IR spectroscopy may be not very helpful. A general confidence exists that sputtering can be a dominant destruction mechanism in some astrophysical scenarios and that chemical processing of ice by ion bombardment is together with gas phase and surface grain reactions fundamental in understanding the chemical history of dense interstellar clouds and the optical properties of some bodies of the solar system (see Pirronello et al.; Strazzulla, these proceedings).

E   Diffusion of molecules in grains

Not only irradiation produces modifications in grain mantles. Also diffusion of radicals and relative species in icy mantles over long time scales will affect their composition. Radical diffusion in ices has been studied in one case by monitoring fluorescence of glyoxal after recombination of two HCO radicals. Diffusion of volatile species are either studied by monitoring gas release from ice (see Bar-Nun, these proceedings) or by IR "in situ" measurements of the ice composition (see Schmidt et al., these proceedings). Prolungation of these types of experiments is suggested in order to reach consistent results and to better understand thermal evolution of ISG and cometary material.

F   Gas-grain interaction

Adsorption and trapping of energy by molecules, atoms and ions on ice mantles and on refractory particles is one of the critical parameter for most of the gas-grain interactions because it determines the residence time on the surface and the amount of the species trapped in the mantle. Existing experiments indicate that amorphous H_2O ice has a very high adsorption and trapping capacity due to an extremely porous structure of ice particles (pore sizes ranging from 10 to 500 Å). Gas to ice ratio as high as 3.3 can be

obtained by flowing gas into amorphous ice (see Bar-Nun, these proceedings). Release
of gas from ice is governed by internal changes in ice.

One of the most exciting points is that extrapolation to 10 K of results
obtained at 77 K lead to predict that amorphous ice condensed at 10 K has micropores
less than 10 Å wide. It is known that in such small pores adsorption energy is greatly
enhanced compared to a flat surface. Presence of micropores in ISG has certainly an
important effect on evaporation and surface reaction processes whose calculation has
been performed only in the case of a flat surface. Consequently, determination of surface
structure of different ice mantles and measurement of adsorption energy of the main
species present in the gas phase is of primary importance.

G   Inferences from PUMA/PIA data concerning comet's dust composition and need for laboratory simulation

The organic material is found to consist mainly of highly unsaturated hydrocarbons without heteroatoms. Whether this matter is mostly aliphatic or aromatic is yet uncertain,
however it is consistent to assume that PAH's may be prominent. Oxygen is found to
be trapped in silicates and water and some oxygen also, possibly, in methanol, methanal
and formic acids. No indications however have been found whatsoever, that these are
O-bearing polymerized species. In simulation experiments when one add CO into the
frozen gases should look wether water is driven out or built into molecular organic (refractory) material. The silicate problem could be attached by dispersing Si atoms in
frozen oxygen-containing gases (e.g. by implantation) and look for amorphous silicate
formation (see Nuth et al., these proceedings). Generally, it is recommended to extend
liquid chromatography (LC) analysis of residues of irradiated ices containing N and/or
O, in order to search for periodically relevant substances being formed by these processes.
In accordance with the results of the carbon working group, Raman measurements on
processed ices are highly desired, in order to determine the number of fractional groups
produced per unit volume, as well as the total order/disorder state of the material.

H   Final statements

Laboratory simulation experiments play a crucial role in understanding processes oc-
curring in space. Individual effects may be of large complexity and even for processes
that seem to be undestood it is urgently needed to study macroscopic effects for realistic
mixtures.
These effects include:

- variations, with dimensions of targets and radiation damage, of physical quantities
  such as density, morphological properties and optical constants;
- mechanism for release of volatile species (sputtering, sublimation, grain-grain collision
  and their relative importance). In particular photoejection from ice could not be
  observed until now. It should be looked at again.

Let us finish such report with some general recommendation for experimentalist who have in mind to apply their results to astrophysical objects:

- it is recommended that researches will often inspect visually their ice samples in order to determine what are the conditions for making different ice structures. Also, an effort should be made to study ice by electron microscopy;
- it is obvious that one has to care to "simulate" as much as possible environmental physical conditions (temperature etc.) but, at least some of us do think that also working with simple mixtures is extremely helpful in order to gain a deep understanding of the processes;
- it is imperative that laboratory results are dose rate independent or lead to a reasonable way to extrapolate from the laboratory simulation to that in space which occur at an enormous lower rate;
- it is very desiderable that different groups use common scales for different irradiations (e.g. eV/molecule or similar);
- it would be very useful to look to similar works made for other purposes (including industrial work). Care must be given to applicability under astrophysical conditions;
- it would be interesting if different groups could use same materials to study any difference depending on the specific physical process investigated (or experimental arrangement i.e. results that are artifacts of experimental set up);
- to publish or circulate a chart of specialized facilities available at different laboratories may facilitate collaboration among groups;
- it is recommended that published works give description of experimental systems and conditions as complete as possible. Experiments should be designed so that these data are known e.g. surface conditions, purity of material, all temperatures and rates of change;
- it is needed to experimental determine how significant the small particle effect is for different experiments e.g. UV and particle irradiations, sputtering, vaporization. This is particularly important for particles smaller than 100 nm.

INDEX OF SUBJECTS